Public Health Risk Assessment
for Human Exposure to Chemicals

ENVIRONMENTAL POLLUTION

VOLUME 6

Editors

Brian J. Alloway, *Department of Soil Science, The University of Reading, U.K.*
Jack T. Trevors, *Department of Environmental Biology, University of Guelph, Ontario, Canada*

Editorial Board

Public Health Risk Assessment
for Human Exposure to Chemicals

by

Kofi Asante-Duah

Human Health Risk Assessment Practice,
Environmental Risk Solutions, inc.,
The ERS Group, California, U.S.A.
and
Anteon Corporation, Environment Division,
San Diego, California, U.S.A.

KLUWER ACADEMIC PUBLISHERS
DORDRECHT / BOSTON / LONDON

Library of Congress Cataloging-in-Publication Data

ISBN 1-4020-0920-8

Published by Kluwer Academic Publishers,
P.O. Box 17, 3300 AA Dordrecht, The Netherlands.

Sold and distributed in North, Central and South America
by Kluwer Academic Publishers,
101 Philip Drive, Norwell, MA 02061, U.S.A.

In all other countries, sold and distributed
by Kluwer Academic Publishers,
P.O. Box 322, 3300 AH Dordrecht, The Netherlands.

Cover image: Illustrative elements of a risk assessment process

Printed on acid-free paper

Printed in the Netherlands.

To the Everlasting and Loving Memories of:
Atta Panin (a.k.a. Ebenezer Asare Duah, Sr.)
Grandma Nana Martha Adwoa Oforiwaa
Grandma Nana Abena Nketia Owusua

&

To Dad – George Kwabena Duah
To Mom – Alice Adwoa Twumwaa
To My Extraordinary Families (of Abaam,
Kade and Nkwantanan) – and all the offsprings

TABLE OF CONTENTS

PREFACE

Risk to human health as a consequence of toxic materials found in modern societies is a matter of grave concern to the world community. In particular, risks to humans that result from chemical exposures from a multiplicity of sources are a complex issue with worldwide implications. The effective management of human exposure to a variety of chemicals present in various sectors of society has therefore become a very important public health policy issue that will remain a growing social challenge for years to come. In fact, with the control and containment of most infectious conditions of the past millennium having been achieved in most developed countries, and with the resultant increase in life expectancies, much more attention seem to have shifted to degenerative health problems. Many of the degenerative health conditions have been linked to thousands of chemicals regularly encountered in human living and occupational/work environments. It is important, therefore, that human health risk assessments are undertaken on a consistent basis – in order to determine the potential impacts of the target chemicals on public health. Overall, risk assessment promises a systematic way for developing appropriate strategies to aid public health risk policy decisions in the arena of human exposures to chemicals.

Risk assessment generally serves as a tool that can be used to organize, structure, and compile scientific information in order to help identify existing hazardous situations or problems, anticipate potential problems, establish priorities, and provide a basis for regulatory controls and/or corrective actions. A key underlying principle of public health risk assessment is that some risks are tolerable – a reasonable and even sensible view, considering the fact that nothing is wholly safe *per se*. In fact, whereas human exposures to large amounts of a toxic substance may be of major concern, exposures of rather limited extent may be trivial and therefore should not necessarily be a cause for alarm. In order to be able to make a credible decision on the cut-off between what really constitutes a 'dangerous dose' and a 'safe dose', systematic scientific tools – such as those afforded by risk assessment – may be utilized. In this regard, therefore, risk assessment seems to represent an important foundation in the development of effectual public health risk management strategies and policies.

This book provides a concise, yet comprehensive overview of the many facets/aspects relating to human health risk assessments in relation to chemical exposure problems. It presents some very important tools and methodologies that can be used to address chemical exposure and public health risk management problems in a consistent, efficient, and cost-effective way. Overall, the book represents a collection and synthesis of the principal elements of the risk assessment process that pertain to human exposures to chemicals found in modern societies. This also includes an elaboration of pertinent risk assessment concepts and techniques/methodologies for performing human health risk assessments. A number of illustrative example problems are interspersed throughout the book, in order to help present the book in an easy-to-follow, pragmatic manner.

Even though the main focus of this title is on risk assessment of the potential human health effects associated with chemical exposures, it is noteworthy that the same principles may be extrapolated to deal with other forms of human exposure problems (such as exposures to radionuclides and pathogens). Thus, the chemical risk assessment framework may be adapted and applied to exposures to other agents – albeit many unique issues may have to be addressed for exposures to the new agent under consideration. The subject matter of this book can indeed be used to aid in the resolution of a variety of environmental contamination and public health risk management problems.

This book should serve as a useful reference for many a professional encountering risk assessment in relation to environmental contamination and public health risk management programs; it offers an understanding of the scientific basis of risk assessment and its applications to public health policy decisions. The specific intended audience include public and occupational health practitioners, and other public health and environmental health professionals; public policy analysts; environmental consulting professionals; consumer product manufacturers; environmental attorneys; environmental and health regulatory agencies; and a miscellany of health, environmental, and consumer advocacy interest groups. The book is also expected to serve as a useful *educational/training resource for both students and professionals in the health-related and environmental fields* – who have to deal with human exposures to chemicals, public health risk assessment issues, and/or environmental health management problems. This book, written for both the novice and the experienced, is an attempt at offering a simplified and systematic presentation of public health risk assessment methods and application tools – that will carefully navigate the user through the major processes involved.

ACKNOWLEDGEMENTS

I am indebted to a number of people for both the direct and indirect support afforded me during the period that I worked on this book project. Many thanks to the Duah family (of Abaam, Kade, and Nkwantanan), as well as to the friends and colleagues who provided much-needed moral and enthusiastic support throughout preparation of the manuscript for this book.

Special thanks go to Dr. Kwabena Duah (Fellow of the College of Public Health Medicine, South Africa) – a public and occupational health practitioner (South Africa and Australia) – for reviewing and providing contributions to various sections of the manuscript for this book. Indeed, as a physician with specialty practice in public and occupational health, environmental health, and general medicine, Dr. Duah's input was most invaluable. Thanks also to several colleagues associated with the Premier [Occupational and Environmental] Health Services Group (an occupational and environmental health firm offering consulting and primary healthcare services) and the Environmental Risk Solutions (ERS) Group (an environmental consulting firm with practice focusing on human health risk assessments) – for providing miscellaneous assistance.

The support of the Publishing, Editorial, and Production staff at Kluwer Academic Publishers (Dordrecht, The Netherlands) in helping to bring this book project to a successful conclusion is very much appreciated – with special mention of Mrs. Betty van Herk (Office of the Publishing Editor) and Dr. Paul Roos (Publishing Editor). Indeed, Mrs. van Herk continuously provided amazingly prompt review comments and guidance throughout the manuscript preparation. Thanks to Max Voigtritter (Vice President, Anteon Corporation, San Diego, California) for providing miscellaneous support. I also wish to thank every author whose work is cited in this book – for having provided some pioneering work to build on.

Finally, it should be acknowledged that this book benefited greatly from review comments of several anonymous individuals, as well as from discussions with a number of professional colleagues. Any shortcomings that remain are, however, the sole responsibility of the author.

Kofi Asante-Duah
San Diego, California, USA
(06-06-2002)

Chapter 1

INTRODUCTION

In the landmark book – *Silent Spring* – from the early 1960s, Rachel Carson wrote (Carson, 1962; 1994): 'For the first time in the history of the world, every human being is now subjected to contact with dangerous chemicals, from the moment of conception until death.' And this statement of some more than four decades ago is not about to change, given our dependency – maybe even obsession – with a 'modern way of life.' Indeed, in everyday living, peoples around the world – directly or indirectly – are exposed to myriad sources and cocktails of chemical hazards. Ultimately, chemical exposure problems may pose significant risks to the general public because of the potential health effects. Risks to human health as a result of exposure to toxic materials present or introduced into our living and work environments are, therefore, a matter of grave concern to modern society. To borrow again from Rachel Carson's *Silent Spring*, 'if we are going to live so intimately with these chemicals – eating and drinking them, taking them into the very marrow of our bones', then at the very least, we should be able to determine the risks that we are exposed to, as well as know how to manage such risks – in order to ensure a quality to our life (Carson, 1962; 1994).

In fact, with the control and containment of most infectious conditions and diseases of the past millennium having been achieved in most developed countries, and with the resultant increase in life expectancies, much more attention seem to have shifted to degenerative health problems. Many of the degenerative health conditions have been linked to thousands of chemicals regularly encountered in human living and occupational/work environments. It is important, therefore, that human health risk assessments are undertaken on a consistent basis – in order to determine the potential impacts of the target chemicals on public health. Overall, risk assessment promises a systematic way for developing appropriate strategies to aid public health risk policy decisions in the arena of human exposures to chemicals.

This book focuses on the application of risk assessment concepts and principles to support responsible and effective public health risk management programs as relates to chemical exposure problems. It offers an understanding of the scientific basis of risk assessment and its applications to public health policy decisions.

1.1. Coming to Terms with the Several Chemicals in Modern Society

It appears that there is no escape from potential chemical exposure problems in any part of the world – especially that resulting from possible environmental contamination, and also from the usage of various consumer products. Indeed, even in the far removed Arctic regions, it has been established that there is contamination of the arctic aquatic food-chain by organochlorine compounds and other anthropogenic chemicals (see, e.g., Barrie *et al.*, 1992; Dewailly *et al.*, 1993; Lockhart *et al.*, 1992; Muir *et al.*, 1992; Thomas *et al.*, 1992).

In general, the key environmental chemicals of greatest concern are believed to be anthropogenic organic compounds. These typically include pesticides – e.g., lindane, chlordane, endrin, dieldrin, toxaphene, and dichlorodiphenyl trichloroethane [DDT]; industrial compounds – e.g., solvents such as trichloroethylene (or, trichloroethene) [TCE] and fuel products derived from petroleum hydrocarbons; and byproducts of various industrial processes – e.g., hexachlorobenzene [HCB], polychlorinated biphenyls [PCBs], polychlorinated dibenzodioxins (or, polychlorodibenzo-*p*-dioxins) [PCDDs], and polychlorinated dibenzofurans (or, polychlorodibenzofurans) [PCDFs] (see, e.g., Dewailly *et al.*, 1993; 1996). Many industries also produce huge quantities of highly toxic waste byproducts that include cyanide ions, acids, bases, heavy metals, oils, dyes, and organic solvents (Table 1.1). Further yet, other rather unsuspecting sources of environmental contaminants are beginning to add to the multitude of chemical exposure problems that contemporary society faces. For instance, low levels of reproductive hormones, birth control pills, steroids, antibiotics, and numerous other prescription and nonprescription drugs, as well as some of their metabolites, have been detected in various water bodies around the world in recent times. Along with the pharmaceuticals, products used in everyday life (such as food additives, cosmetics, fragrances, plasticizers, cleaners, detergents, disinfectants, insect repellants, pesticides, fire retardants, etc.) are turning up in aquatic environments (Erickson, 2002; NRC, 1999). Indeed, it is probably reasonable to assume that pollutants from pharmaceuticals and other everyday products have been in the human environments for as long as they have been in use – albeit only recently have analytical methods been developed to detect them at the low levels typically found in the environment.

A general review of various chemical materials and their usage in social contexts reveals that hazards from several of the commonly encountered 'social chemicals' could be problematic with respect to their potential human health impacts; this is illustrated by a limited number of the select examples enumerated below.

- *Arsenic.* A poison famous from murder mysteries, arsenic [As] has been used in insecticides (among other uses, such as in alloying agents and wood preservatives) – and these have resulted in extensive environmental contamination problems. Also, there has been a number of medicinal, agricultural, and industrial uses for arsenic compounds; for example, arsenic has been used extensively in medicine (viz., Fowler's Solution) for the treatment of leukemia, psoriasis, and asthma, as well as in the formulation of anti-parasitic drugs.

  It is also noteworthy that arsenic is a naturally-occurring element distributed throughout the environment. As a consequence, arsenic is introduced into waters through the dissolution of natural minerals and ores – and thus concentrations in groundwater in some areas are elevated as a result of releases from local rocks. Still, industrial effluents also contribute arsenic to waters in some areas.

Table 1.1. Examples of typical potentially hazardous waste-streams from selected industrial sectors

Sector/source	Typical hazardous waste-stream
Agricultural and food production	Acids and alkalis; fertilizers (e.g., nitrates); herbicides (e.g., dioxins); insecticides; unused pesticides (e.g., aldicarb, aldrin, DDT, dieldrin, parathion, toxaphene)
Airports	Hydraulic fluids; oils
Auto/vehicle servicing	Acids and alkalis; heavy metals; lead-acid batteries (e.g., cadmium, lead, nickel); solvents; waste oils
Chemical/pharmaceuticals	Acids and alkalis; biocide wastes; cyanide wastes; heavy metals (e.g., arsenic, mercury); infectious and laboratory wastes; organic residues; PCBs; solvents
Domestic	Acids and alkalis; dry-cell batteries (e.g., cadmium, mercury, zinc); heavy metals; insecticides; solvents (e.g., ethanol, kerosene)
Dry-cleaning/laundries	Detergents (e.g., boron, phosphates); dry-cleaning filtration residues; halogenated solvents
Educational/research institutions	Acids and alkalis; ignitable wastes; reactives (e.g., chromic acid, cyanides; hypochlorites, organic peroxides; perchlorates, sulfides); solvents
Electrical transformers	PCBs
Equipment repair	Acids and alkalis; ignitable wastes; solvents
Leather tanning	Inorganic chemicals (e.g., chromium, lead); solvents
Machinery manufacturing	Acids and alkalis; cyanide wastes; heavy metals (e.g., cadmium, lead); oils; solvents
Medical/health services	Laboratory wastes; pathogenic/infectious wastes; radionuclides; solvents
Metal treating/manufacture	Acids and alkalis; cyanide wastes; heavy metals (e.g., antimony, arsenic, cadmium, cobalt); ignitable wastes; reactives; solvents (e.g., toluene, xylenes)
Military training grounds	Heavy metals
Mineral processing/extraction	High-volume/low-hazard wastes (e.g., mine tailings); red muds
Motor freight/railroad terminals	Acids and alkalis; heavy metals; ignitable wastes (e.g., acetone; benzene; methanol); lead-acid batteries; solvents
Paint manufacture	Heavy metals (e.g., antimony, cadmium, chromium); PCBs; solvents; toxic pigments (e.g., chromium oxide)
Paper manufacture/printing	Acids and alkalis; dyes; heavy metals (e.g., chromium, lead); inks; paints and resins; solvents
Petrochemical industry/gasoline stations	Benzo-a-pyrene (BaP); hydrocarbons; oily wastes; lead; phenols; spent catalysts
Photofinishing/photographic industry	Acids; silver; solvents
Plastic materials and synthetics	Heavy metals (e.g., antimony, cadmium, copper, mercury); organic solvents
Shipyards and repair shops	Heavy metals (e.g., arsenic, mercury, tin); solvents
Textile processing	Dyestuff heavy metals and compounds (e.g., antimony, arsenic, cadmium, chromium, mercury, lead, nickel); halogenated solvents; mineral acids; PCBs
Timber/wood preserving industry	Heavy metals (e.g., arsenic); non-halogenated solvents; oily wastes; preserving agents (e.g., creosote, chromated copper arsenate, pentachlorophenol)

Inorganic arsenic can occur in the environment in several forms; in natural waters – and thus in drinking-water – it is mostly found as trivalent arsenite, As(III) or pentavalent arsenate, As(V). Organic arsenic species – quite common in seafood – are far less harmful to human health, and are also readily eliminated by the body. Indeed, drinking water poses the greatest threat to public health from arsenic – severe health effects having been observed in populations drinking arsenic-rich water over long periods. Exposure at work, as well as mining and industrial emissions may also be significant in some locations.

Overall, human exposure to arsenic can result in serious health effects; for instance, large doses can cause gastrointestinal disorders – and even small quantities may be carcinogenic. Following long-term exposure, the first changes are usually observed in the skin – namely, pigmentation changes, and then thickening (hyperkeratosis). Cancer is a late phenomenon, and usually takes more than ten years to develop. Also, some studies have reported hypertensive and cardiovascular disease, diabetes, and reproductive effects. On the other hand, absorption of arsenic through the skin is believed to be minimal, and thus hand-washing, bathing, laundry, etc. with water containing arsenic do not appear to pose significant human health risk. In any case, the relationship between arsenic exposure and other health effects is not quite as clear-cut. According to a 1999 study by the US National Academy of Sciences, long-term exposure to arsenic in drinking water causes cancer of the skin, lungs, urinary bladder, and may cause kidney and liver cancer. The study also found that arsenic harms the central and peripheral nervous systems, as well as heart and blood vessels, and causes serious skin problems; it also may cause birth defects and reproductive problems. In particular, recent studies appear to strengthen the evidence of a link between bladder and lung cancer and exposure to arsenic in drinking water. Indeed, even very low concentrations of arsenic in drinking water are believed to be associated with a higher incidence of cancer. Also, recent research by the US EPA's Office of Research and Development has shown that arsenic can induce an interaction of arsenic compounds with DNA, causing genetic alterations. The study found that methylated trivalent arsenic derivatives (which can be produced by the body in an attempt to detoxify arsenic) produce reactive compounds that cause DNA to break.

- *Asbestos.* A known human carcinogen, asbestos has found a wide range of uses in various consumer products. Indeed, processed asbestos has typically been fabricated into a wide variety of materials that have been used in consumer products (such as cigarette filters, wine filters, hair dryers, brake linings, vinyl floor tiles, and cement pipe), and also in a variety of construction materials (e.g., asbestos-cement pipe, flooring, friction products, roofing, sheeting, coating and papers, packing and gaskets, thermal insulation, electric insulation, etc.). Notwithstanding the apparent useful commercial attributes, asbestos has emerged as one of the most complex, alarming, costly, and tragic environmental health problems (Brooks *et al.*, 1995). Its association with lung cancer has been proven – with synergistic effect observed with cigarette smoke exposures.

There are two general sub-divisions of asbestos: the serpentine group – containing only chrysotile (which consists of bundles of curly fibrils); and the amphibole group – containing several minerals (which tend to be more straight and rigid). Because asbestos is neither water-soluble nor volatile, the form of concern with respect to human exposure relates to the microscopic fibers (usually reported as, or measured in the environment in units of fibers per m^3 or fibers per cc). For asbestos fibers to cause any disease in a potentially exposed population, they must gain access to the potential receptor's body. Since they do not pass through the intact skin, their main entry routes are by inhalation or ingestion of contaminated air or water (Brooks *et al.*, 1995) – with the inhalation pathway apparently being the most critical in typical exposure scenarios. In fact, for asbestos exposures, inhalation is expected to be the only significant exposure pathway. Consequently, potential human exposure and intake is derived based on estimates of the asbestos concentration in air, the rate of contact with the contaminated air, and the duration of exposure. Subsequently, the intake can be integrated with the toxicity index for asbestos to determine the potential risks associated with any exposures; this then forms a basis for developing appropriate public health risk management actions.

- *Lead.* Known to be neurotoxic, as well as a cause of anaemia, among others, lead [Pb] has been used in water supply systems, gasolines, automobile batteries, and paints for a long time in modern human history; this has resulted in extensive releases into the environment. The typical sources of environmental lead contamination include industry (such as metal smelters and lead-recycling facilities), paints, and exhaust from motor vehicles that used leaded gasoline. Domestic water supply systems have also been a major source of human exposure to lead. Further elaboration on these is provided in Section 1.2.4.

  Inorganic lead is indeed one of the most ubiquitous toxic substances; it has been used since antiquity, but its use seems to have increased exponentially during the 20th Century (Levallois *et al.*, 1991; Harrison and Laxen, 1981). As a result of past and current industrial uses, lead has in fact become a common environmental pollutant globally, and is often more problematic in economically disadvantaged and minority-populated areas or regions.

  Lead has come to be recognized as a primary public health hazard globally (see, e.g., Needleman and Gatsonis, 1990; Pirkle *et al.*, 1985; Schwartz, 1994). Lead toxicity derives from the fact that it is absorbed through respiratory or digestive routes, and then preferentially binds to RBCs for distribution to the body tissues. Common observable human health effects include nausea and irritability at low levels, and brain damage at large doses. Of special interest is the storage of lead in the human bone, where its half-life may be in excess of twenty years. Also, the threat of lead poisoning in children and pregnant women is of particular public health concern; indeed, lead poisoning can cause a number of adverse human health effects – but this is particularly detrimental to the neurological development of children. Further discussion of the effects of lead is provided in Section 1.2.4.

- *Mercury.* A nervous system toxin, mercury [Hg], is a significant environmental pollutant in several geographical regions/areas (although far less common than the more ubiquitous lead) because of its use in measuring instruments (e.g., thermometers and manometers); in medicines (as antiseptics); in dental practice; in lamps; and in fungicides. Major sources of Hg to the human environment are the release of elemental Hg from manometers used to measure the flow of natural gas through pipelines and distribution systems, electrochemical industries, and certain fungicides (Henke *et al.*, 1993; Stepan *et al.*, 1995). Potential sources of airborne Hg releases include combustion of fossil fuels, chlor-alkali plants, waste incineration, mining and smelting of Hg ores, and industrial processes involving the use of Hg (ATSDR, 1999; Porcella, 1994). Inorganic Hg may [also] be present in soil due to atmospheric deposition of Hg released from both natural and anthropogenic sources as elemental or inorganic Hg vapor, or as inorganic Hg adsorbed to particulate matter.

It is noteworthy that, Hg can be present in both organic and inorganic forms in the environment. Indeed, Hg released into the environment will persist for a long time – and the Hg can change between the organic and inorganic forms. One form of organic Hg – namely, methylmercury – can produce a buildup in certain fish. Thus, even very low levels of Hg in the ocean and lakes can contaminate these fish to the point of being a significant environmental and public health concern.

The form of Hg and the manner of human exposure determine the nature and/or type of the consequential health effects. Long-term exposure to either organic or inorganic Hg can permanently damage the brain, kidneys, and developing fetuses. Commonly observable human health effects from exposure to large doses of organic Hg compounds include brain damage, often fatal.

• *Organochlorine Compounds.* Most organochlorine compounds – including the chlorinated aromatic hydrocarbons, such as PCBs (that have been widely used in electrical transformers) and DDT (that has been widely used as a powerful pesticide/insecticide) – have proven to be notoriously persistent in the environment. PCBs and DDT are indeed persistent lipophilic chlorinated organic compounds that have been used rather extensively globally -- as noted in the further discussion below. It is also noteworthy that, in various organisms, DDT is slowly transformed to the even more stable and persistent DDE (dichlorodiphenyl dichloroethylene). Thus, such chemicals qualify for classification as part of the group often referred to as the persistent organic pollutants [POPs].

PCBs found use in heat exchange and dielectric fluid; as stabilizers in paints, polymers, and adhesives; and as lubricants in various industrial processes. In the past, PCBs had been used in the manufacture of electrical transformers and capacitors due to their exhibiting low flammability, high heat capacity, and low electrical conductivity – and are virtually free of fire and explosion hazards. PCBs also found several 'open-ended applications' (referred to as such due to relative ease with which the PCB may enter the environment during use, when compared to a 'closed system' for transformer/capacitor use) in products such as plasticizers, surface coatings, ink and dye carriers, adhesives, pesticide extenders, carbonless copy paper, dyes, etc. For instance, they gained widespread use in plasticizers because PCBs are permanently thermoplastic, chemically stable, non-oxidizing, non-corrosive, fire resistant, and are excellent solvents; and they were used in laminating adhesive formulations involving polyurethanes and polycarbonates to prepare safety and acoustical glasses. PCBs have been used in adhesive formulas to improve toughness and resistance to oxidative and thermal degradation when laminating ceramics and metals; and were used in paints and varnishes to impart weatherability, luster, and adhesion. PCBs have also been used in 'nominally closed systems' (due to relative ease with which the PCB may enter the environment during use, when compared to a 'closed system' for transformer/capacitor use) as hydraulic fluids, heat transfer fluids, and lubricants.

DDT, which belongs to the chlorinated insecticide family, was used extensively from the early 1940s to about the early 1970s for agricultural and public health purposes. (It is noteworthy that DDT is still used to a degree of concern in some parts of the world, especially in the developing nations – albeit its use has been banned or curtailed in most industrialized nations.)

POPs have indeed become environmental disaster stories, especially in view of their potential to cause severe health effects. For instance, some PCB congeners and DDT isomers possess an endocrine-disrupting capacity, and are believed to contribute to breast cancer risk and various reproductive and developmental disorders (Colborn *et al.*, 1993; Davis *et al.*, 1993; Dewailly *et al.*, 1994; 1996; Falck *et al.*, 1992; Wolff *et al.*, 1993). Indeed, there are several adverse health effects associated with both PCBs and DDT – as, for instance, tests on animals show that PCBs can harm reproduction and growth, as well as can cause skin lesions and tumors. Furthermore, when PCB fluid is partially burned (as may happen in the event of a transformer fire), PCDDs and PCDFs are produced as byproducts – and these byproducts are indeed much more toxic than the PCBs themselves. For instance, dioxin is associated with a number of health risks, and has been shown to cause cancer of the liver, mouth, adrenal gland, and lungs in laboratory animals; and tests on rats have shown that furans can cause anemia and other blood problems.

Overall, the high lipophilicity and the resistance to biodegradation of most organochlorine compounds allow the bioaccumulation of these chemicals in fatty tissues of organisms and their biomagnification through food chains (Dewailly *et al.*, 1996). As a consequence of humans being located at the top of most food chains, therefore, relatively high levels of these compounds have been found in human adipose tissues, blood lipids, and breast milk fat.

The above list – illuminating the 'two-edged sword' nature of a variety of 'social chemicals' – could be continued for several different families of both naturally occurring and synthetic groups of chemicals or their derivatives. All these types of situations represent very important public health risk management problems that call for appropriate decisions on what toxic insults are tolerable, and also on what levels of exposure may indeed pose significant danger – i.e., 'which/what dose makes the poison?' Invariably, human exposure to chemicals and the consequential health problems are generally a logical derivative of human activities and/or lifestyles. Even so, modern society is not about to abandon the hazard-causing activities and materials – albeit most chemical products are often used in a regulated manner.

1.2. The Nature of Chemical Hazards and Human Response from Exposure to Chemical Substances

There generally are varying degrees of hazards associated with different chemical exposure problem situations. Such variances may be the result of both chemical-specific and receptor-specific factors and/or conditions. Thus, chemical exposure problems may pose different levels of risk, depending on the type of chemicals and extent of contacting by the receptor; the degree of hazard posed by the contacted substance will generally be dependent on several factors, including:

- physical form and chemical composition;
- quantities contacted;
- reactivity;
- toxicity effects; and
- local conditions and environmental setting (e.g., temperature, humidity, and light)

Also, it is noteworthy that the biological effects of two or more toxic substances can be different in nature and degree, in comparison to those of the individual substances acting alone (Williams and Burson, 1985). Chemical interactions between substances may indeed affect their toxicities – 'positively' or 'negatively' – in that, both/all substances may act upon the same physiologic function, or all substances may compete for binding to the same physiologic receptor. In situations where both/all substances have the same physiologic function, their total effects may be simply *additive* (i.e., the simple arithmetic sum of the individual effects), or they may be *synergistic* (i.e., the situation when the total effect is greater than the simple arithmetic sum of the effects of each separately). Under some circumstances, the outcome is a *potentiation* effect – that occurs when an 'inactive' or 'neutral' substance enhances the action of an 'active' one; and in yet other situations, it is one of *antagonism* – when an 'active' substance decreases the effect of another 'active' one.

In general, it is very important to comprehensively/adequately characterize the nature and behavior of all chemicals of potential concern – with careful consideration given to the above-stated and related factors. Ultimately, depending on the numbers and types of chemicals involved, as well as the various receptor-specific factors, significantly different human response could result from any given chemical hazard and/or exposure situation.

1.2.1. CLASSIFICATION OF CHEMICAL TOXICITY

Human response to chemical exposures is as much dependent on the toxicity of the contacted substance as it is on the degree of exposure – among other factors. Chemical toxicity may be characterized differently – particularly, according to the duration and location of exposure to an organism, and/or in accordance with the timing between exposure to the toxicant and the first appearance of symptoms associated with toxicity. The categories commonly encountered in public health risk assessments are identified and contrasted below (Brooks *et al.*, 1995; Davey and Halliday, 1994; Hughes, 1996).

- *Acute vs. Chronic toxicity. Acute toxicity* involves the sudden onset of symptoms that last for a short period of time – usually less than 24 hours, whereas *chronic toxicity* results in symptoms that are of long, continuous duration. In general, the cellular damage that produces the symptoms associated with acute toxicity is usually reversible, whereas there tends to be a permanent outcome from chronic toxicity due to the irreversible cellular changes that have occurred in the organism. Indeed, if cellular destruction and related loss of function are severe, then death of the organism may result.

- *Local vs. Systemic toxicity. Local toxicity* occurs when the symptoms resulting from exposure to a toxicant are restricted or limited to the site of initial exposure, whereas *systemic toxicity* occurs when the adverse effects occur at sites far removed from the initial site of exposure. The latter effects are those elicited after absorption and distribution of the toxicant from its entry point to a distant site. Indeed, toxicants are often absorbed at one site, and then are distributed to distant regions of the receptor through transport within the organism via the blood or lymphatic circulatory systems. In general, it tends to be easier to attribute a toxic response in the case of local toxicity (because the response occurs at the site of first contact between the biological system and the toxicant), in comparison to systemic toxicity.

- *Immediate vs. Delayed toxicity.* *Immediate toxicity* results when symptoms occur rapidly (usually within seconds to minutes) following the exposure of an organism to a toxicant, whereas *delayed toxicity* generally results long after exposure and therefore sometimes adds to the difficulty in establishing a cause-and-effect relationship. (These effects have also been referred to as acute and chronic, respectively.) Indeed, the relationship between causative agents or toxicants and the pathologic symptoms or toxicity is relatively more easily established in the case of 'immediate toxicity'.

In general, an understanding of the time-dependent behavior of a toxicant as related to its absorption, distribution, storage, biotransformation, and elimination is necessary to explain how such toxicants are capable of producing 'acute' or 'chronic' toxicity, 'local' or 'systemic' toxicity, and 'immediate' or 'delayed' toxicity (Hughes, 1996). Consequently, *toxicokinetics* (which is the study of the processes of absorption, distribution, storage, biotransformation, and elimination in relation to toxicants as they interact with living organisms) becomes a very important area of investigation for human exposures to chemicals. Also, *toxicodynamics* (which examines the mechanisms by which toxicants produce unique cellular effects within an organism) is another important area of study in this respect. In any case, whether reversible or irreversible cellular injury occurs upon exposure of an organism to a given toxicant will depend on the duration of exposure as well as the specific toxicokinetic properties of the toxicant (Hughes, 1996).

It is noteworthy that, the terms 'acute' and 'chronic' as applied to toxicity may also be used to describe the duration of exposure – namely, 'acute exposure' and 'chronic exposure'. Indeed, it has been established that acute and chronic exposure to many toxicants will parallel acute and chronic toxicity – albeit, in some cases, acute exposure can lead to chronic toxicity (Hughes, 1996).

1.2.2. FACTORS INFLUENCING CHEMICAL TOXICITY TO HUMANS AND HUMAN RESPONSE TO CHEMICAL TOXICANTS

Within the human body, a chemical may be metabolized, or it may be stored in the fat (as typical of some fat-soluble substances such as DDT that accumulate in the body and become more concentrated as they pass along the food-chain), or excreted unchanged. Metabolism will probably make some chemicals more water-soluble and thus more easily excreted – albeit, sometimes, metabolism increases toxicity (WHO, 1990). In general, the severity of adverse effects resulting from exposures to chemical substances depends on several factors – particularly those annotated in Box 1.1.

Ultimately, the potential for adverse health effects on populations contacting hazardous chemicals can involve any organ system. The target and/or affected organ(s) will depend on several factors – especially the specific chemicals contacted; the extent of exposure (i.e., dose or intake); the characteristics of the exposed individual (e.g., age, gender, body weight, nutritional status, psychological status, genetic make-up, immunological status, susceptibility to toxins, hypersensitivities); the metabolism of the chemicals involved; time of the day during exposure and weather conditions (e.g., temperature, humidity, barometric pressure, season); and the presence or absence of confounding variables such as other diseases (Brooks *et al.*, 1995; Derelanko and Hollinger, 1995; Grisham, 1986; Hughes, 1996).

Box 1.1. Factors potentially influencing human response to toxic chemicals

♦ Nature of toxic chemical (i.e., the types, behavior and effects of the chemical substance and its
  metabolites)
  ‣ Physical/chemical properties of the agent
  ‣ Chemical potency
  ‣ Mechanism of action
  ‣ Interactions between chemicals in a mixture
  ‣ Absorption efficiency (i.e., how easily the chemical is absorbed)
♦ Exposure characteristics
  ‣ Dose (because large dose may mean more immediate effects)
  ‣ Route of exposure
  ‣ Levels and duration of exposure
  ‣ Timing and frequency of exposure
  ‣ Storage efficiency (i.e., accumulation and persistence of chemical in the body)
  ‣ Time of day during exposure (as hormones and enzyme levels are known to fluctuate during the
    course of a day – i.e., circadian rhythms)
  ‣ Environmental factors relating to weather conditions (since temperature, humidity, barometric pressure,
    season, etc., potentially affect absorption rates)
♦ Individual susceptibility
  ‣ Age (since the elderly and children are more susceptible to toxins, and therefore may show different
    responses to a toxicant)
  ‣ Gender (since each sex has hormonally controlled hypersensitivities – and thus females and males
    may exhibit different responses to a toxicant)
  ‣ Body weight (which is inversely proportional to toxic responses/effects)
  ‣ Nutritional status (because, in particular, a lack of essential vitamins and minerals can result in
    impaired cellular function and render cells more vulnerable to toxicants and vice versa – e.g., levels
    of nutrients like iron, calcium, and magnesium can protect against cadmium absorption and retention
    in the human body)
  ‣ Hormonal status (e.g., associated with menopause and pregnancy in women)
  ‣ Psychological status (because stress increases vulnerability)
  ‣ Genetics (because different metabolic rates, related to genetic background, affects receptor responses)
  ‣ Immunological status and presence of other diseases (because health status influences general
    metabolism and may also affect an organism's interaction with toxicants)
  ‣ Anatomical variability (i.e., variations in anatomical parameters between genders, and between
    healthy people *vs.* those with pre-existing 'obstructive' disease conditions)
♦ Hazard controls
  ‣ Source reduction
  ‣ Administrative/institutional and engineering controls
  ‣ Personal protective equipment/clothing
  ‣ Safe work practices
♦ Medical intervention
  ‣ Screening
  ‣ Treatment

1.2.3. DISTRIBUTION AND STORAGE OF TOXICANTS IN THE HUMAN BODY

Distribution of toxicants (following exposure and absorption) occurs when a toxicant is absorbed and then subsequently enters the lymph or blood supply for transport to other regions of the human body. The lymphatic system is a part of the circulatory system and drains excess fluid from the tissues (Davey and Halliday, 1994; Hughes, 1996). Several factors affect the distribution of toxicants to tissues in the human body – most importantly the following:

- Physical and chemical properties/characteristics of the toxicant
- Concentration gradient (between the amount of the toxicant in the blood as compared to the tissue)
- Volume of blood flowing through a specific tissue or organ in the human body
- Affinity of toxicants for specific tissues (i.e., tissue specificity or preference of the toxicant)
- Presence of special structural barriers to slow down toxicant entrance.

Further elaboration of these major factors that affect the distribution of toxicants to human body tissues can be found elsewhere in the literature (e.g., Davey and Halliday, 1994; Hughes, 1996).

Storage results when toxicants accumulate in specific tissues of the human body, or become bound to circulating plasma proteins (Hughes, 1996). The common storage sites/locations for toxicants in the human body tissues include circulating plasma proteins, bones, liver, kidneys, and fat.

1.2.4.   SCOPE OF CHEMICAL HAZARD PROBLEMS: LEAD EXPOSURES AS AN EXAMPLE

Lead is a naturally occurring element that humans have used for a variety of purposes since about the beginning of modern civilization – and various human activities have resulted in the extensive spread of lead throughout the environment. Consequently, lead can now be found in the human physiological system of just about every individual – to the extent that several people have lead levels that are within an order of magnitude of levels associated with adverse health effects (Budd et al., 1998; Flegal and Smith, 1992, 1995). Indeed, lead exposure is an international/global issue – since no contemporary society seems to be completely immune to the presence of lead in their environments. Also, both children and adults are susceptible to health effects from lead exposure – albeit the typical exposure pathways and effects are somewhat different.

Most human exposure to lead occurs through ingestion or inhalation; the general public is less likely to encounter lead that readily enters the human body through the skin (i.e., via dermal exposure). In fact, lead exposure in the general population (including children) occurs primarily through ingestion, although inhalation also contributes to lead body burden and may indeed be the major contributor for workers in lead-related occupations. Almost all inhaled lead is absorbed into the body, whereas between 20% and 70% of ingested lead is absorbed – with children generally absorbing a higher percentage than adults (ATSDR, 1999).

It is noteworthy that, once absorbed into the body, lead may be stored for long periods in mineralizing tissue (viz., teeth and bones) – and then released again into the bloodstream, especially in times of calcium stress (e.g., during pregnancy, lactation, osteoporosis), or calcium deficiency; this would constitute an 'endogenous' exposure. In fact, lead poses a substantial threat to pregnant women and their developing fetuses – because blood lead readily crosses the placenta, putting the developing fetus at risk (especially with respect to the neurologic development of the fetus, since there is no blood-brain barrier at this stage). In general, the mother's blood Pb level serves as an important indicator of risk to the fetus.

The major 'exogenous' sources and associated pathways of lead exposure (excerpted from the US Agency for Toxic Substances and Diseases Registry literature) follows – with further details provided elsewhere (e.g., ATSDR, 1999). In general, occupational lead exposures may occur in the following workers: lead mining, refining, smelting, and manufacturing industry employees; plumbers and pipe fitters; auto mechanics/repairers; glass manufacturers; shipbuilders; printers; plastic manufacturers; law-enforcement officers and military personnel; steel welders or cutters; construction workers; rubber product manufacturers; fuel station attendants; battery manufacturers and recyclers; bridge reconstruction workers; firing range instructors. Environmental lead exposures to the general population (including both children and adults) may occur via lead-containing paint; leaded gasoline; soil/dust near lead industries, roadways, lead-painted homes; plumbing leachate (from pipes or solder); and ceramic ware. Hobbies and related activities are additional sources of lead exposure – that may include glazed-pottery making; target shooting at firing ranges; lead soldering (e.g., electronics); painting; preparing lead shot or fishing sinkers; stained-glass making; car or boat repair; and home remodeling. Other potential sources of lead exposure may occur from use of folk remedies; cosmetics; and tobacco smoking. Further elaboration on some of the major sources is provided below.

- *Lead-Based Paints.* Lead-based paint (LBP) is a primary source of environmental exposure to lead in several places. For example, according to the US Centers for Disease Control and Prevention (CDC), between 83% and 86% of all homes built before 1978 in the United States have LBP in them – and the older the house, the more likely it is to contain LBP and to have a higher concentration of lead in the paint. It is not surprising, therefore, that in 1993 the America Academy of Pediatrics identified LBP as the major source of lead exposure for children.

  In general, as LBP deteriorates, peels off, chips away, is removed (e.g., during renovation activities), or pulverizes because of friction (e.g., in window sills), house dust and surrounding soil may become contaminated with lead (ATSDR, 1999). Subsequently, the lead released into the human environment can then enter the human body through normal hand-to-mouth activities and inhalation. Children are particularly at increased risk from the ingestion of paint chips – and children with pica behavior are at an even greater risk.

- *Automobile Emissions.* Prior to lead being phased out and then banned (in most places around the world) as a gasoline additive, automobile emissions were a major source of exposure to lead. Much of the lead released into the air (especially from automobiles in the past) and in recent times from industrial discharges is deposited onto the land or surface water. In fact, although some industries continue to discharge lead into the air, lead inhalation is no longer the major exposure pathway of significant concern for most developed economies; however, the same cannot be said about most of the developing economies – where lead inhalation exposures remain a significant public health concern. Also, in some other countries, leaded gasoline is still used, and the resulting emissions pose a major public health threat.

  In general, much of the lead discharged into the air is ultimately brought back to the ground surface or surface waters through wet or dry deposition (ATSDR, 1999). Past and present atmospheric emissions have, therefore, contributed to the extensive amounts of lead in soils globally – and areas of high traffic flow or near industrial release sources are likely to have greater concentrations of lead in soils and dust than the more remote areas.

- *Occupational Worker Exposures.* Workers (and indirectly, their families) in up to a hundred or more types of industries may experience occupational exposures to lead. Workers in the lead mining, smelting, refining, and manufacturing industries typically experience the highest and most prolonged occupational exposures to lead. Others at increased risk from lead exposures include workers in brass/bronze foundries, rubber products and plastics industries, soldering, steel welding/cutting operations, battery manufacturing plants, and other manufacturing industries (ATSDR, 1999). Increased risk for occupational lead exposures also occur among construction workers, bridge maintenance and repair workers, municipal waste incinerator workers, pottery/ceramics industry employees, radiator/auto repair mechanics, and people working with lead solder. Furthermore, many so-called 'cottage industries' are actually located in the home or in residential areas – in which case both the workers and families in the homes (or even the neighborhoods) are potentially at risk from direct exposures.

   The major exposure pathways for industrial workers are inhalation and ingestion of lead-laden dust and fumes. It is noteworthy that, occupational exposures can also produce secondary exposures in a worker's family if, for instance, a worker brings home lead-contaminated dust on the skin, clothes, or shoes. Of course, workers can prevent such secondary exposures by showering and/or changing clothing before returning to their homes.

- *Consumer Products.* Drinking water, food, and alcohol can become significant sources of environmental exposure to lead. For instance, lead may occur in drinking water through leaching from lead-containing pipes, faucets, and solder found in plumbing of older buildings; leaching rates accelerate when water is acidic or hot, or when it has been standing in the pipes for extended periods (e.g., overnight). Indeed, faucet fixtures have been shown by a number of researchers to be a significant source of lead exposure (see, e.g., Samuels and Meranger, 1984; Schock and Neff, 1988; Gardels and Sorg, 1989; Lee et al., 1989; Maas and Patch, 1990; Patch et al., 1998). It is noteworthy that, in their study conducted for some US residential water supply systems, Patch et al. (1998) determined that: lead concentrations caused by faucets are significantly greater than lead concentrations that occur in the plumbing line just behind the faucets; bathroom faucets leach more lead than kitchen faucets; lead concentrations increase with standing times (for the water); newer faucets leach more lead than older faucets; and faucets manufactured primarily with sand-casting methods yield significantly higher lead concentrations than those manufactured with other methods.

   Lead may also contaminate food during production, processing, and packaging. Production sources may include soil lead uptake by root vegetables or atmospheric lead deposition onto leafy vegetables; processing and packaging sources of lead in consumer diets may include lead-soldered food cans, and some plastic food wrappers printed with lead-containing pigments. Other sources of food contamination include certain ceramic tableware, lead-glazed pottery, leaded-crystal glassware, certain so-called 'natural' calcium supplements, and bright red and yellow paints on bread bags. Yet additional sources of lead exposure include wine and homemade alcohol that is distilled and/or stored in leaded containers (ATSDR, 1999).

Next, people using paints, pigments, facial cosmetics, or hair coloring with lead or lead acetate also increase their risk from lead exposures. For instance, certain lead-containing cosmetics (e.g., 'surma' and 'kohl') are quite popular in some Asian countries. Also, certain folk remedies may result in significant lead exposures – as, for instance, the ingestion of certain home remedy medicines may expose people to lead or lead compounds. General examples of these types of consumer products include certain Mexican folk remedies; lead-containing remedies used by some Asian communities; and certain Middle Eastern remedies and cosmetics. Lastly, smoking cigarettes or even the breathing of second-hand smoke may increase a person's exposure to lead – because tobacco smoke contains small amounts of lead (ATSDR, 1999).

- *Recreational and Related Activities.* Certain hobbies, home activities, and car repairs (e.g., radiator repair) can contribute to a person's lead exposures. Some of the more common hobbies include glazed-pottery making; artistic painting; stained-glass making; glass or metal soldering; target shooting; electronics soldering; and construction of bullets, slugs, or fishing sinkers; and house renovation involving scraping, remodeling, or otherwise disturbing lead-based paint (ATSDR, 1999).

- *Proximity to Active Release Sources.* People living near hazardous waste sites, lead smelters/refineries, battery recycling/crushing centers or other industrial lead sources may be more easily exposed to lead and other lead-containing chemicals. For instance, industrial and mining activities may result in the release of lead and lead compounds into the air and soil; the releases will invariably be within the exposure setting of the neighboring communities. Local community residents may then be exposed to emissions from these sources through ingestion and/or inhalation of lead-contaminated dust or soils. The typical sources may range in size from large mines and hazardous waste sites to small garages working with old car batteries. Indeed, even abandoned industrial lead sites (such as old mines or lead smelters) may continue to pose significant potential public health hazards.

Once it enters the human body, the absorption and biologic fate of lead depends on a variety of factors. An especially important determinant is the physiologic characteristics of the exposed person – including nutritional status, health, and age. Children and pregnant women, for example, can absorb up to 70% of ingested lead, whereas a general adult typically absorbs up to 20%; most inhaled lead in the lower respiratory tract is absorbed (ATSDR, 1999). The chemical form of lead, or lead compounds, entering the body is also an important factor; organic lead compounds (far rarer since the discontinuation of leaded gasoline additives) are metabolized in the liver, whereas inorganic lead (the most common form of lead) does not undergo such transformation.

Most of the lead that is absorbed into the body is excreted either by the kidney (in urine) or through biliary clearance (ultimately, in the feces). The percentage of lead excreted and the timing of excretion depend on a number of factors. Studies indicate that adults excrete the majority of an absorbed fraction of lead. Ultimately, adults may retain only 1% of absorbed lead, but children tend to retain more than adults; in infants from birth to 2 years, approximately one-third of the total amount of lead is retained

(ATSDR, 1999). Absorbed lead that is not excreted is exchanged primarily among three compartments – namely, blood; soft tissue (liver, kidneys, lungs, brain, spleen, muscles, and heart); and mineralizing tissues (bones and teeth), which typically contain the vast majority of the lead body burden.

Once in the bloodstream, lead is primarily distributed among the three compartments of blood, soft tissue, and mineralizing tissue; the bones and teeth of adults contain more than 95% of the total lead in the body (ATSDR, 1999). In times of stress, the body can mobilize stored lead, thereby increasing the level of lead in the blood. Although the blood generally carries only a small fraction of the total lead body burden, it serves as the initial receptacle of absorbed lead and distributes lead throughout the body, making it available to other tissues (or for excretion). In general, the body tends to accumulate lead over a lifetime and normally releases it very slowly; thus, both past and current elevated exposures to lead increase a person's risks for lead effects. To facilitate public health risk management decisions on lead exposure problems, blood lead level measurements become important because it is about the most widely used measure of lead exposure.

1.3. 'The Dose Makes the Poison' – So, What Dose is Safe Enough?

Current level of knowledge shows that many metals may be considered essential to normal cellular activity and evolutionary development. However, in excess, they may cause toxic responses – as, for example, are noted below for the select list of essential and medically important metals (Berlow et al., 1982; Hughes, 1996).

- Aluminum [Al] – finds medical uses in antacids, and also in dialysis fluids. However, it has an associated toxic effect of dialysis dementia with excesses.
- Cobalt [Co] – found in vitamin B12 is an essential metal, but which could cause polycythemia and cardiomyopathy in excesses. Like iron (Fe^{2+}) in hemoglobin, Co^{2+} serves to hold the large vitamin molecule together and to make it function properly.
- Copper [Cu] – facilitates the synthesis of hemoglobin, but may cause microcytic anemia when present in excessive amounts. Indeed, Cu is required for a variety of roles in the human body, several of which are connected to the use of iron. Although the total amount of Cu in the body is rather small, its deficiency may result in weak blood vessels and bones as well as nerve damage.
- Gold [Au] – finds medical uses in pharmaceuticals (rheumatoid arthritis), but excesses could result in nephropathies.
- Iron [Fe] – important to the formation of RBCs (viz., erythropoiesis), but may cause liver or cardiovascular damage in excesses. In the human body, the iron-containing molecule (called hemoglobin) carries oxygen from the lungs to the rest of the human body. Indeed, small amounts of Fe are found in every tissue cell in molecules that use oxygen. It is noteworthy that, although the actual need for iron is very low (approximately 1 to 1.5 mg/day for a normal person), about ten times as much must be taken in human foods because only a small fraction of the iron passing through the human body is absorbed.
- Lithium [Li] – finds medical uses in pharmaceuticals (depression), but excesses may result in nephropathies and cardiopathies.

- Manganese [Mn] – is an enzyme potentiator, but may cause CNS (central nervous system) disorders and manganese pneumonitis in excess. Indeed, Mn has many essential functions in every cell. However, Mn is also highly neurotoxic and the effects are largely irreversible; consequently, the recommended exposure limits have been lowered drastically in a number of countries in recent years. It is noteworthy that, with its increased industrial use and emissions into the general environment, the harmful effects of Mn cannot be overlooked – and close monitoring is important.
- Molybdenum [Mo] – is an enzyme cofactor, but may cause anemia and diarrhea in excesses. Indeed, Mo is part of several important enzymes.
- Selenium [Se] – is an enzyme cofactor, but subject to cause neuropathies, dermatopathies, decreased fertility, and teratogenesis in excesses.
- Zinc [Zn] – is essential (as Zn^{2+}) for the normal growth of genital organs, wound healing, and general growth of all tissues. It is also associated with the hormone insulin, which is used to treat diabetes. Even so, excess of this essential nutrient is not recommended. It is noteworthy that oysters are believed to be an unusually rich source of Zn.

It is noteworthy that even some of the more 'suspicious' chemicals (e.g. arsenic and chromium) are believed to be essential nutrients in small amounts but are extremely toxic in slightly larger amounts. In general, even the essential elements can be toxic at concentrations that are too high, and yet a deficiency of these same metals can also be harmful to the health of most living organisms – including humans. Thus, it is quite important to make a distinction between the therapeutic and toxic properties of chemicals – recognizing that these properties are sometimes, but not always, indistinguishable except by dose.

Indeed, the 16th century Swiss philosopher and physician-alchemist, Paracelsus, indicated once upon a time that only the dose of a substance determines its toxicity. This notion makes it even more difficult to ascertain the levels that constitute hazardous human exposure to chemicals. But careful application of risk assessment and risk management principles and tools should generally help remove some of the fuzziness in defining the cut-off line between what may be considered a 'safe level' and what apparently is a 'dangerous level' for most chemicals.

1.4.   Managing the Chemical Exposure Problem: The Need for Public Health Risk Assessment

Risk assessment is a tool used to organize, structure, and compile scientific information in order to help identify existing hazardous situations or problems, anticipate potential problems, establish priorities, and provide a basis for policy decisions about regulatory controls and/or corrective actions. A key underlying principle of public health risk assessment is that some risks are tolerable – a reasonable and even sensible view, considering the fact that nothing is wholly safe *per se*. In fact, whereas human exposures to large amounts of a toxic substance may be of major concern, exposures of rather limited extent may be trivial and therefore should not necessarily be a cause for alarm. In order to be able to make a credible decision on the cut-off between what really constitutes a 'dangerous dose' and a 'safe dose', systematic scientific tools – such as those afforded by risk assessment – may be utilized. In this regard, therefore, risk assessment seems to represent an important foundation in the development of effectual public health risk management strategies and policies for populations subjected to toxic chemical insults and assaults.

Indeed, several groups of peoples around the world are exposed to a barrage of chemical constituents on a daily basis – through their use of a variety of consumer products, and via exposure to ambient environmental contaminants. Because of the several health and socioeconomic implications associated with chemical exposure problems, it is important to generally use systematic and technically sound methods of approach in the relevant scientific evaluations. Usually, risk assessments – which allow receptor exposures to be estimated by measurements and/or models – assist in the determination of potential health problems associated with the use of specific consumer products. The exposure assessment component of this process tends to be particularly complicated by the huge diversity in usage and composition of consumer products, and also by the variability of the types and sources of environmental contaminants in the human living and work environments (van Veen, 1996; Vermeire *et al.*, 1993). It is noteworthy that, the huge diversity in consumer products usage and composition typically results in intermittent exposures to varying amounts and types of products that also contain varying concentrations of chemical compounds.

In the application of the various risk assessment tools, one should also become aware of the fact that developments in other fields of study – such as data management systems – are likely to greatly benefit the public health analyst. In fact, an important aspect of public health risk management with growing interest relates to the coupling of environmental/public health data with Geographic Information System (GIS) – to allow effectual risk mapping of a study area with respect to the location and proximity of risk to identified or selected populations. The GIS can process geo-referenced data and provide answers to questions such as the distribution of selected phenomena and their temporal changes; the impact of a specific event on populations; or the relationships and systematic patterns of chemical exposures vis-à-vis observed health trends in a region; etc. Indeed, it has been suggested that, as a planning and policy tool, the GIS technology could be used to 'regionalize' a risk analysis process. Once risks have been mapped using GIS, it may then be possible to match estimated risks to risk reduction strategies, and also to delineate spatially the regions where resources should be invested, as well as the appropriate public health risk management strategies to adopt for various geographical dichotomies.

Overall, the general objectives of a public health risk management program usually will consist of the following typical tasks:

- Determine if a hazardous substance exists and/or may be contacted by humans
- Estimate the potential threat to public health, as posed by the chemical substances of concern
- Determine if immediate response action is required to abate potential problems
- Identify possible remedy or corrective action strategy(s) for the situation
- Provide for public health informational needs of the population-at-risk in the potentially affected community.

Risk assessment provides one of the best mechanisms for completing the tasks involved here. Indeed, a systematic and accurate assessment of risks associated with a given chemical exposure problem is crucial to the development and implementation of a cost-effective corrective action plan. Consequently, risk assessment should generally be considered as an integral part of most public health risk management programs that are directed at controlling the potential effects of chemical exposure problems. The application of risk assessment can indeed provide for prudent and technically feasible and scientifically justifiable decisions about corrective actions that will help protect public health in a most cost-effective manner.

Finally, it should be recognized that, there are several direct and indirect legislative issues that affect public health risk assessment programs in different regions of the world. Differences in legislation amongst different nations (or even within a nation) tend to result in varying types of public health risk management strategies being adopted or implemented. Indeed, legislation remains the basis for the administrative and management processes in the implementation of most public health policy agendas. Despite the good intents of most regulatory controls, however, it should be acknowledged that, in some cases, the risk assessment seems to be carried out simply to comply with the prevailing legislation – and may not necessarily result in any significant hazard or risk reduction.

1.5.   Suggested Further Reading

American Academy of Pediatrics, 1993. Lead poisoning: from screening to primary prevention, *Pediatrics*, 92(1): 176 – 183

American Academy of Pediatrics, 1995. Treatment guidelines for lead exposure in children, *Pediatrics*, 96(1): 155 – 160

Barry, PSI, 1975. A comparison of concentrations of lead in human tissue, *British Journal of Industrial Medicine*, 32:119 – 139

Barry, PSI, 1981. Concentrations of lead in the tissues of children, *British Journal of Industrial Medicine*, 38: 61 – 71

Bates, DV, 1994. *Environmental Health Risks and Public Policy*, University of Washington Press, Seattle, Washington

British Medical Association, 1998. *Health & Environmental Impact Assessment (An Integrated Approach)*, British Medical Association (BMA), England, UK

Christen, K., 2001. The arsenic threat worsens, *Environmental Science & Technology*, 35(13): 286A – 291A

Francis, BM, 1994. *Toxic Substances in the Environment*, J. Wiley, New York

Freeze, RA, 2000. *The Environmental Pendulum*, University of California Press, Berkeley, CA

Gilliland, FD, KT Berhane, *et al.*, 2002. Dietary magnesium, potassium, sodium, and children's lung function, *American Journal of Epidemiology*, 155(2): 125 – 131

Mitchell, P. and D. Barr, 1995. The nature and significance of public exposure to arsenic: a review of its relevance to South West England, *Environmental Geochemistry and Health*, 17: 57 – 82

Pocock, SJ, M. Smith, and P. Baghurst, 1994. Environmental lead and children's intelligence: a systematic review of the epidemiological evidence, *British Medical Journal*, 309: 1189 – 97

USEPA, 2001. *Providing Solutions for a Better Tomorrow – reducing the risks associated with lead in soil*, EPA/600/F-01/014, Office of Research and Development, US Environmental Protection Agency (USEPA), Washington, DC

Chapter 2

HUMAN EXPOSURE TO CHEMICALS

Human exposure to chemicals is virtually an inevitable part of life in this day and age. Such exposures may occur via different human contact sites and target organs, and also under a variety of exposure scenarios. The contact sites represent the physical areas of chemical contacting, and the target organ are the internal body organs that tend to transport, process, and/or store the absorbed chemicals; an exposure scenario is a description of the activity that brings a human receptor into contact with a chemical material, product, or medium. Chemical exposure investigations, consisting of the planned and managed sequence of activities carried out to determine the nature and distribution of hazards associated with potential chemical exposure problems, can be designed and used to address human exposure and response to chemical toxicants.

Several characteristics of the chemicals of concern as well as the human contact sites will provide an indication of the critical features of exposure; these will also provide information necessary to determine the chemical's distribution, uptake, residence time, magnification, and breakdown to new chemical compounds. In particular, the physical and chemical characteristics of the chemicals as well as the target organs involved can significantly affect the intake, distribution, half-life, metabolism, and excretion of such chemicals by potential receptors. This chapter looks at the major human contact sites, target organs, and exposure scenarios that can be expected to become key players in the assessment of human exposure to and response from chemical hazards.

2.1. An Overview of Human Contact Sites and Target Organs Most Susceptible to Chemical Exposures

The major routes of both intentional and accidental exposure of chemicals to humans (and indeed other living organisms) include the following (Brooks *et al.*, 1995; Homburger *et al.*, 1983; Hughes, 1996):

- the skin – i.e., the percutaneous route;
- the lungs – i.e., the inhalation-respiration pulmonary route; and
- the mouth – i.e., the oral route

Minor routes of exposure may consist of rectal, vaginal, and parenteral (i.e., intravenous or intramuscular, a common means for the administration of drugs or toxic substances in test subjects) (Homburger *et al.*, 1983). Indeed, the manner in which a chemical substance is taken up and/or enters the complex physiologic system of an organism is very much dependent on the physical and chemical properties of the contacted substance. For instance, the pulmonary system is most likely to take in vapor-phase and very fine, respirable particulate matter; non-respirable particulates usually enter the body via the oral route; and absorption through the skin is possible for most physical forms, but especially from contacts with liquids and adhering solid materials.

In general, upon human exposure to chemical substances, the contacted material is often absorbed into the receptor bloodstream via three primary routes – i.e., inhalation, oral ingestion, and dermal/skin contact. The three corresponding primary physiological routes of absorption associated with the human body are comprised of the respiratory system; the digestive system; and the percutaneous (i.e., through the skin). Thus, an awareness of these anatomical and physiological characteristics associated with each route of absorption is important as a first step in understanding how toxicants enter the human body.

2.1.1. FUNDAMENTALS OF HUMAN PHYSIOLOGY

The most important physiological elements/organs crucial to the study of human exposure to chemicals are annotated below – and discussed in greater details elsewhere (e.g., Berlow *et al.*, 1982; Berne and Levy, 1993; Brum *et al.*, 1994; Davey and Halliday, 1994; Dienhart, 1973; Frohse *et al.*, 1961; Guyton, 1968, 1971, 1982, 1986; Hughes, 1996; Scanlon and Sanders, 1995; Willis, 1996).

- *The Skin.* The skin is a highly organized, heterogeneous, and multi-layered organ of the human body. It serves as a protective layer that impedes the entry of harmful agents and chemicals into the human body. Indeed, the skin is more than just an inert barrier, since it supports a multitude of life functions – and it should be viewed as a dynamic, living tissue whose permeability characteristics are susceptible to change.

   The skin, which is the largest organ in the body, consists of two primary layers – the nonvascular *epidermis* layer, and the highly vascularized *dermis* layer – and is also separated from deeper body tissues by a *subcutaneous* layer, called the hypodermis (Figure 2.1). The outermost layer of the epidermis – called the stratum corneum – is thought to provide the major barrier to the absorption into the circulation of most substances deposited on the skin surface. Below this layer lies the viable epidermis that contains enzymes that metabolize certain penetrating substances – albeit enzymes may also be active in the stratum corneum.

   The *vascular system*, representing the blood stream, is of concern for the distribution of absorbed chemical substances; this extends through the dermis and subcutaneous layers, but not the epidermis. Consequently, the skin functions as a barrier to the entry of many toxic substances into the human body. When toxicants become localized in the epidermis, local toxicity (rather than systemic toxicity) is the likely result; this is because the epidermis is avascular (i.e., having no blood vessels) – and without a transport mechanism, toxicants cannot be distributed to other areas of the body where systemic toxicity may result (Hughes, 1996).

Several routes of absorption are possible through the skin – the most common being the cutaneous adsorption of a toxicant, followed by passive diffusion through the epidermis into the dermis where the toxicant might enter a blood vessel. Indeed, passage into the dermis is enhanced if the toxicant enters a sweat gland or hair follicle; since these structures originate in the dermis and penetrate through the epidermis, this route effectively bypasses the protective barrier provided by the epidermis (Hughes, 1996). By far, the greatest area of the skin is composed of the epidermal cell layer, and most toxicants absorbed through the skin do so through epidermal cells – albeit, despite their much smaller total areas, cells in the follicular walls and in sebaceous glands are much more permeable than epidermal cells.

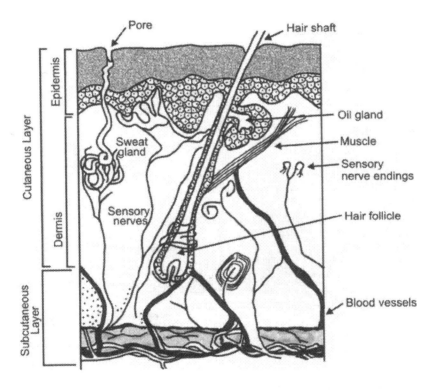

Figure 2.1. Illustrative sketch of the general structure of the human skin
(as a dermal contact exposure route for chemical materials)

The permeability coefficient (K_p) is a key parameter in estimating dermal absorption – albeit the extent of absorption of a compound in humans is often dependent on the anatomical site to which the compound is applied. The permeability of the skin to a toxic substance is indeed a function of both the substance and the skin. It is noteworthy that the K_p values can only be calculated from steady-state absorption rates that usually occur only after prolonged exposure (minutes to hours) to an infinite dose. Calculation of exposure to aqueous solutions of chemicals during swimming and bathing are instances where permeability constants can be used to approximate percutaneous absorption (USEPA, 1992).

- *The Respiratory System.* The human respiratory system is comprised of a series of organs and body parts – most importantly the mouth, the nose, the trachea, and the lungs (Figure 2.2). Each region of the respiratory system contributes a unique functional component that prohibits or limits the ability of toxicants to enter the body. In any case, the respiratory system, with its close anatomical and physiological association with the cardiovascular system, constitutes one of the prime sites for absorption and distribution of toxicants (Hughes, 1996). The pulmonary system is indeed the site of entry for numerous toxicants in the human living and work environments.

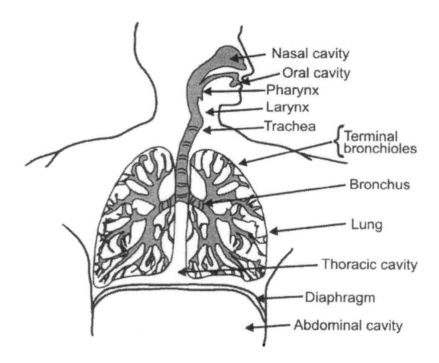

Figure 2.2. Illustrative sketch of the general structure of the human respiratory system (as an inhalation exposure route for chemical materials)

Indeed, the lungs represent the site of respiration in the human body. In general, inhaled air enters the lungs, where it encounters a huge area of tissue that allows the exchange of gas in the lungs with gas in the blood. If the lung tissue is damaged, the alveoli walls may be destroyed (causing emphysema) or scar tissue may form in the bronchioles (causing chronic bronchitis). (The alveoli are the small air sacs in the lungs through which oxygen passes from the lungs into the bloodstream – partly absorbed into red blood cells and carried to the rest of the body; carbon dioxide passes from the bloodstream into the lungs – to be exhaled.)

Damage to the lungs may be caused by various factors – including recurrent infections, severe asthma, smoking, and air pollution problems. Indeed, certain air pollutants have a direct effect on the ability of the human body to transport oxygen; for example, lead poisoning interferes with the body's ability to manufacture hemoglobin (which carries oxygen in the red blood cells) – and this can produce severe chronic anemia. It is noteworthy that, the 'suspended particles' in air pollution (i.e., soot, dust, and smoke) tend to present a unique sort of problem. Such particles tend to collect on the walls of the bronchial tubes and interfere with the ability of the lungs to get rid of irritants – due to interference with gas exchanges. Other particles – for example, asbestos and some other industrial fibers and particulates – have the ability to cause cancer. In general, only particulate matter of size $\leq 10\mu m$ (referred to as PM10 or PM-10) can usually be transported through the upper respiratory system into the lungs.

- *The Digestive System.* The general features of the human gastrointestinal tract – including the mouth, pharynx, esophagus, stomach, small intestines, large intestine, rectum, and the anus – are shown in Figure 2.3. In general, the mouth receives and chews food; the esophagus carries the food to the stomach; the stomach liquefies the food and begins digestion; the small intestine does the major job of breaking down the food molecules into smaller units – which can then be absorbed into the bloodstream; and the large intestine removes water and forms the feces from waste food matter.

  The *small intestine* is the most important organ for absorbing food (and of course toxic chemicals as well, if present) along the gastrointestinal tract. Although absorption into the bloodstream can occur in the stomach (which is the muscular sac that stores food and other materials taken through the mouth), this entry route is generally considered minor relative to that which occurs in the small intestine. For materials that remain undigested and/or unabsorbed in the body, the *large intestine* serves as the final major organ of the gastrointestinal tract whose function is to store and concentrate feces to be excreted later.

- *The Circulatory System.* The distribution and removal of chemicals after they are absorbed, or after entering the human body is a very important aspect of toxicological studies. The distribution of chemical toxins occurs through the circulatory or vascular system (whereas removal may occur through the kidneys). The human circulatory system, therefore, represents a very important route of distribution.

- *The Liver.* The liver may be considered as a filter for the blood and also as a control system for the levels of chemicals (including important nutrients such as sugars), as well as a place where toxic substances can be transformed via detoxification reactions. The liver, therefore, represents an organ system most important in facilitating chemical transformations in the human body.

- *The Kidneys.* When blood passes through the kidneys, substances not needed by the body (including toxic substances and their metabolites) are generally separated and excreted in the urine. The kidneys, therefore, constitute the organ facilitating excretions from the body. Indeed, the kidneys contribute a large share of the work required to eliminate toxic substances from the human body.

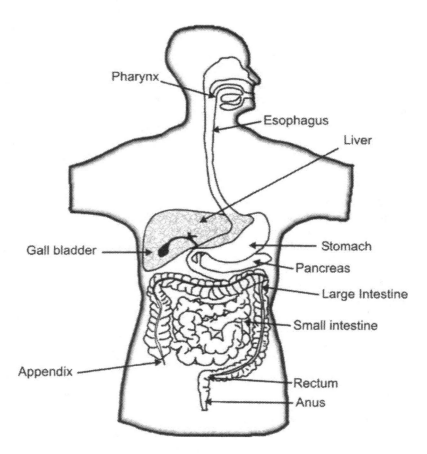

Figure 2.3. Illustrative sketch of the general structure of the human digestive system
(as an ingestion exposure route for chemical materials)

In general, chemical contacting or exposure may occur via the first three of the above-listed physiological elements, whereas the transport and fate of the chemicals in the human body (i.e., pertaining to the distribution and removal of chemicals entering humans) will generally be dictated by the latter three. These organ systems do indeed represent primary routes of chemical absorption by the human body.

2.1.2. TARGET ORGAN TOXICITY

Target organ toxicity is defined as the adverse effects or disease states manifested in specific organs in the human body. The major disease states from toxicity in human body organs include the following (Hughes, 1996):

- Hematotoxicity (i.e., blood cell toxicity) – occurs when too many or too few of the different blood cell components (i.e., erythrocytes, leukocytes, and thrombocytes) are present, or structural anomalies occurring in blood components interfere with normal functioning. Hematotoxins alter the general characteristics of blood cells to produce symptoms.
- Hepatotoxicity (i.e., toxic effects in the liver) – occurs when liver toxicants (typically characterized as being cytotoxic or cholestatic) enter the liver. Cytotoxic mechanisms affect hepatocytes and are responsible for different types of liver injury, and cholestatic mechanisms affect the flow of bile.
- Nephrotoxicity (i.e., toxic effects in the kidney) – occurs when nephrotoxins are present. The pathologies associated with nephrotoxicity are dependent on the anatomical region of the nephron affected by the toxicant.
- Neurotoxicity (i.e., toxic effects to the nervous systems) – occurs when toxicants interrupt the normal mechanisms of neuronal communication. Neurotoxins are known to alter neurons in the nervous system; they interfere with the communication ability of neurons, impeding receptor or motor neuron signaling and central nervous system (CNS) functioning.
- Dermatotoxicity (i.e., adverse effects produced by toxicants in the skin) – occurs when dermatotoxins are present at skin contact sites. Toxic skin reactions are diverse and may involve any one or several combinations of the skin components.
- Pulmonotoxicity (i.e., disease states in the respiratory system resulting from inhalation of toxicants) – occurs when pulmonotoxins enter the respiratory system. Ultimately, concern is crucial if/when toxic responses results in a decreased ability for the lung to exchange oxygen and carbon dioxide across the lung membrane walls.

Indeed, toxicity is unique for each organ, since each organ is an assemblage of tissues, and each tissue is a unique assemblage of cells. Consequently, under the influence of a chemical toxicant, each organ will manifest different disease states (from toxicity) that depend on the structural and functional characteristics of the cells present (Brooks *et al.*, 1995; Davey and Halliday, 1994; Hughes, 1996).

In general, human exposure to chemical constituents present in consumer products and/or in the environment can produce several adverse effects and/or specific diseases. For example, human exposures to certain chemicals may result in such diseases as allergic reaction, anemia, anxiety, asthma, blindness, bronchitis, various cancers, contact dermatitis, convulsions, embryotoxicity, emphysema, pneumonoconiosis, heart disease, hepatitis, obstructive lung disease, memory impairment, nephritis, and neuropathy. In fact, human exposures to chemicals can cause various severe health impairment or even death if intake occurs in sufficiently large amounts. Also, there are those chemicals of primary concern that can cause adverse impacts, even from limited exposures.

2.2. General Types of Human Exposures

Human populations may become exposed to a variety of chemicals via several different exposure routes – represented primarily by the inhalation, ingestion, and dermal exposure routes (Figure 2.4). Indeed, human chemical uptake occurs mainly through the skin (from dermal contacts), via inhalation (from vapors/gases and particulate matter), and/or by ingestion (through oral consumptions). The likely types and significant categories of human exposures to a variety of chemical materials that could affect public health risk management decisions are annotated below (Al-Saleh and Coate, 1995; Corn, 1993; OECD, 1993).

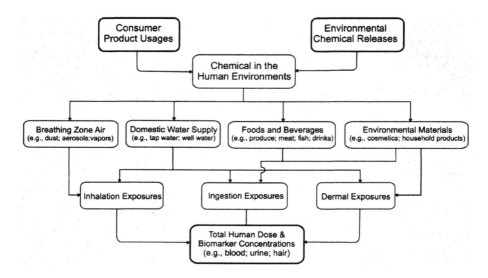

Figure 2.4. Major types of human exposures to chemicals:
a simplified 'total' human exposure conceptual model

- *Skin Exposures.* The major types of dermal exposures that could affect public health risk management decisions consist of dermal contacts with chemicals present in consumer products or in the environment, and also dermal absorption from contaminated waters. Dermal exposures that results from the normal usage of consumer products may be expressed by the following form of generic relationship:

$$\text{Dermal Exposure} = \frac{\{[\text{CONC}] \times [\text{PERM}] \times [\text{AREA}] \times [\text{EXPOSE}]\}}{[\text{BW}]} \quad (2.1)$$

where: CONC is the concentration of material (in the medium of concern); PERM is the skin permeability constant; AREA is the area of exposed skin (in contact with the medium); EXPOSE is the exposure duration (i.e., duration of contact); BW is the average body weight.

In general, fat-soluble chemical substances, and to some extent the water-soluble chemicals are absorbed through intact skins. Overall, skin characteristics such as sores and abrasions may facilitate skin/dermal uptakes. Environmental factors such as temperature and humidity may also influence skin absorption of various chemicals. Furthermore, the physical state (i.e., solid, liquid or gas), acidity (i.e., pH), as well as the concentration of the active ingredient of the contacted substance will generally affect the skin absorption rates/amounts.

- *Oral Exposures.* Ingestion takes place when chemical-containing food materials, medicines, etc. are consumed via the mouth or swallowed. The major types of chemical ingestion exposures that could affect public health risk management decisions consist of the oral intake of contaminated materials (e.g., soils intake by children exercising pica behavior), food products (e.g., plant products, fish, animal products, and mother's milk), and waters. Ingestion exposures that results from the normal usage of consumer products may be expressed by the following form of generic relationship:

$$\text{Oral Exposure} = \frac{\{[\text{CONC}] \times [\text{CONSUME}] \times [\text{ABSORB}] \times [\text{EXPOSE}]\}}{[\text{BW}]} \qquad (2.2)$$

where: CONC is the concentration of material (i.e., the concentration of the contaminant in the material ingested – e.g., soil, water, or food products such as crops, and dairy/beef); CONSUME is the consumption amount/rate of material; ABSORB is the per cent (%) absorption (i.e., the gastrointestinal absorption of the chemical in solid or fluid matrix); EXPOSE is the exposure duration; BW is the average body weight.

The total dose received by the potential receptors from chemical ingestions will, in general, be dependent on the absorption of the chemical across the gastro-intestinal (GI) lining. The scientific literature provides some estimates of such absorption factors for various chemical substances. For chemicals without published absorption values and for which absorption factors are not implicitly accounted for in toxicological parameters, absorption may conservatively be assumed to be 100%.

- *Inhalation Exposures to Volatiles.* Exposures to volatile chemical materials that results from the normal usage of consumer products may be expressed by the following form of generic relationship:

Inhalation Exposure to Volatiles

$$= \frac{\{[\text{VAPOR}] \times [\text{INHALE}] \times [\text{RETAIN}] \times [\text{EXPOSE}]\}}{[\text{BW}]} \qquad (2.3)$$

where: VAPOR is the vapor phase concentration of material (i.e., the concentration of chemical in the inhaled air); INHALE is the inhalation rate (of the exposed individual); RETAIN is the lung retention rate (i.e., the amount retained in the lungs); EXPOSE is the exposure duration (i.e., the length of exposure of the exposed individual); BW is the average body weight (of the exposed individual).

It is noteworthy that, showering – which represents one of the most common and universal human activities – generally encompasses a system that promotes release of VOCs from water (due to high turbulence, high surface area, and small droplets of water involved). In fact, contemporary studies have shown that risks from inhalation while showering can be comparable to – if not greater than – risks from drinking contaminated water (Jo *et al.*, 1990a, 1990b; Kuo *et al.*, 1998;

McKone, 1987; Richardson *et al.*, 2002; Wilkes *et al.*, 1996). Thus, this exposure scenario represents a particularly important one to evaluate in a public health risk assessment, as appropriate. In this case, the concentration of any contaminants in the shower air is assumed to be in equilibrium with the concentration in the water.

Also, considering the fact that the degree of dilution in the indoor air of a building is generally far less than the situation outdoors, contaminant vapors entering/infiltrating into a building structure may represent a significant risk to occupants of the building. In fact, the migration of subsurface contaminant vapors into buildings can become a very important source of human exposure via the inhalation route. As appropriate, therefore, a determination of the relative significance of vapor transport and inhalation as a critical exposure scenario should be given serious consideration during the processes involved in the characterization of a chemical exposure problem, and in establishing environmental quality criteria and/or public health goals. Risk assessment methods can generally be used to make this determination – i.e., as to whether or not vapor transport and inhalation represent a significant exposure scenario worth focusing on in a given study. For example, a risk characterization scenario involving exposure of populations to vapor emissions from cracked concrete foundations/floors can be determined on such basis, in order for responsible risk management and/or mitigative measures to be adopted.

● *Inhalation Exposures to Particulate Matter.* Exposures to inhalable chemical particulates that results from the normal usage of consumer products may be expressed by the following form of generic relationship:

Inhalation Exposure to Particulates

$$= \frac{\{[PARTICLE] \times [RESPIRABLE] \times [INHALE] \times [ABSORB] \times [EXPOSE]\}}{[BW]}$$

(2.4)

where: PARTICLE is the total aerosol or particulate concentration of material; RESPIRABLE is the % of respirable material; INHALE is the inhalation rate; ABSORB is the % absorbed; EXPOSE is the exposure duration; BW is the average body weight.

It is noteworthy that, in general, only particulate matter of size $\leq 10\mu m$ (referred to as PM-10 or PM10) can usually be transported through the upper respiratory system into the lungs.

In addition to the above major exposure situations, it must be acknowledged that accidental exposures may also occur via the same routes (i.e., from dermal contact, oral ingestion, and/or inhalation). Furthermore, chemical vapors or aerosols may be absorbed through the lungs.

Indeed, the analysis of potential human receptor exposures to chemicals found in our everyday lives and in the human living and work environments often involves several complex issues. In all cases, however, the exposures are generally evaluated via the calculation of the average daily dose (ADD) and/or the lifetime average daily dose (LADD). Typically, the carcinogenic effects (and sometimes the chronic non-carcinogenic effects) associated with a chemical exposure problem involve estimating the LADD; for non-carcinogenic effects, the ADD is usually used. The ADD differs from the LADD in that the former is not averaged over a lifetime; rather, it is the average of the daily dose pertaining to the actual number of days of exposure. The maximum daily dose (MDD) will typically be used in estimating acute or subchronic exposures. Details of the requisite algorithms for estimating potential human exposures and intakes are elaborated in Chapter 6.

2.3. The Nature of Chemical Exposure Problems

Public and/or consumer exposures may occur via the inhalation, dermal, and/or oral routes. A wide variety of *potential* exposure patterns can be anticipated from human exposures to chemicals.   As an illustrative example, a select list of typical and commonly encountered exposure scenarios in relation to environmental contamination problems will include the following (Asante-Duah, 1998; HRI, 1995):

- Inhalation Exposures
  - Indoor air, resulting from potential receptor exposure to contaminants (both volatile constituents and fugitive dust) present in indoor ambient air.
  - Indoor air, resulting from potential receptor exposure to volatile chemicals in domestic water that may volatilize inside a house (e.g., during hot water showering) and contaminate indoor air.
  - Outdoor air, resulting from potential receptor exposure to contaminants (both volatile constituents and fugitive dust) present in outdoor ambient air.
  - Outdoor air, resulting from potential receptor exposure to volatile chemicals in irrigation water, or other surface water bodies, that may volatilize and contaminate outdoor air.
- Ingestion Exposures
  - Drinking water, resulting from potential receptor oral exposure to contaminants in domestic water used for drinking or cooking.
  - Swimming, resulting from potential receptor exposure (via incidental ingestion) to contaminants in surface water bodies.
  - Incidental soil ingestion, resulting from potential receptor exposure to contaminants in dust and soils.
  - Crop consumption, resulting from potential receptor exposure to contaminated foods (such as vegetables and fruits produced in household gardens that used contaminated soils, groundwater, or irrigation water in the cultivation process).
  - Dairy and meat consumption, resulting from potential receptor exposure to contaminated foods (such as locally grown livestock that may be contaminated from domestic water supplies, or from feeding on contaminated crops, or from contaminated air and soils).
  - Seafood consumption, resulting from potential receptor exposure to contaminated foods (such as fish and shellfish harvested from contaminated waters, or that have been exposed to contaminated sediments, and consequently have bioaccummulated toxic levels of chemicals in the edible portions).
- Dermal Exposures
  - Showering, resulting from potential receptor exposure (via skin absorption) to contaminants in domestic water supply.

- ♦ Swimming, resulting from potential receptor exposure (via skin absorption) to contaminants in surface water bodies.
- ♦ Direct soils contact, resulting from potential receptor exposure to contaminants present in outdoor soils.

These types of exposure scenarios will typically be evaluated as part of an exposure assessment component of a public health risk management program. It should be emphasized, however, that this listing is by no means complete, since new exposure scenarios are always possible for case-specific situations – albeit this demonstrates the multiplicity and inter-connections of numerous pathways via which populations may be exposed to chemical constituents.

Whereas the above-listed exposure scenarios may not all be relevant for every chemical exposure problem, a number of other exposure scenarios may have to be evaluated for the particular local conditions of interest. In any case, once the complete set of potential exposure scenarios has been fully determined, the range of critical exposure pathways can be identified. In general, careful consideration of the types and extent of potential human exposures, combined with hazard assessment and exposure-response information, is necessary for the conduct of a credible human health risk assessment. The hazard assessment for a consumer product or component thereof is associated with any human health effects, and the exposure-response assessments involve an examination of the relationship between the degree of exposure to a product or component and the magnitude of any specific adverse effect(s). Additionally, the exposure assessment (which is very critical to determining potential risks) requires realistic data to determine the extent of possible skin, inhalation, and ingestion exposures to products and components (Corn, 1993). Ultimately, the resulting information generated can then be used to support the design of cost-effective public health risk management programs.

2.3.1. HUMAN EXPOSURES TO AIRBORNE CHEMICAL TOXICANTS

Airborne pollutants can generally be transported over long distances – and this could result in the deposition of pollutants very far away from where they were produced or used. For example, high levels of pesticides (such as DDT, chlordane, and toxaphene) have been found to be present in beluga whales from the Arctic, where such chemicals were not used. Thus, airborne chemical toxicants can very well impact population groups that are geographically widely dispersed. As a consequence, air pollution presents one of the greatest risks to human health globally. The long list of health problems caused or aggravated by air pollution includes respiratory problems, cancer, and eye irritations (Holmes *et al.*, 1993).

Chemical concentration in air – represented by the ground-level concentration (GLC) – is a function of the source emission rate and the dilution factor at the points of interest (usually the potential receptor location). Of particular interest are air emissions from chemical release sources that often represent a major source of human exposure to toxic or hazardous substances. The emissions of critical concern relate to volatile organic chemicals (VOCs), semi-VOCs (SVOCs), particulate matter, and other chemicals associated with wind-borne particulates such as metals, PCBs, dioxins, etc. Volatile chemicals may be released into the gaseous phase from such sources as landfills, surface impoundments, contaminated surface waters, open/ruptured chemical tanks or containers, etc. Also, there is the potential for subsurface gas movements into underground structures such as pipes and basements, and eventually into indoor air. Additionally, toxic chemicals adsorbed to soils may be transported to the ambient air as particulate matter or fugitive dust. Finally, several consumer products and materials in the human living and work environments will tend to release potentially hazardous chemicals into the human breathing zone/space.

Chemical release sources can indeed pose significant risks to public health as a result of possible airborne release of particulate matter laden with toxic chemicals, and/or volatile emissions. In fact, even very low-level air emissions could pose significant threats to exposed individuals, especially if toxic or carcinogenic contaminants are involved. Consequently, there is increased concern and attention on the assessment of public health risks associated with chemical releases into air. Of particular concern, it should be recognized that certain air pollutants have a direct effect on the ability of the human body to transport oxygen (Berlow et al., 1982). For example, lead poisoning interferes with the body's ability to manufacture hemoglobin (which carries oxygen in the red blood cells) – and this can produce severe chronic anemia; carbon monoxide replaces oxygen on hemoglobin molecules – and thus reduces the efficiency with which the blood transfers oxygen to the cells. Also, some toxic gases (such as the oxides of sulfur and nitrogen, and also ozone) that are often found in the smog of cities as a result of industrial pollution can present major health hazards; nitrogen and sulfur oxides form very strong acids when they dissolve in the water of lining membranes – and these gases can cause damage to the bronchial tubes and alveoli.

2.3.2. WATER POLLUTION PROBLEMS AND HUMAN EXPOSURES TO CHEMICALS IN WATER

Historically, surface waters were among the first environmental media to receive widespread attention with regards to environmental/chemical pollution problems. This attention was due in part to the high visibility and extensive public usage of surface waters, as well as for their historical use as 'waste receptors' (Hemond and Fechner, 1994). Surface water contamination may also result from contaminated runoff and overland flow of chemicals (from leaks, spills, etc.) and chemicals adsorbed to mobile sediments. In addition, it must be recognized that groundwater resources are just about as vulnerable to environmental/chemical contamination; groundwater contamination may result from the leaching of toxic chemicals from contaminated soils, or the downward migration of chemicals from lagoons and ponds. Indeed, it is worth mentioning the fact that groundwater is extensively used by public water supply systems in several places around the world; thus, it is always important to give very close attention to the seemingly 'hidden' groundwater pollution problems in chemical exposure evaluation programs. In general, human exposure to chemicals in contaminated water problems should address both intakes resulting from water ingestion and also from water dermal contacting and inhalation of the volatile constituents.

Another major 'hidden' concern of water quality management that should not be overlooked relates to the issue of eutrophication – i.e., the nutrient enrichment of the water and the bottom of surface water bodies. Indeed, human-made eutrophication has been considered one of the most serious global water quality problems of surface water bodies during the past few decades. Increasing discharges of domestic and industrial wastewater, the intensive use of crop fertilizers, the rise in airborne pollution, and the natural mineralization of streamflows can be seen as some of the primary causes of this undesirable phenomenon. Typical symptoms of eutrophication include, among other things, sudden algal blooms, water coloration, floating water-plants and debris, and excretion of toxic substances that causes taste and odor problems in the production of drinking water, and sometimes fish kills. These symptoms can result in limitations of water use for domestic, agricultural, industrial, or recreational purposes. In addition, the nitrates coming from fertilizer applications tend to eventually become drinking water hazards, especially because the nitrate ion (NO_3^-) is reduced to the nitrite ion (NO_2^-) in the human body following the consumption of the nitrate-containing water – and nitrite destroys the ability of hemoglobin to transport oxygen to the cells. Indeed, high nitrate concentrations in drinking water are particularly dangerous to small infants.

2.3.3. CONTAMINATED SOIL PROBLEMS AND HUMAN EXPOSURES TO CHEMICALS ON LAND

Contaminated land issues are a complex problem with worldwide implications, and risk to public health arising especially from soils at contaminated lands is a matter of grave concern. Contaminated soils may arise in a number of ways – many of which are the result of manufacturing and other industrial activities or operations. In fact, much of the soil contamination problems encountered in a number of places are the result of waste generation associated with various forms of industrial activities. In particular, the chemicals and allied products manufacturers are generally seen as the major sources of industrial hazardous waste generation that culminates in contaminated soil problems. These industries generate several waste types, such as organic waste sludge and still bottoms (containing chlorinated solvents, metals, oils, etc.); oil and grease (contaminated with polychlorinated biphenyls [PCBs], polyaromatic hydrocarbons [PAHs], metals, etc.); heavy metal solutions (of arsenic, cadmium, chromium, lead, mercury, etc.); pesticide and herbicide wastes; anion complexes (containing cadmium, copper, nickel, zinc, etc.); paint and organic residuals; and several other chemicals and byproducts that have the potential to contaminate lands.

In addition to the above situations, several different physical and chemical processes can also affect contaminant migration from contaminated soils. Thus, contaminated soils can potentially impact several other environmental matrices. For instance, atmospheric contamination may result from emissions of contaminated fugitive dusts and volatilization of chemicals present in soils; surface water contamination may result from contaminated runoff and overland flow of chemicals (from leaks, spills, etc.) and chemicals adsorbed to mobile sediments; groundwater contamination may result from the leaching of toxic chemicals from contaminated soils, or the downward migration of chemicals from lagoons and ponds; etc. Consequently, human exposures to chemicals at contaminated lands may occur in a variety and multiplicity of ways – including via the following more common pathways:

- Direct inhalation of airborne vapors and also respirable particulates
- Deposition of airborne contaminants onto soils, leading to human exposure via dermal absorption or ingestion
- Ingestion of food products that have been contaminated as a result of deposition onto crops or pasture lands, and subsequent introduction into the human food chain
- Ingestion of contaminated dairy and meat products from animals consuming contaminated crops or waters

- Deposition of airborne contaminants onto waterways, uptake through aquatic organisms, and eventual human consumption of impacted aquatic foods
- Leaching and runoff of soil contamination into water resources, and consequential human exposures to contaminated waters in a water supply system.

Contaminated lands therefore represent a potentially long-term source for human exposure to a variety of chemical toxicants.

2.3.4. HUMAN EXPOSURES TO CHEMICALS IN FOODS

Food products represent a major source of human exposure to chemicals, even if in incrementally minute amounts. For example, a number of investigations have shown that much of the seafood coming from most locations contains detectable levels of environmental pollutants (such as Pb, Cr, PCBs, dioxins and pesticides). Also, chemicals such as tartrazine, a previously revered food preservative that was widely used in some countries, has now been determined to cause allergies in significant numbers of human populations. Consequently, there is a clear move away from the use of such chemicals – as, for example, is demonstrated by the fact that 'chips' and indeed many other food items sold in South Africa now proudly display on the packaging that the products are 'tartrazine-free'.

In fact, because of the potential human exposure to the variety of toxic/hazardous chemicals, it is very important to understand the potential human health risks associated with these exposures and the potential public health implications of such chemicals being present in the food sources. In general, human dietary exposure to chemicals in food (and indeed similar consumable products) depends both on food consumption patterns and the residue levels of a particular chemical on/in the food or consumer product, generally expressed by the following conceptual relationship (Driver *et al.*, 1996; Kolluru *et al.*, 1996):

$$\text{Dietary Exposure} = f(\text{Consumption, Chemical concentration}) \tag{2.5}$$

Typically, multiplying the average consumption of a particular food product by the average chemical concentration on/in that food provides the average ingestion rate of that chemical from the food product. In reality, however, estimation of dietary exposure to chemicals – such as pesticides or food additives – becomes a more complex endeavor, especially because of the following likely reasons (Driver *et al.*, 1996; Kolluru *et al.*, 1996):

- Occurrence of a particular chemical in more than one food item
- Variation in chemical concentrations in food products and other consumer items
- Person-to-person variations in the consumption of various food products
- Variation in dietary profiles across age, gender, ethnic groups, and geographic regions
- Fraction of consumable food product actually containing the chemical of concern (e.g., treated with a given pesticide)
- Possible reductions or changes in chemical concentrations or composition due to transformation during transport, storage, and food preparation.

In fact, the inherent variability and uncertainty in food consumption and chemical concentration data tend to produce a high degree of variability in dietary exposure and risk of a given chemical. For instance, the dietary habits of a home gardener may result in an increase or decrease in exposure – possibly attributable to their unique consumption rates, and also the contaminated fractions involved. In general, individual consumers may indeed ingest significantly different quantities of produce and, depending on their fruit/vegetable preferences, may also be using more of specific crops that are efficient accumulators of contaminants/chemicals or otherwise. Consequently, both food consumption and chemical concentration data are best represented or characterized by dynamic distributions that reflect a wide range of values, rather than by a single value. The distribution of dietary exposures and risks is determined by using both the distribution of food consumption levels and the distribution of chemical concentrations in food (see, e.g., Brown et al., 1988; Driver et al., 1996; NRC, 1993; Rodricks and Taylor, 1983; USEPA, 1986).

2.4. Public Health and Socio-Economic Implications of Chemical Exposure Problems

Human populations are continuously in contact with varying amounts of chemicals present in air, water, soil, food, and other consumer products. Such human exposures to chemical constituents can indeed produce several adverse health effects in the target receptors. In general, any chemical intake can cause severe health impairment or even death, if taken in sufficiently large amounts. Also, there are those chemicals of primary concern that can cause adverse impacts, even from limited exposures.

Invariably, the mere presence of a chemical exposure source within a community or a human population habitat zone can lead to potential receptor exposures – possibly resulting in both short- and long-term effects on a variety of populations within the 'zone of influence.' Indeed, historical records (Table 2.1) have clearly demonstrated the dangers that may result from the presence of chemical exposure situations within or near residential communities and human work environments (Alloway and Ayres, 1993; BMA, 1991; Brooks et al., 1995; Canter et al., 1988; Gibbs, 1982; Grisham, 1986; Hathaway et al., 1991; Kletz, 1994; Levine, 1982; Long and Schweitzer, 1982; Meyer et al., 1995; Petts et al., 1997).

Ultimately, human exposures to chemicals can result in a reduction of life-expectancy – and possibly a period of reduced quality of life (e.g., as caused by anxiety from exposures, diseases, etc.). The presence of toxic chemicals can therefore create potentially hazardous situations and pose significant risks of concern to society at large. In general, potential health and socio-economic problems are averted by carefully implementing substantive corrective action and/or risk management programs appropriate for the specific chemical exposure problem. Methods for identifying and linking all the multiple chemical sources to the human receptor exposures (as discussed throughout this book) are often used to facilitate the development of a sound public health risk management program.

Table 2.1. Selected typical examples of potential human exposures to hazardous chemicals/materials

Chemical hazard location	Source/nature of problem	Contaminants of concern	Nature of exposure settings and scenarios, and observed effects
Love Canal, Niagara Falls, New York, USA	Section of an abandoned excavation for a canal was used as industrial waste landfill. Site received over 20 000 tonnes of chemical wastes containing more than 80 different chemicals.	Various carcinogenic and volatile organic chemicals – including hydrocarbon residues from pesticide manufacture.	Section of an abandoned excavation for a canal that lies within suburban residential setting had been used as industrial waste landfill. Problem first uncovered in 1976. Industrial waste dumping occurred from the 1940s through the 1950s; this subsequently caused entire blocks of houses to be rendered uninhabitable. Potential human exposure routes included direct contact and also various water pathways. Several apparent health impairments – including birth defects and chromosomal abnormalities – observed in residents living in vicinity of the contaminated site.
Chemical Control, Elizabeth, New Jersey, USA	Fire damage to drums of chemicals – resulting in leakage and chemical releases.	Various hazardous wastes from local industries.	The Chemical Control site was adjacent to an urban receptor community; site located at the confluence of two rivers. Leaked chemicals from fire-damaged drums contaminated water (used for fire fighting) – that subsequently entered adjacent rivers. Plume of smoke from fire deposited ash on homes, cars, and playgrounds. Potential exposures mostly via inhalation of airborne contaminants in the plume of smoke from the fire that blew over surrounding communities.
Bloomington, Indiana, USA	Industrial wastes entering municipal sewage system. Sewage material was used for garden manure/fertilizer.	PCBs	PCB-contaminated sewage sludge used as fertilizer – resulting in crop uptakes. Also discharges and runoff into rivers resulted in potential fish contamination. Direct human contacts and also exposures via the food chain (as a result of human ingestion of contaminated food).
Times Beach, Missouri, USA	Dioxins (tetrachlorinated dibenzo(*p*)dioxin, TCDD) in waste oils sprayed on public access areas for dust control.	Dioxins (TCDD)	Waste oils contaminated with dioxins (TCDD) were sprayed in several public areas (residential, recreational, and work areas) for dust control of dirt roads, etc. in the late 1960s and early 1970s. Problem was deemed to present extreme danger in 1982 – i.e., from direct contacts, inhalation, and probable ingestion of contaminated dust and soils.

(continues overleaf)

Chemical hazard location	Source/nature of problem	Contaminants of concern	Nature of exposure settings and scenarios, and observed effects
Triana, Alabama, USA	Industrial wastes dumped in local stream by a pesticide plant.	DDT and other compounds.	High DDT metabolite residues detected in fish consumed by community residents. Potential for human exposure via food chain – i.e., resulting from consumption of fish.
Woburn, Massachusetts, USA	Abandoned waste lagoon with several dumps.	Arsenic compounds, various heavy metals, and organic compounds.	Problem came to light in 1979 when construction workers discovered more than 180 large barrels of waste materials in an abandoned lot alongside a local river. Potential for leachate to contaminate groundwater resources, and also for surface runoff to carry contamination to surface water bodies. High levels of carcinogens found in several local wells – which were then ordered closed. Potential human receptors and ecosystem exposure via direct contacts and water pathways indicated. Inordinately high degree of childhood leukemia observed. This apparent excess of childhood leukemia was linked to contaminated well water in the area. In general, leukemia and kidney cancers in the area were found to be higher than normal.
Santa Clarita, California, USA	Runoff from an electronics manufacturing industry resulted in contamination of drinking water.	Trichloroethylene (TCE) and various other volatile organic compounds (VOCs).	TCE and other VOCs contaminated drinking water in this community (due to runoff from industrial facility). Excess of adverse reproductive outcomes, and excess of major cardiac anomalies among infants suspected.
Three Mile Island, Pennsylvania, USA	Overheating of nuclear power station in March 1979.	Radioactive materials	Small amount of radioactive materials escaped into atmosphere. Emission of radioactive gases – and potential for radioactivity exposures. Unlikely that anyone was harmed by radioactivity from incident. Apparently, the discharge of radioactive materials was too small to cause any measurable harm.
Flixborough, England, UK	Explosion in nylon manufacturing factory in June 1974.	Mostly hydrocarbons	Hydrocarbons processed in reaction vessels/reactors (consisting of oxidation units, etc.). Destruction of plant in explosion, causing death of 28 men on site and extensive damage and injuries in surrounding villages. Explosive situation – i.e., vapor cloud explosion.

(continues overleaf)

Chemical hazard location	Source/nature of problem	Contaminants of concern	Nature of exposure settings and scenarios, and observed effects
Chernobyl, Ukraine (then part of the former USSR)	Overheating of a water-cooled nuclear reactor in April 1986.	Radioactive materials	Nuclear reactor blew out and burned, spewing radioactive debris over much of Europe. General concern relates to exposure to radioactivity. About 30 people reported killed immediately, or died within a few months that may be linked to the accident. It has further been estimated that several thousands more may/could die from cancer during the next 40 years or so as a result of incident.
Seveso (near Milan), Italy	Discharge containing dioxin contaminated a neighboring village over a period of approximately 20 minutes in July 1976.	Dioxin and caustic soda.	Large areas of land contaminated – with part of it being declared uninhabitable. Mostly dermal contact exposures (resulting from vapor-phase/gas-phase deposition on the skin) – especially from smoke particles containing dioxin falling onto skins, etc. About 250 people developed the skin disease, chloracne, and about 450 were burned by caustic soda.
Lekkerkirk (near Rotterdam), The Netherlands	Residential development built on land atop layer of household demolition waste and covered with relatively thin layer of sand. Housing project spanned 1972 – 1975. Problem of severe soil contamination was discovered in 1978. Evacuation of residents commenced in the summer of 1980.	Various chemicals – comprised mainly of paint solvents and resins (containing toluene, lower boiling point solvents, antimony, cadmium, lead, mercury, and zinc).	Rising groundwater carried pollutants upward from underlying wastes into the foundations of houses. This caused deterioration of plastic drinking water pipes, contamination of the water, noxious odors inside the houses, and toxicity symptoms in garden crops. Several houses had to be abandoned, while the waste materials were removed and transported by barges to Rotterdam for destruction by incineration. Polluted water was treated in a physico-chemical purification plant.
Union Carbide Plant, Bhopal, India	Leak of methyl isocyanate from storage tank in December 1984.	Methyl isocyanate (MIC)	Leak of over 25 tonnes of MIC from storage tank occurred at Bhopal, India. In general, exposure to high concentrations of MIC can cause blindness, damage to lungs, emphysema, and ultimately death. MIC vapor discharged into the atmosphere – and then spread beyond plant boundary, killing well over 2,000 people and injuring several tens of thousands more.

(continues overleaf)

Chemical hazard location	Source/nature of problem	Contaminants of concern	Nature of exposure settings and scenarios, and observed effects
Kamioka Zinc Mine, Japan	Contaminated surface waters.	Cadmium	Water containing large amounts of cadmium discharged from the Kamioka Zinc Mine into river used for drinking water, and also for irrigating paddy rice. Ingestion of contaminated water and consumption of rice contaminated by crop uptake of contaminated irrigation water. Long-term exposures resulted in kidney problems for population.
Minamata Bay and Agano River at Niigata, Japan	Effluents from wastewater treatment plants entering coastal waters near a plastics-manufacturing factory.	Mercury – giving rise to the presence of the highly toxic methylmercury.	Accumulation of methylmercury in fish and shellfish. Human consumption of contaminated seafood – resulting in health impairments, particularly severe neurological symptoms.

2.4.1. THE GENERAL NATURE OF HUMAN HEALTH EFFECTS FROM CHEMICAL EXPOSURES

Several health effects may arise if/when people are exposed to certain chemicals introduced into the human environments. The following represent the major broad categories of human health effects that could be anticipated from exposure to chemicals typically found in contemporary societies (Andelman and Underhill, 1988; Asante-Duah, 1998; Bertollini *et al.*, 1996; Brooks *et al.*, 1995; Grisham, 1986; Hathaway *et al.*, 1991; Lippmann, 1992):

- Carcinogenicity (i.e., capable of causing cancer in humans and/or laboratory animals)
- Heritable genetic and chromosomal mutation (i.e., capable of causing mutations in genes and chromosomes that will be passed on to the next generation)
- Developmental toxicity and teratogenesis (i.e., capable of causing birth defects or miscarriages, or damage to developing foetus)
- Reproductive toxicity (i.e., capable of damaging the ability to reproduce)
- Neurotoxicity (i.e., capable of causing harm to the nervous system)
- Alterations of immunobiological homeostasis
- Congenital abnormalities.

Furthermore, most of the chemicals of concern will usually possess either of the following toxicity attributes:

- Acute toxicity (i.e., capable of causing adverse effects, and possibly death, from even short-term exposures)
- Chronic toxicity (i.e., capable of causing long-term damage, other than cancer).

In general, several different symptoms and human health effects may be produced from exposure to various potentially toxic chemicals commonly found in consumer products and/or encountered in the human environments.

Table 2.2 lists some typical symptoms, health effects, and other biological responses that could be produced from a wide range of toxic chemicals commonly encountered in the human environment. Indeed, a number of 'social chemicals' are known or suspected to cause cancer; several others may not have carcinogenic properties – but are, nonetheless, of significant concern due to their systemic toxicity effects.

Table 2.2. Some typical health effects resulting from chemical exposures: a listing for selected toxic chemicals in the human environments

Chemical	Typical health effects/symptoms and toxic manifestations/responses
Arsenic and compounds	Acute hepatocellular injury, anemia, angiosarcoma, cirrhosis, developmental disabilities, embryotoxicity, heart disease, hyperpigmentation, peripheral neuropathies
Antimony	Heart disease
Asbestos	Asbestosis (scarring of lung tissue)/fibrosis (lung and respiratory tract)/lung cancer, mesothelioma, emphysema, irritations, pneumonia/pneumoconioses
Benzene	Aplastic anemia, CNS depression, embryotoxicity, leukemia and lymphoma, skin irritant
Beryllium	Granuloma (lungs and respiratory tract)
Cadmium	Developmental disabilities, kidney damage, neoplasia (lung and respiratory tract), neonatal death/fetal death, pulmonary edema
Carbon tetrachloride	Narcosis, hepatitis, renal damage, liver tumors
Chromium and compounds	Asthma, cholestasis (of liver), neoplasia (lung and respiratory tract), skin irritant
Copper	Gastrointestinal irritant, liver damage
Cyanide	Asthma, asphyxiation, hypersensitivity, pneumonitis, skin irritant
Dichlorodiphenyl trichloroethane (DDT)	Ataxic gait, convulsions, human infertility/reproductive effects, kidney damage, neurotoxicity, peripheral neuropathies, tremors
Dieldrin	Convulsions, kidney damage, tremors
Dioxins and furans (PCDDs/PCDFs)	Hepatitis, neoplasia, spontaneous abortion/fetal death; bioaccumulative
Formaldehyde	Allergic reactions; gastrointestinal upsets; tissue irritation
Lead and compounds	Anemia, bone marrow suppression, CNS symptoms, convulsions, embryotoxicity, neoplasia, neuropathies, kidney damage, seizures; biomagnifies in food chain
Lindane	Convulsions, coma and death, disorientation, headache, nausea and vomiting, neurotoxicity, paresthesias
Lithium	Gastroenteritis, hyperpyrexia, nephrogenic diabetes, Parkinson's disease
Manganese	Bronchitis, cirrhosis (liver), influenza (metal-fume fever), pneumonia, neurotoxicity
Mercury and compounds	Ataxic gait, contact allergen, CNS symptoms; developmental disabilities, neurasthenia, kidney and liver damage, Minamata disease; biomagnification of methyl mercury
Methylene chloride	Anesthesia, respiratory distress, death
Naphthalene	Anemia

(continues overleaf)

Chemical	Typical health effects/symptoms and toxic manifestations/responses
Nickel and compounds	Asthma, CNS effects, gastrointestinal effects, headache, neoplasia (lung and respiratory tract)
Nitrate	Methemoglobinemia (in infants)
Organo-chlorine pesticides	Hepatic necrosis, hypertrophy of endoplasmic reticulum, mild fatty metamorphosis
Pentachlorophenol (PCP)	Malignant hyperthermia
Phenol	Asthma, skin irritant
Polychlorinated biphenyls (PCBs)	Embryotoxicity/infertility/fetal death, dermatoses, hepatic necrosis, hepatitis, immune suppression
Silver	Blindness, skin lesions, pneumonoconiosis
Toluene	Acute renal failure, ataxic gait, neurotoxicity/CNS depression, memory impairment
Trichloroethylene (TCE)	CNS depression, deafness, liver damage, paralysis, respiratory and cardiac arrest, visual effects
Vinyl chloride	Leukemia and lymphoma, neoplasia, spontaneous abortion/fetal death, tumors, death
Xylene	CNS depression, memory impairment
Zinc	Corneal ulceration, esophagus damage, pulmonary edema

Source: Compiled from various sources – including, Blumenthal, 1985; Chouaniere, et al., 2002; Grisham, 1986; Hughes, 1996; Lave and Upton, 1987; Rowland and Cooper, 1983; and personal communication with Dr. Kwabena Duah, Australia (2002).

2.5. Assessing Public Health Risks

In order to arrive at cost-effective public health risk management decisions, answers will typically have to be generated for several pertinent questions when one is confronted with a potential chemical exposure problem (Box 2.1). In general, when it is suspected that a potential hazard exists at a particular locale, then it becomes necessary to further investigate the situation – and to fully characterize the prevailing or anticipated hazards. This activity may be accomplished by the use of a well-designed exposure investigation program. Ultimately, a thorough investigation – culminating a risk assessment – that establishes the nature and extent of receptor exposures may become necessary, in order to arrive at appropriate and realistic corrective action and/or risk management decisions.

Box 2.1. Major issues important to making cost-effective public health risk management decisions for chemical exposure problems

♦ What is the nature of the chemical exposure(s)?
♦ What are the sources of, and the 'sinks' or receptors for, the chemicals of potential concern?
♦ What population groups are potentially at risk?
♦ What are the likely and significant exposure pathways and scenarios that connect chemical source(s) to potential receptors?
♦ What is the current extent of receptor exposures?
♦ What is the likelihood of health and environmental effects resulting from the chemical exposure?
♦ What interim measures, if any, are required as part of a risk management and/or risk prevention program?
♦ What corrective action(s) may be appropriate to remedy the prevailing situation?
♦ What level of residual chemical exposures will be tolerable or acceptable for the target receptors?

Chemical exposure characterizations typically will consist of the planned and managed sequence of activities carried out to determine the nature and distribution of hazards associated with the specific chemical exposure problem. The activities involved usually are comprised of several specific tasks – broadly listed to include the following:

- Problem definition (including identifying study objectives and data needs)
- Identification of the principal hazards
- Design of sampling and analysis programs
- Collection and analysis of appropriate samples
- Recording or reporting of laboratory results for further evaluation
- Logical analysis of sampling data and laboratory analytical results
- Interpretation of study results (consisting of enumeration of the implications of, and decisions on corrective action).

In order to get the most out of the chemical exposure characterization, this activity must be conducted in a systematic manner. Indeed, systematic methods help focus the purpose, the required level of detail, and the several topics of interest – such as physical characteristics of the potential receptors; contacted chemicals; extent and severity of possible exposures; effects of chemicals on populations potentially at risk; probability of harm to human health; and possible residual hazards following implementation of risk management and corrective action. Ultimately, the data derived from the exposure investigation may be used to perform a risk assessment – which then becomes a key element in the public health risk management decision process.

2.6. Suggested Further Reading

Ashford, NA and CS Miller, 1998. *Chemical Exposures: Low Levels and High Stakes,* J. Wiley & Sons, New York, NY

Craun, GF, FS Hauchman, and DE Robinson (eds.), 2001. *Microbial Pathogens and Disinfection By-products in Drinking Water – Health Effects and Management of Risks,* ILSI Press, Washington, DC

Francis, BM, 1994. *Toxic Substances in the Environment,* J. Wiley, New York

Falck, F., A. Ricci, MS Wolff, J. Godbold, and P. Deckers, 1992. Pesticides and polychlorinated biphenyl residues in human breast lipids and their relation to breast cancer, *Archives of Environmental Health,* 47: 143 – 146

Goldman, M., 1996. Cancer risk of low-level exposure, *Science,* 271(5257): 1821 – 1822

Hwang, J-S and C-C Chan, 2002. Effects of air pollution on daily clinic visits for lower respiratory tract illness, *American Journal of Epidemiology,* 155(1): 1 – 10

Kent, C., 1998. *Basics of Toxicology,* J. Wiley & Sons, New York, NY

Klaassen, CD (ed.), 1996. *Casarett and Doull's Toxicology: The Basic Science of Poisons,* 5th edition. McGraw-Hill, New York

Lanphear, B. and R. Byrd, 1998. Community characteristics associated with elevated blood lead levels in children, *Pediatrics,* 101(2): 264 – 71

Mitchell, P. and D. Barr, 1995. The nature and significance of public exposure to arsenic: a review of its relevance to South West England, *Environmental Geochemistry and Health,* 17: 57 – 82

Moeller, DW, 1997. *Environmental Health,* Revised Edition, Harvard University Press, Cambridge, MA

NRC (National Research Council), 1993. *Pesticides in the Diets of Infants and Children,* National Academy Press: Washington, DC

Pocock, SJ, M. Smith, and P. Baghurst, 1994. Environmental lead and children's intelligence: a systematic review of the epidemiological evidence, *British Medical Jour.,* 309: 1189 – 97

Sherman JD. 1994. *Chemical Exposure and Disease,* Princeton Scientific Publishing Co., New Jersey

Silkworth, JB and JF Brown, Jr., 1996. Evaluating the impact of exposure to environmental contaminants on human health, *Clinical Chemistry,* 42: 1345 – 1349

van Veen, MP, 1996. A general model for exposure and uptake from consumer products, *Risk Analysis,* 16(3): 331 – 338

Vermeire, TG, P. van der Poel, R. van de Laar, and H. Roelfzema, 1993. Estimation of consumer exposure to chemicals: application of simple models, *Science of the Total Environment,* 135: 155 – 176

White RF, RG Feldman, PH Travers, 1990. Neurobehavioral effects of toxicity due to metals, solvents, and insecticides, *Clinical Neuropharmacology,* 13: 392 – 412

Zartarian, VG and JO Leckie, 1998. Dermal exposure: the missing link, *Environmental Science & Technology,* 32(5): 134A – 137A

Chapter 3

PRINCIPLES AND CONCEPTS IN RISK ASSESSMENT

In its application to chemical exposure problems, the risk assessment process is used to compile and organize the scientific information that is necessary to support environmental and public health risk management decisions. The approach is used to help identify potential problems, establish priorities, and provide a basis for regulatory actions. Indeed, it is apparent that the advancement of risk analysis in regulatory decision-making has promoted rational policy deliberations over the past several decades. Yet, as real-world practice indicates, risk analyses have often been as much the source of controversy in regulatory considerations as the facilitator of consensus (ACS and RFF, 1998).

Indeed, risk assessment can appropriately be regarded as a valuable tool for public health and environmental decision-making – albeit there tends to be disagreement among experts and policy makers about the extent to which its findings should influence decisions about risk. To help produce reasonable/pragmatic and balanced policies in its application, it is essential to explicitly recognize the character, strengths, and limitations of the analytical methods that are involved in the use of risk analyses techniques in the decision-making process.

In general, risk assessment methods commonly encountered in the literature of environmental and public health management, and/or relevant to the management of chemical exposure problems require a clear understanding of several fundamental issues. This chapter discusses key fundamental principles and concepts that will facilitate the application and interpretation of risk assessment information – and thus make it more suitable in public health risk management decisions.

3.1. Fundamental Principles of Chemical Hazard, Exposure, and Risk Assessments

Hazard is that object with the potential for creating undesirable adverse consequences; exposure is the situation of vulnerability to hazards; and risk is considered to be the probability or likelihood of an adverse effect due to some hazardous situation. Indeed, it is the likelihood to harm as a result of exposure to a hazard that distinguishes risk from hazard. For example, a toxic chemical that is hazardous to human health does not constitute a risk unless human receptors/populations are exposed to such a substance –

as conceptually illustrated by the Venn diagram representation shown in Figure 3.1. Thus, a complete assessment of potential hazards posed by a substance or an object involves, among several other things, a critical evaluation of available scientific and technical information on the substance or object of concern, as well as the possible modes of exposure. In addition, potential receptors will have to be exposed to the hazards of concern before any risk could be said to exist. In fact, the availability of an adequate and complete information set is an important prerequisite for producing sound hazard, exposure, and risk assessments.

From the point of view of human exposure to chemicals, risk can be defined as the probability that public health could be affected to various degrees (including an individual or group suffering injury, disease, or even death) under specific set of circumstances. Potential risks are estimated by considering the probability or likelihood of occurrence of harm; the intrinsic harmful features or properties of specified hazards; the population-at-risk (PAR); the exposure scenarios; and the extent of expected harm and potential effects. The integrated assessment of hazards, exposures and risks are indeed a very important contributor to any decision that is aimed at adequately managing a given hazardous situation.

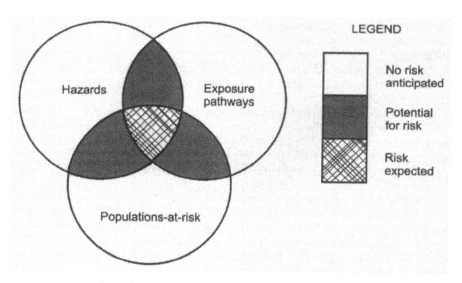

Figure 3.1. When do hazards actually represent risks?

3.1.1. THE NATURE OF CHEMICAL HAZARD, EXPOSURE, AND RISK

Hazard is defined as the potential for a substance or situation to cause harm, or to create adverse impacts on populations and/or property. It represents the undetermined loss potential, and may comprise of a condition, a situation, or a scenario with the potential for creating undesirable consequences. The degree of chemical hazard will usually be determined from the type of exposure scenario and the potential effects or responses resulting from any exposures.

There is no universally accepted single definition of risk. In any case, risk may be considered as the probability or likelihood of an adverse effect, or an assessed threat to persons due to some hazardous situation. It is a measure of the probability and severity of adverse consequences from an exposure of potential receptors to hazards – and may simply be represented by the measure of the frequency of an event.

Procedures for analyzing hazards and risks may typically be comprised of several steps (Figure 3.2), consisting of the following general elements:

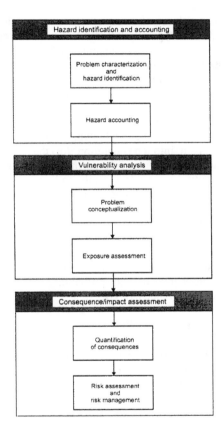

Figure 3.2. Basic steps in the analyses of hazards and risks

- Hazard Identification and Accounting
  - identify hazards (including nature/identity of hazard, location, etc.)
  - identify initiating events (i.e., causes)
  - identify resolutions for hazard
  - define exposure setting
- Vulnerability Analysis
  - identify vulnerable zones or locales
  - identify concentration/impact profiles for affected zones

- ♦ determine populations potentially at risk (such as human populations, and critical facilities)
- ♦ define exposure scenarios
- Consequences/Impacts Assessment
  - ♦ determine risk categories for all identifiable hazards
  - ♦ determine probability of adverse outcome (from exposures to hazards)
  - ♦ estimate consequences (including severity, uncertainties, etc.).

Some or all of these elements may have to be analyzed in a comprehensive manner, depending on the nature and level of detail of the hazard and/or risk analysis that is being performed. The analyses typically fall into two broad categories – endangerment assessment (which may be considered as contaminant-based, such as human health and environmental risk assessment associated with chemical exposures), and safety assessment (which is system failure-based, such as probabilistic risk assessment of hazardous facilities or installations). Ultimately, the final step will be comprised of developing risk management and/or risk prevention strategies for the problem situation.

It is noteworthy that, human exposures to radiological contaminants may be evaluated in a manner similar to the chemical exposure problems – albeit certain unique issues may have to be taken into consideration for the radiological exposures. In general, the radiological exposures may occur through medical and dental X-rays; naturally-occurring radioactive materials in soils and groundwater; ambient air; and various food sources and indeed several other consumer product sources.

3.1.2. BASIS FOR MEASURING RISKS

Risk represents the assessed loss potential, often estimated by the mathematical expectation of the consequences of an adverse event occurring. It is defined by the product of the two components of the probability of occurrence (p) and the consequence or severity of occurrence (S), viz.:

$$Risk = p \times S \tag{3.1}$$

Risk – interpreted as the probability of a harmful event to humans or to the environment that is caused by a chemical, physical, or biological agent – can also be described by the following conceptual relationship:

$$Risk = [f(I) \times f(P)] - f(D) \tag{3.2}$$

where $f(I)$ represents an 'intrinsic risk' factor that is a function of the characteristic nature of the agent or the dangerous properties of the hazard; $f(P)$ is a 'presence' factor that is a function of the quantity of the substance or hazard released into the human environment, and of all the accumulation and removal methods related to the chemical and physical parameters of the product, as well as to the case-specific parameters typical of the particular environmental setting; and $f(D)$ represents a 'defense' factor that is a function of what society can do in terms of both protection and prevention to minimize the harmful effects of the hazard. Probably, the most important factor in this equation is $f(D)$, that may include both the ordinary defense mechanisms for hazard abatement as well as some legislative measures. In fact, the level of risk is dependent on the degree of hazard as well as on the amount of safeguards or preventative measures against adverse effects; consequently, risk can also be defined by the following simplistic conceptual relationships:

$$\text{Risk} = \frac{[\text{Hazard}]}{[\text{Preventative Measures}]} \qquad (3.3)$$

or

$$\text{Risk} = f\{\text{Hazard, Exposure, Safeguards}\} \qquad (3.4)$$

where 'Preventative Measures' or 'Safeguards' is considered to be a function of exposure – or rather inversely proportional to the degree of exposure. 'Preventative Measures' or 'Safeguards' represent the actions that are generally taken to minimize potential exposure of target populations to the specific hazards.

Invariably, the estimation of risks involves an integration of information on the intensity, frequency, and duration of exposure for all identified exposure routes associated with the exposed or impacted group(s). For instance, risk may represent the probability for a chemical to cause adverse impacts to potential receptors as a result of exposures over specified time periods. Ultimately, the risk measures give an indication of the probability and severity of adverse effects (Figure 3.3) – and this is generally established with varying degrees of confidence according to the importance of the decision involved.

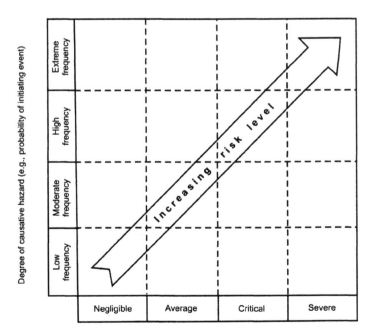

Figure 3.3. General conceptual categories of risk measures

In general, measures used in risk analysis take various forms, depending on the type of problem, degree of resolution appropriate for the situation on hand, and the analysts' preferences. Thus, the risk parameter may be expressed in quantitative terms – in which case it could take on values from zero (associated with certainty for no-adverse effects) to unity (associated with certainty for adverse effects to occur). In several other cases, risk is only described qualitatively – such as by use of descriptors like 'high', 'moderate', 'low', etc.; or indeed, the risk may be described in semi-quantitative/semi-qualitative terms. In any case, the risk qualification or quantification process will normally rely on the use of several measures, parameters and/or tools as reference yardsticks (Box 3.1). Individual lifetime risk (represented by the probability that the individual will be subjected to an adverse effect from exposure to identified hazards) is about the most commonly used measure of risk.

Box 3.1. Typical/common measures, parameters, and/or tools that form the basis for risk qualification or quantification

---

- ♦ Probability distributions (based on probabilistic analyses)
- ♦ Expected values (based on statistical analyses)
- ♦ Economic losses or damages
- ♦ Public health damage
- ♦ Risk profile diagrams (e.g., iso-risk contours plotted on area map, to produce an iso-risk contour map)
- ♦ Incidence rate (defined by the ratio of [number of new cases over a period of time]:[population at risk])
- ♦ Prevalence rate (defined by the ratio of [number of existing cases at a point in time]:[total population])
- ♦ Relative risk (i.e., risk ratio) (defined by a ratio such as [incidence rate in exposed group]:[incidence rate in non-exposed group])
- ♦ Attributable risk (i.e., risk difference) (defined by an arithmetic difference, such as [incidence among an exposed group] – [incidence among the non-exposed group])
- ♦ Margin of safety (defined by the ratio of [the highest dose level that does *not* produce an adverse effect]:[the anticipated human exposure])
- ♦ Individual lifetime risk (equal to the product of exposure level and severity, e.g., [dose x potency])
- ♦ Population or societal risk (defined by the product of the individual lifetime risk and the population exposed)
- ♦ Frequency-consequence diagrams (also known as F-N curves for fatalities, to define societal risk)
- ♦ Quality of life adjustment (or quality adjusted life expectancy, QALE)
- ♦ Loss of life expectancy (given by the product of individual lifetime risk and the average remaining lifetime)

3.1.3. WHAT IS RISK ASSESSMENT?

Several somewhat differing definitions of risk assessment have been published in the literature by various authors to describe a variety of risk assessment methods and/or protocols (e.g., Asante-Duah, 1998; Cohrssen and Covello, 1989; Conway, 1982; Cothern, 1993; Covello *et al.*, 1986; Covello and Mumpower, 1985; Crandall and Lave, 1981; Davies, 1996; Glickman and Gough, 1990; Gratt, 1996; Hallenbeck and Cunningham 1988; Kates, 1978; Kolluru *et al.*, 1996; LaGoy, 1994; Lave, 1982; Neely, 1994; NRC, 1982, 1983, 1994; Richardson, 1990, 1992; Rowe, 1977; Turnberg, 1996; USEPA 1984; Whyte and Burton, 1980). In a generic sense, risk assessment may be considered to be a systematic process for arriving at estimates of all the significant risk factors or parameters associated with an entire range of 'failure modes' and/or exposure scenarios in connection with some hazard situation(s). It entails the evaluation of all pertinent scientific information to enable a description of the likelihood, nature, and extent of harm to human health as a result of exposure to chemicals present in the human environments.

Risk assessment is indeed a scientific process that can be used to identify and characterize chemical exposure-related human health problems. In its application to the management of chemical exposure problems, the process encompasses an evaluation of all the significant risk factors associated with all feasible and identifiable exposure scenarios that are the result of specific chemicals being introduced into the human environments. It may, for instance, involve the characterization of potential adverse consequences or impacts to human receptors that are potentially at risk from exposure to chemicals found in consumer products and/or in the environment.

In general, the public health risk assessment process seeks to estimate the likelihood of occurrence of adverse effects resulting from exposures of human receptors to chemical, physical, and/or biological agents present in the human living and work environments. The process consists of a mechanism that utilizes the best available scientific knowledge to establish case-specific responses that will ensure justifiable and defensible decisions – as necessary for the management of hazardous situations in a cost-efficient manner. The process is also concerned with the assessment of the importance of all identified risk factors to the various stakeholders whose interests are embedded in a candidate problem situation (Petak and Atkisson, 1982).

3.1.4. THE NATURE OF RISK ASSESSMENTS

Risk assessment methods may fall into several general major categories – typically under the broad umbrellas of hazard assessment, exposure assessment, consequence assessment, and risk estimation (Covello and Merkhofer, 1993; Norrman, 2001). The *hazard assessment* may consist of monitoring (e.g., source monitoring and laboratory analyses), performance testing (e.g., hazard analysis and accident simulations), statistical analyses (e.g., statistical sampling and hypotheses testing), and modeling methods (e.g., biological models and logic tree analyses). The *exposure assessment* may be comprised of monitoring (e.g., personal exposures monitoring, media contamination monitoring, biologic monitoring), testing (e.g., laboratory tests and field experimentation), dose estimation (e.g., as based on exposure time, material disposition in tissue, and bioaccumulation potentials), chemical fate and behavior modeling (e.g., food-chain and multimedia modeling), exposure route modeling (e.g., inhalation, ingestion, and dermal contact), and populations-at-risk modeling (e.g., general populuation *vs.* sensitive groups). The *consequence assessment* may include health surveillance, hazard screening, animal tests, human tests, epidemiologic studies, animal-to-human extrapolation modeling, dose-response modeling, pharmacokinetic modeling, ecosystem monitoring, and ecological effects modeling. The *risk estimation* will usually take such forms as relative risk modeling, risk indexing (e.g., individual risk *vs.* societal risk), nominal *vs.* worst-case outcome evaluation, sensitivity analyses, and uncertainty analyses. Detailed listings of key elements of the principal risk assessment methods are provided elsewhere in the literature (e.g., Covello and Merkhofer, 1993; Norrman, 2001).

Most of the techniques available for performing risk assessments are structured around decision analysis procedures – in order to facilitate comprehensible solutions for even complicated problems. In fact, the risk assessment process can be used to provide a 'baseline' estimate of existing risks that can be attributed to a given agent or hazard, as well as to determine the potential reduction in exposure and risk under various mitigation scenarios. In general, some risk assessments may be classified as *retrospective* – focusing on injury after the fact (e.g., nature and level of risks at a given contaminated site); or it may be considered as *predictive* – such as evaluating possible future harm to human health or the environment (e.g., risks anticipated if a newly developed food additive is approved for use in consumer food products). In the investigation of chemical exposure problems, the focus of most public health risk assessments tends to be on a determination of potential or anticipated risks to the populations potentially at risk.

Risk assessment is indeed a powerful tool for developing insights into the relative importance of the various types of exposure scenarios associated with potentially hazardous situations. But as Moeller (1997) points out, it has to be recognized that a given risk assessment provides only a snapshot in time of the estimated risk of a given toxic agent at a particular phase of our understanding of the issues and problems. To be truly instructive and constructive, therefore, risk assessment should preferably be conducted on an iterative basis – being continually updated as new knowledge and information become available.

3.1.5. RECOGNITION OF UNCERTAINTY AS AN INTEGRAL COMPONENT OF RISK ASSESSMENTS

A major difficulty in decision-making resides in the uncertainties of system characteristics for the situation at hand. Uncertainty is the lack of confidence in the estimate of a variable's magnitude or probability of occurrence; scientific judgment becomes an important factor in problem-solving under uncertainty, and decision analysis provides a means of representing the uncertainties in a manner that allows informed discussion. The presence of uncertainty means, in general, that the best outcome obtainable from an evaluation and/or analysis cannot be guaranteed. Nonetheless, as has been pointed out by Bean (1988), decisions ought to be made even in an uncertain setting, otherwise several aspects of environmental (and indeed public health) management actions could become completely paralyzed. Indeed, there are inevitable uncertainties associated with risk estimates, but these uncertainties do not invalidate the use of the risk estimates in the decision-making process. However, it is important to identify and define the confidence levels associated with the evaluation; depending on the specific level of detail of a risk assessment, the type of uncertainty that dominates at each stage of the analysis can be different (see Chapter 8).

In view of the fact that risk assessment may constitute a very important part of the environmental and public health management decision-making process, it is essential that all the apparent sources of uncertainty be well documented. Indeed, the need to be explicit about uncertainty issues in risk analysis has long been recognized – and this remains a recurrent theme for policy analysts and risk management practitioners. In general, the uncertainty can be characterized via sensitivity analysis and/or probability analysis techniques – with the technique of choice usually being dependent on the available input data statistics. Broadly speaking, sensitivity analyses require data on the range of values for each exposure factor in the scenario; and probabilistic analyses require data on the range and probability function (or distribution) of each exposure factor within the scenario.

Uncertainty analysis can indeed be performed qualitatively or quantitatively – with sensitivity analysis often being a useful adjunct to the uncertainty analysis. Sensitivity analysis entails the determination of how rapidly the output of a given analysis changes with respect to variations in the input data; thus, in addition to presenting the best estimate, the evaluation will also provide a range of likely estimates in the form of a sensitivity analysis. In fact, it is recommended that a sensitivity analysis should generally become an integral part of a detailed risk evaluation process. Through such analyses, uncertainties can be assessed properly, and their effects on given decisions accounted for in a systematic way. In this manner, the risk associated with given decision alternatives may be delineated, and then appropriate corrective measures can be taken accordingly. Further discussion of this topic appears later on in Chapter 8 of this title.

3.2. Fundamental Concepts in Risk Assessment Practice

In order to adequately evaluate the risks associated with a given hazard situation, several concepts are usually employed in the processes involved. Some of the fundamental concepts and definitions that will facilitate a better understanding of the risk assessment process and application principles, and that may also affect risk management decisions, are introduced below in this section.

3.2.1. QUALITATIVE *VERSUS* QUANTITATIVE RISK ASSESSMENT

The general types of risk assessment often encountered in practice may range from an evaluation of the potential effects of toxic chemical releases known to be occurring, to evaluations of the potential effects of releases due to events whose probability of occurrence is uncertain (Moeller, 1997). Regardless, the attributable risk for any given problem situation can be expressed in qualitative, semi-quantitative, or quantitative terms. For instance, in conveying qualitative conclusions regarding chemical hazards, narrative statements incorporating 'weight-of-evidence' or 'strength-of-evidence' conclusions, in lieu of alpha-numeric designations alone, may be used. In other situations, pure numeric parameters are used; and yet in other circumstances, a combination of both numeric parameters and qualitative descriptors are used in the risk presentation/discussion.

In public health risk assessments, quantitative tools are often used to better define exposures, effects, and risks in the broad context of risk analysis. Such tools will usually employ the plausible ranges associated with default exposure scenarios, toxicological parameters, and indeed other assumptions and policy positions. Although the utility of numerical risk estimates in risk analysis has to be appreciated, these estimates should be considered in the context of the variables and assumptions involved in their derivation – and indeed in the broader context of biomedical opinion, host factors, and actual exposure conditions. Consequently, directly or indirectly, qualitative descriptors also become part of the quantitative risk assessment process. For instance, in evaluating the assumptions and variables relating to both toxicity and exposure conditions for a chemical exposure problem, the risk outcome may be provided in qualitative terms – albeit the risk levels are expressed in quantitative terms.

Risk Categorization
Oftentimes in risk studies, it becomes necessary to put the degree of hazards or risks into different categories for risk management purposes. A typical risk categorization scheme for potential chemical exposure problems may involve a grouping of the

'candidate' problems on the basis of the potential risks attributable to various plausible conditions – such as high-, intermediate- and low-risk problems, conceptually depicted by Figure 3.4. Such a classification can indeed facilitate the development and implementation of a more efficient public health risk management or corrective action program.

In general, the high-risk problems will prompt the most concern – requiring immediate and urgent attention or corrective measures. A problem may be designated as 'high-risk' when exposure represents a real or imminent threat to human health; in this case, an immediate action will generally be required to reduce the threat. Indeed, to ensure the development of adequate and effectual public health risk management or corrective action strategies, potential chemical exposure problems may need to be categorized in a similar or other appropriate manner during the risk analysis.

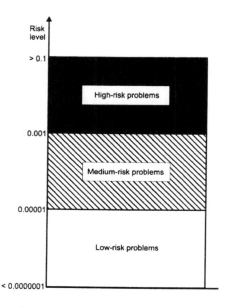

Figure 3.4. A conceptual representation of typical risk categories for chemical exposure problems

3.2.2. CONSERVATISMS IN RISK ASSESSMENTS

Many of the parameters and assumptions used in hazard, exposure, and risk evaluation studies tend to have high degrees of uncertainties associated with them. Thus, it is common practice for safe design and analysis, to model risks such that risk levels determined for management decisions are preferably over-estimated. Such conservative estimates (also, often cited as 'worst-case', or 'plausible upper bound' estimates) used in risk assessment are based on the premise that pessimism in risk assessment (with resultant high estimates of risks) is more protective of public health and/or the environment. For example, in performing risk assessments, scenarios have often been developed that will reflect the worst possible exposure pattern; this notion of *'worst-case scenario'* in the risk assessment refers to the event or series of events resulting in the greatest exposure or potential exposure. Also, quantitative cancer risk assessments are typically expressed as plausible upper bounds rather than estimates of central tendency; but when several plausible upper bounds are added together, the question arises as to whether the overall result is still plausible (Bogen, 1994; Burmaster and Harris, 1993; Cogliano, 1997). In any case, although it is believed that the overall risk depends on the independence, additivity, synergistic/antagonistic interactions among the carcinogens, and the number of risk estimates – as well as on the shapes of the underlying risk distributions – sums of upper bounds still provide useful information about the overall risk. On the other hand, gross exaggeration of actual risks could lead to poor decisions being made with respect to the oftentimes very limited resources available for general risk mitigation purposes. Thus, after establishing a worst-case scenario, it is often desirable to also develop and analyze more realistic or 'nominal' scenarios, so that the level of risk posed by a hazardous situation can be better bounded – by selecting 'best' or 'most likely' sets of assumptions for the risk assessment. But in deciding on what realistic assumptions are to be used in a risk assessment, it is imperative that the analyst chooses parameters that will, at worst, result in erring on the side of safety.

Lately, a number of investigators have been elaborating on a variety of ways and means of making risk assessments more realistic – rather than the dependence on wholesale compounded conservative assumptions (see, e.g., Anderson and Yuhas, 1996; Burmaster and von Stackelberg, 1991; Cullen, 1994; Maxim, *in* Paustenbach, 1988). Among other things, there is the need to undertake sensitivity analyses – including the use of multiple assumption sets that reflect a wider spectrum of exposure scenarios. This is important because regulations based on the so-called upper-bound estimate or worst-case scenario may address risks that are almost nonexistent and impractical. Indeed, risk assessment using extremely conservative biases do not provide risk managers with the quality information needed to formulate efficient and cost-effective management strategies. Also, using plausible upper-bound risk estimates or worst-case scenarios may lead to spending scarce and limited resources to regulate or control insignificant risks – whiles more serious risks are probably being ignored. Thus, conservatism in individual assessments may not be optimal or even conservative in a broad sense if some sources of risk are not addressed simply because others receive undue attention. Consequently, the overall recommendation is to strive for accuracy rather than conservatism.

3.2.3. INDIVIDUAL *VERSUS* GROUP RISKS

In the application of risk assessment to environmental and public health risk management programs, it often becomes important to distinguish between individual and societal risks, in order that the most appropriate one can be used in the analysis of case-specific problems. In fact, more broadly, three types of risks can be identified for most situations – namely:

- risks to individuals
- risks to the general population
- risks to highly exposed or highly sensitive subgroups of a population

The latter two categories may then constitute the 'societal' or 'group' risk – representing population risks associated with more than one person or the individual.

Individual risks are considered to be the frequency at which a given individual could potentially sustain a given level of adverse consequence from the realization or occurrence of specified hazards. *Societal risk*, on the other hand, relates to the frequency and the number of individuals sustaining some given level of adverse consequence in a given population due to the occurrence of specified hazards. The population risk provides an estimate of the extent of harm for the population or population segment being addressed.

Individual risk estimates represent the risk borne by individual persons within a population – and are more appropriate in cases where individuals face relatively high risks. However, when individual risks are not inequitably high, then it becomes important during resources allocation, to consider possible society-wide risks that might be relatively higher. Indeed, risk assessments almost always deal with more than a single individual. Frequently, individual risks are calculated for some or all of the persons in the population being studied, and are then put into the context of where they fall in the distribution of risks for the entire population.

It is noteworthy that, at an individual level, the choice of whether or not to accept a risk is primarily a personal decision. However, on a societal level (where values tend to be in conflict and decisions often produce 'winners' and 'losers'), the decision to accept or reject a risk is much more difficult (Cohrssen and Covello, 1989). In fact, no numerical level of risk will likely receive universal acceptance, but also eliminating all risks is an impossible task – especially for our modern society in which people have become so accustomed to numerous 'hazard-generating' luxuries of life. Indeed, for many activities and technologies of today, some level of risk has to be tolerated in order to gain the benefits of the activity or technology. Consequently, levels of risk that may be considered as tolerable or relatively 'safe enough' should typically be identified/defined – at least on the societal level – to facilitate rational risk management and related decision-making tasks. In this process, it must be acknowledged that individuals at the high end of a risk distribution are often of special interest to risk managers – i.e., when considering various actions to mitigate the risk; these individuals often are either more susceptible to the adverse health effect than others in the population or are highly exposed individuals, or both.

3.2.4. CONSIDERATION OF RISK PERCEPTION ISSUES

The general perception of risks tends to vary amongst individuals and/or groups, and may even change with time. Risk perception may therefore be considered as having both spatial and temporal dimensions. In general, the public often views risk differently from the risk estimates developed by technical experts. Indeed, this notion ties in very well with the concept that public perception of risk is a function of hazard and the so-called 'outrage' factors; the 'outrage' component describes a range of factors, other than the actual likelihood of a hazard, that contribute to an enhanced perception of risk (Sandman, 1993; Slovic, 1993; 1997). Perhaps these 'outrage' factors explain why multiple hazards of similar magnitude can be perceived as having vastly differently levels of risk. In any case, whereas public outrage is not tangible, it is real – and must therefore be addressed to ensure program success.

In general, risks that are involuntary (e.g., environmental risks) or 'novel' seem to arouse more concern from the target/affected populations than those that are voluntary (e.g., associated with use of certain cosmetics and other consumer products) or 'routine'; thus, the latter tends to be more acceptable to the affected individuals (van Leeuwen and Hermens, 1995). Consequently, 'natural' toxins and contaminants in foods may be considered acceptable (even though they may cause illness), whereas food additives (used in foodstuffs to assist in preservation) may not be as much acceptable to some people (Richardson, 1986). Also, perceptions about risk tend to be influenced by the sources of information; styles of presentation; personal background and educational levels; cultural contexts; and the dimensions of a particular risk problem. For instance, cultural explanations for risk management controversies – in regards to the ways people differ in their thinking about risk (or risk acceptability for that matter) – have gained increasing recognition/acceptance over the past decade and a half or so (Earle and Cvetkovich, 1997). In fact, several value judgments become an important component of the consequential decision-making process – with the value judgments involving very complex social processes. A fairly well established hierarchy of risk 'tolerability' has indeed emerged in recent times that involve several issues/factors – including those enumerated in Box 3.2 (Cassidy, 1996; Cohrssen and Covello, 1989; Lowrance, 1976).

Invariably, issues relating to risk perception become a very important consideration in environmental and public health management decisions – especially because, in some situations, the perception of a group of people may alter the priorities assigned to the reduction of competing risks. In fact, the differences between risk perception and risk estimation could have crucial consequences on the assessment, management, and communication of risks. This is because the particular risks estimated in a given risk assessment may not necessarily be consistent with the perceptions or concerns of those individuals most directly affected.

Box 3.2. Key factors affecting the 'tolerability' of risk by individuals and society

♦ Voluntariness (i.e., Voluntary *vs.* Involuntary exposures)
♦ Response time (i.e., Delayed *vs.* Immediate effects)
♦ Source (i.e., Natural *vs.* Human-made risks)
♦ Controllability (i.e., Controllable *vs.* Uncontrollable)
♦ Perception of personal control
♦ Familiarity with the type of hazard (i.e., Old/Known *vs.* New/Unknown hazards or risks)
♦ Perceptions about potential benefits (i.e., Exposure is an essential *vs.* Exposure is a luxury)
♦ Nature of hazard and/or consequences (i.e., Ordinary *vs.* Catastrophic)
♦ Perception of the extent and type of risk
♦ Perceptions about comparative risks for other activities
♦ Reversibility of effects (i.e., Reversible *vs.* Irreversible)
♦ Perceptions about available choices (i.e., No alternatives available *vs.* Availability of alternatives)
♦ Perceptions about equitability/fairness of risk distribution
♦ Continuity of exposure (i.e., Occasional *vs.* Continuous)
♦ Visual indicators of risk factors or levels (i.e., Tangible *vs.* Intangible risks)

3.3.   Risk Acceptability and Risk Tolerance Criteria

The concept of an 'acceptable risk level' relates to a very important issue in risk assessment – albeit the desirable or tolerable level of risk is not always attainable. Regardless, with maintenance of public health and safety being a crucial goal for public health risk management, it should be acknowledged that reasons such as budgetary constraints alone could not be used as justification for establishing an acceptable risk level on the higher side of a risk spectrum. Meanwhile, it must also be noteworthy that risk acceptability (i.e., the level of risk that society can allow for a specified hazard situation) usually will have a spatial and temporal variability to it.

An important concept in risk management is that there are levels of risk that are so great that they must not be allowed to occur at all cost, and yet there are other risk levels that are so low that they are not worth bothering with even at insignificant costs; these are known, respectively, as *de manifestis* and *de minimis* levels (Kocher and Hoffman, 1991; Suter, 1993; Travis *et al.*, 1987; Whipple, 1987). Risk levels between these bounds are typically balanced against costs, technical feasibility of mitigation actions, and other socioeconomic, political and legal considerations – in order to determine their acceptability or tolerability. It is noteworthy that, the concept of *de manifestis* risk is generally not seen as being controversial – because, after all, some hazard effects are clearly unacceptable. However, the *de minimis* risk concept tends to be controversial – in view of the implicit idea that some exposures to and effects of pollutants or hazards are acceptable (Suter, 1993). In any case, it is still desirable to use these types of criteria to eliminate obviously trivial risks from further risk management actions – considering that society cannot completely eliminate or prevent all human and environmental health effects associated with chemical exposure problems.

Indeed, virtually all social systems have target risk levels – whether explicitly indicated or not – that represent tolerable limits to danger that the society is prepared to accept in consequence of potential benefits that could accrue from a given activity. This tolerable limit is often designated as the *de minimis* or 'acceptable' risk level. Thus, in the general process of establishing 'acceptable' risk levels, it is possible to use *de minimis* levels below which one need not be concerned (Rowe, 1983). Current regulatory requirements are particularly important considerations in establishing such acceptable risk levels.

3.3.1. THE *DE MINIMIS* OR 'ACCEPTABLE' RISK

Risk is *de minimis* if the incremental risk produced by an activity is sufficiently small so that there is no incentive to modify the activity (Cohrssen and Covello, 1989; Covello *et al.*, 1986; Fischhoff *et al.*, 1981; Whipple, 1987). These are risk levels judged to be too insignificant to be of any social concern or to justify use of risk management resources to control them, compared with other beneficial uses for the often limited resources available in practice. In simple terms, the *de minimis* principle assumes that extremely low risks are trivial and need not be controlled. A *de minimis* risk level would therefore represent a cutoff, below which a regulatory agency could simply ignore alleged problems or hazards.

The concept of *de minimis* or *acceptable* risk is essentially a threshold concept, in that it postulates a threshold of concern below which there would be indifference to changes in the level of risk. In fact, considerable controversy exists concerning the concept of 'acceptable' risk in the risk/decision analysis literature. This is because, in practice, acceptable risk is the risk associated with the most acceptable decision – rather than being acceptable in an absolute sense. It has been pointed out (Massmann and Freeze, 1987) that, acceptable risk is decided in the political arena and that 'acceptable' risk really means 'politically acceptable' risk.

In general, the selection of a *de minimis* risk level is contingent upon the nature of the risks, the stakeholders involved, and a host of other contextual variables (such as other risks being compared against). This means that *de minimis* levels will be fuzzy (in that they can never be precisely specified), and relative (in that they will depend on the special circumstances). Also, establishing a *de minimis* risk level is often extremely difficult because people perceive risks differently. Moreso, the cumulative burden of risks could make a currently insignificant risk become significant in the future. Consequently, stricter *de minimis* standards will usually become necessary in dealing with newly introduced risks that affect the same population groups.

There are several approaches to deriving the *de minimis* risk levels, but that which is selected should be justifiable based on the expected socioeconomic, environmental, and public health impacts. A common approach in placing risks in perspective is to list many risks (which are considered similar in nature), along with some quantitative measures of the degree of risk. Typically, risks below the level of one-in-a-million (i.e., 10^{-6}) chance of premature death will often be considered insignificant or *de minimis* by regulatory agencies in most nations, since this compares favorably with risk levels from several 'normal' human activities – e.g., 10^{-3} for smoking a pack of cigarette/day, or rock climbing, etc.; 10^{-4} for heavy drinking, home accidents, driving motor vehicles, farming; 10^{-5} for truck driving, home fires, skiing, living downstream of a dam, use of contraceptive pills, etc.; 10^{-6} for diagnostic X-rays, fishing, etc.; and 10^{-7} for drinking about 10 liters of diet soda containing saccharin, etc. (Paustenbach, 1988; Rowe, 1977, 1983; Whipple, 1987). In considering a *de minimis* risk level, however, the possibility of multiple *de minimis* exposures with consequential large aggregate risk should not be overlooked. In fact, Whipple (*in* Paustenbach, 1988) suggests the use of a *de minimis* probability idea that will help develop a generally workable *de minimis* policy.

In summary, *de minimis* is a lower bound on the range of acceptable risk for a given activity. When properly utilized, a *de minimis* risk concept can help prioritize risk management decisions in a socially responsible and beneficial way. It may also be used to define the threshold for regulatory involvement. Indeed, it is only after deciding on an acceptable risk level that an environmental or public health risk management program can be addressed in a most cost-effective manner. Ultimately, in order to make a determination of the best environmental or public health risk management strategy to adopt for a given problem situation, a pragmatic and realistic acceptable risk level ought to haven been specified *a priori*.

3.4.   General Attributes of Risk Assessment

The conventional paradigm for risk assessment is *predictive*, which deals with localized effects of a particular action that could result in adverse effects. However, there also is increasing emphasis on assessments of the effects of environmental and public health hazards associated with 'in-place' chemical exposure problems. This assessment of past pollutions and exposures, with possible on-going consequences, generally fall under the umbrella of what has been referred to as *retrospective* risk assessment (Suter, 1993). The impetus for a retrospective risk assessment may be a source, observed effects, or evidence of exposure. Source-driven retrospective assessments result from observed pollution or exposures that requires elucidation of possible effects (e.g., hazardous waste sites, spills/accidental releases, consumer product usage, etc.); effects-driven retrospective assessments result from the observation of apparent effects in the field that requires explanation (e.g., localized public health indicators, fish or bird kills, declining populations of a species, etc.); and exposure-driven retrospective assessments are prompted by evidence of exposure without prior evidence of a source or effects (e.g., the case of a scare over mercury found in the edible portions of dietary fish). In all cases, however, the principal objective of the risk assessment is to provide a basis for actions that will minimize the impairment of the environment and/or of public health, welfare and safety.

In general, risk assessment – which seems to be one of the fastest evolving tools for developing appropriate strategies in relation to environmental and public health management decisions – seeks to answer three basic questions:

- What could potentially go wrong?
- What are the chances for this to happen?
- What are the anticipated consequences, if this should indeed happen?

A complete analysis of risks associated with a given situation or activity will generate answers to these questions. Subsequently, a decision has to be made as to whether any existing risk is sufficiently high to represent a public health concern, and if so, to determine the nature of risk management actions. Appropriate mitigative activities can then be initiated by implementing the necessary corrective action and risk management decisions.

Invariably, tasks performed during a risk assessment will help answer the infamous questions: 'how safe is safe enough?' and/or 'how clean is clean enough?' Subsequently, risk management enters the process to help address the follow-up question of: 'what can be done about the prevailing situation?'

3.4.1. THE PURPOSE

The overall goal in a risk assessment is to identify potential 'system failure modes' and exposure scenarios – and this is achieved via the fulfillment of several general objectives (Box 3.3). This process is intended to facilitate the design of methods that will help reduce the probability of 'failure', and also the attending public health, socioeconomic, and environmental consequences of any 'failure' and/or exposure events. Its purpose is to provide, insofar as possible, a complete information set to risk managers – so that the best possible decision can be made concerning a potentially hazardous situation. Indeed, as Whyte and Burton (1980) succinctly indicate, a major objective of risk assessment is to help develop risk management decisions that are more systematic, more comprehensive, more accountable, and more self-aware of appropriate programs than has often been the case in the past. The risk assessment process provides a framework for developing the risk information necessary to assist risk management decisions; information developed in the risk assessment will typically facilitate decisions about the allocation of resources for safety improvements and hazard/risk reduction. Also, the analysis will generally provide decision-makers with a more justifiable basis for determining risk acceptability, as well as aid in choosing between possible corrective measures developed for risk mitigation programs.

In general, the information generated in a risk assessment is often used to determine the need for, and the degree of mitigation required for chemical exposure problems. For instance, risk assessment techniques and principles are frequently utilized to facilitate the development of effectual site characterization and corrective action programs for contaminated lands scheduled for decommissioning and subsequent re-development for residential housing. In addition to providing information about the nature and magnitude of potential health and environmental risks associated with the contaminated land problem, the risk assessment also provides a basis for judging the need for any type of remedial action (Asante-Duah, 1998). Furthermore, risk assessment can be used to compare the risk reductions afforded by different remedial or risk control strategies. Indeed, the use of risk assessment techniques in contaminated land cleanup plans in particular, and corrective action programs in general, is becoming increasingly important in several places. This is because the risk assessment serves as a useful tool for evaluating the effectiveness of remedies at contaminated sites, and also for establishing cleanup objectives (including the determination of cleanup levels) that will produce efficient, feasible, and cost-effective remedial solutions. Typically, its general purpose is to gather sufficient information that will allow for an adequate and accurate characterization of the potential risks associated with the project site. In these types of situations, the risk assessment process is used to determine whether the level of risk at a contaminated site warrants remediation, and then to further project the amount of risk reduction necessary to protect public health and the environment. On this basis, an appropriate corrective action plan can then be developed and implemented for the case site and/or the impacted area.

Box 3.3. An annotation of typical general objectives of a risk assessment

- Determining if potentially hazardous situations exist – i.e., determine 'baseline' risks and the possible need for corrective action.
- Providing a consistent process for evaluating and documenting public and environmental health threats associated with a potential hazardous situation.
- Estimation of the potential threat to public health and/or the environment that is posed by a facility or hazardous situation – e.g., evaluation of health impacts of emission from industrial facilities and other sources; evaluation of health impacts of chemicals migrating from hazardous waste sites; etc.
- Evaluation of potential risks of new facilities and development projects.
- Estimation of potential health risks associated with use of several chemicals and consumer products – to ensure the development and implementation of acceptable public health policies.
- Pre-marketing safety evaluation of new chemicals (such as pesticides, food additives, drugs, etc.) and other consumer products.
- Post- marketing safety evaluation of existing chemicals (such as pesticides, food additives, drugs, etc.) and other consumer products.
- Determination of the relative size of different problem situations – in order to facilitate priority setting, where necessary.
- Preliminary project scoping – in order to identify possible data gaps in an exposure and risk evaluation problem.
- Determining if there is a need for an immediate response action.
- Identifying possible corrective action strategies.
- Providing basis for comparing and choosing between several remedial action alternatives.
- Providing a basis for determining the levels of chemicals that can remain at a given locale, and still be adequately protective of public health and the environment.
- Providing for the risk management informational needs of property owners and general community.
- Evaluation of product liability and toxic tort claims.

3.4.2. THE ATTRIBUTES

The risk assessment process typically utilizes the best available scientific knowledge and data to establish case-specific responses to hazard-receptor interactions. Depending on the scope of the analysis, the methods used in estimating risks may be either qualitative or quantitative. Thus, the process may be one of data analysis or modeling, or a combination of the two. Indeed, the type and degree of detail of any risk assessment depends on its intended use. Its purpose will shape the data needs, the protocol, the rigor, and related efforts. Overall, the processes involved in any risk assessment usually require a multidisciplinary approach – often covering several areas of expertise in most situations.

In general, the process of quantifying risks does, by its very nature, give a better understanding of the strengths and weaknesses of the potential hazards being examined. It also shows where a given effort can do the most good in modifying a system – in order to improve their safety and efficiency. The major attributes of risk assessment that are particularly relevant to environmental and public health risk management programs include:

- Identification and ranking of all existing and anticipated potential hazards
- Explicit consideration of all current and possible future exposure scenarios
- Qualification and/or quantification of risks associated with the full range of hazard situations, system responses, and exposure scenarios
- Identification of all significant contributors to the critical pathways, exposure scenarios, and/total risks
- Determination of cost-effective risk reduction policies, via the evaluation of risk-based remedial action alternatives and/or the adoption of efficient risk management and risk prevention programs
- Identification and analysis of all significant sources of uncertainties.

As previously noted in Section 3.1.5, there are inherent uncertainties associated with all risk assessments. This is due in part to the fact that the analyst's knowledge of the causative events and controlling factors usually is limited, and also because, to a reasonable extent, the results obtained depend on the methodology and assumptions used. Furthermore, risk assessment can impose potential delays in the implementation of appropriate corrective measures – albeit the overall gain in program efficiency is likely to more than compensate for any delays. Even so, as Moeller (1997) points out, unless care is exercised and all interacting factors considered, then the outcome could be a risk assessment directed at single issues, followed by ill-conceived management strategies – and this can create problems worse than those the management strategies were designed to correct. In fact, the single-issue approach can also create public myopia by excluding the totality of [feasible] alternatives and consequences necessary for a more informed public/stakeholder choice – and thus the need for a more comprehensive evaluation that considers other feasible alternative management strategies.

3.4.3. RISK ASSESSMENT *VERSUS* RISK MANAGEMENT

Risk assessment has been defined as 'the characterization of the potential adverse health effects of human exposures to environmental hazards' (NRC, 1983). In a risk assessment, the extent to which a group of people has been or may be exposed to a certain chemical is determined; the extent of exposure is then considered in relation to the kind and degree of hazard posed by the chemical – thereby allowing an estimate to be made of the present or potential risk to the target population. Depending on the problem situation, different degrees of detail may be required for the process; however, the continuum of acute to chronic hazards and exposures should be fully investigated in a comprehensive assessment, so that the complete spectrum of risks can be defined for subsequent risk management decisions.

The risk management process – that utilizes prior-generated risk assessment information – is used in making a decision on how to protect public health. Examples of risk management actions include: deciding on how much of a chemical a company may discharge into a river; deciding on which substances may be stored at a hazardous waste disposal facility; deciding on the extent to which a hazardous waste site must be cleaned up; setting permit levels for chemical discharge, storage, or transport; establishing levels for air pollutant emissions; and determining the allowable levels of contamination in drinking water or food products.

Overall, risk assessment is generally conducted to aid in risk management decisions. Whereas risk assessment focuses on evaluating the likelihood of adverse effects, risk management involves the selection of a course of action in response to an identified risk – and the latter is based on many other factors (e.g., social, legal, political, or economic) in addition to the risk assessment results. Essentially, risk assessment provides *information* on the health risk, and risk management is the *action* taken based on that information.

3.5. Risk Assessment as a Diagnostic Tool

It has long been recognized that, nothing is wholly safe or dangerous *per se,* but that the object involved, and the manner and conditions of use determine the degree of hazard or safety. Consequently, it may rightly be concluded that there is no escape from all risk, no matter how remote, but that there only are choices among risks (Daniels, 1978). In that spirit, risk assessment is usually designed to offer an opportunity to understanding a system better – by adding an orderliness and completeness to a problem evaluation. It must be acknowledged, however, that risk assessment has usefulness only if it is used properly. Also, the risk analyst must be cognizant of the fact that hazard perception and risk thresholds – which can have significant impact on the ultimate risk decision – tend to be quite different in different regions or locations.

Risk assessment is often considered an integral part of the diagnostic assessment of chemical exposure problems. In its application to the investigation of chemical exposure problems, the risk assessment process encompasses an evaluation of all the significant risk factors associated with all feasible and identifiable exposure scenarios. It includes a characterization of potential adverse consequences or impacts to the populations potentially at risk from the chemical exposure. Procedures typically used in the risk assessment process will usually be comprised of the following tasks:

- Identification of the sources of chemical exposures
- Determination of the chemical exposure routes
- Identification of populations potentially at risk
- Determination of the specific chemicals of potential concern
- Determination of frequency of potential receptor exposures to chemicals
- Evaluation of chemical exposure levels
- Determination of receptor response to chemical exposures
- Estimation of likely impacts or damage resulting from receptor exposures to the chemicals of potential concern.

Potential risks are estimated by considering the probability or likelihood of occurrence of harm; the intrinsic harmful features or properties of specified hazards; the populations potentially at risk; the exposure scenarios; and the extent of expected harm and potential effects.

In most applications, risk assessment is used to provide a baseline estimate of existing risks that are attributable to a specific agent or hazard; the baseline risk assessment consists of an evaluation of the potential threats to human health and the environment in the absence of any remedial or response action. The process can also be used to determine the potential reduction in exposure and risk under various corrective action scenarios, as well as to support remedy selection in risk mitigation/abatement or control programs.

3.5.1. BASELINE RISK ASSESSMENTS

Baseline risk assessments involve an analysis of the potential adverse effects (current or future) caused by receptor exposures to hazardous substances, in the absence of any actions to control or mitigate these exposures (i.e., under an assumption of 'no-action'). Thus, the baseline risk assessment provides an estimate of the potential risks to the populations-at-risk that follows from the receptor exposure to the hazards of concern, when no mitigative actions have been considered. Because this type of assessment identifies the primary threats associated with the situation, it also provides valuable input to the development and evaluation of alternative risk management and mitigative options. In fact, baseline risk assessments are usually conducted to evaluate the need for, and the extent of corrective action required for a hazardous situation; that is, they provide the basis and rationale as to whether or not remedial action is necessary. Overall, the results of the baseline risk assessment are generally used to:

- Document the magnitude of risk at a given locale, as well as the primary causes of the risk.
- Help determine whether any response action is necessary for the problem situation.
- Prioritize the need for remedial action, where several problem situations are involved.
- Provide a basis for quantifying remedial action objectives.
- Develop and modify remedial action goals.
- Support and justify 'no further action' decisions – as appropriate, by documenting the likely insignificance of the threats posed by the hazard source(s).

In general, baseline risk assessments are designed to be case-specific – and therefore may vary in both detail and the extent to which qualitative and quantitative analyses are used. The level of effort required to conduct a baseline risk assessment depends largely on the complexity and particular circumstances associated with the hazard situation under consideration.

3.5.2. COMPARATIVE RISK ASSESSMENTS

Comparative risk assessment (CRA) has become an important aspect of risk analysis. In essence, CRA is directed at developing risk rankings and priorities that would put various kinds of hazards on an ordered scale from small to large (ACS and RFF, 1998; NRC, 1989). ACS and RFF (1998) identify two principal forms of CRA – namely:

- *Specific risk comparisons* – that involve side-by-side evaluations of distinct risks, on the basis of likelihood and severity of effects. This form of CRA is comprised of a side-by-side evaluation of the risk (on an absolute or relative basis) associated with exposures to a few substances, products, or activities. Such comparisons may involve similar risk agents (e.g., the comparative cancer risks of two chemically similar pesticides) or widely different agents (e.g., the cancer risk from a particular pesticide compared with the risk of death or injury from automobile travel). Specific risk comparisons can be particularly useful when one is considering the relative importance of risks within the context of similar products, activities, or risk management actions. A more popular application has been in the area of risk communication – where such comparisons have been helpful in facilitating non-technical audiences' understanding of the significance of varying risk levels (as for example, weighing the expected risks of new products or technologies against those that are already accepted or tolerated). Paired comparisons of reasonably similar risks represent the most straightforward application of comparative risk analysis; such evaluations may be conducted simply, based on estimated risk levels and the extent of anticipated harm. For example, a pair of chemical pesticides might be compared with respect to their expected chronic health effects, adjusted for likelihood.
- *Programmatic comparative risk assessment* – which seeks to make macro-level (i.e., 'big-picture') comparisons among many and widely different types of hazards/risks. This is usually carried out in order to provide information for setting regulatory and budgetary priorities for hazard reduction. In this kind of comparison, risk rankings are based on the relative magnitude of risk (i.e., which hazards pose the greatest threat) or on relative risk reduction opportunities (i.e., the amount of risk that can be avoided with available technologies and resources). In fact, by its nature, programmatic CRA spans many, dissimilar risks and provides a ready forum for value debates over what is important in gauging the seriousness of a hazard and in establishing priorities. Arguably, the major strength of programmatic CRA is the opportunity it provides for discussion and debate among various important points of view – especially those from technical experts, policy makers, and the public.

Indeed, methods for appropriately carrying out these kinds of analyses remain controversial, as is the concept of using relative risk comparisons to establish priorities for hazard reduction. The challenges are particularly difficult when comparisons across widely different risks are involved. Additionally, all risk comparisons become considerably more complex when the views of differing individuals are brought into focus – many of whom may disagree on matters such as the relevant attributes for comparisons, the trade-off relationships to be assumed, and the way uncertainty should be included in the analysis. Furthermore, sizable uncertainties – such as relates to the nature of health effects, the level of exposures, or various other factors – can make it difficult to combine the various attributes of a hazard into a single risk measure and, thereby, blunt the precision of the comparison process (ACS and RFF, 1998). In general, risk comparisons are especially useful in situations requiring the comparison of the risks of alternative options, and also to gauge the importance of different causes of the same hazard.

3.5.3. PUBLIC HEALTH RISK ASSESSMENTS

A public health assessment typically consists of a review of available information about hazardous substances in the human environments – followed by an evaluation of whether exposure to these substances might cause any harm to people. It must be emphasized here that, a public health assessment is *not* the same thing as a 'medical examination' or a 'community health study' – albeit it can sometimes lead to those types of evaluation, as well as to other public health risk management actions. Overall, all forms of public health assessment consider the following key issues:

- Levels (or concentrations) of hazardous substances present
- Likelihood that people might be exposed to chemicals present in a locale
- Exposure pathways and routes (such as breathing air, drinking or contacting water, contacting or eating soil, or eating food) via which people might exposed to the chemicals of concern
- Nature of harm the substances might cause to people (i.e., the chemical toxicity)
- Potential health impacts on populations working and/or living near the chemical source(s)
- Other dangers that could potentially exacerbate the likely effects of the chemical exposure problems.

The following three primary sources of information may be used to make the above determinations:

- *Environmental data* – such as information about the chemical constituents and how people could come in contact with them;
- *Health data* – including available information on community-wide morbidity and mortality rates, or the local incidences of illness, disease, and death in comparison with national and/or other regional/provincial/state incident rates; and
- *Community concerns* – such as reports from the public about how a hazardous chemical exposure is affecting the community's health and/or quality of life.

Ultimately, the public health assessment may be used to identify health studies or other public health risk management actions, such as community environmental health education, that might be needed. The information generated this process provides a basis for actions by policy makers – usually consisting of actions to prevent or reduce population exposures to hazardous substances.

Conducting an Exposure Investigation
An exposure investigation is one approach used to develop better characterization of past, current, and possible future human exposures to hazardous substances in the human living and work environments, and to more thoroughly evaluate existing and possible health effects related to those exposures. Information is typically gathered in three main ways during an exposure investigation, viz.:

- *Biomedical testing* – as for example, urine or blood samples can serve as an important source of information that may be gathered and evaluated during an exposure investigation. Biomedical samples can show current (and sometimes past) exposures to a chemical constituent.

- *Environmental testing* – associated with contaminated environmental media (for example, contamination of soil, water, or air) can serve as an important source of information gathered and evaluated during an exposure investigation. Investigators may focus environmental testing on where people live and/or work, or indeed any place where they might come in contact with the substances under investigation.

- *Exposure-dose reconstruction analyses* – involving the use of environmental sampling information and computer models to estimate the constituent levels that people may have been exposed to in the past, or could become exposed to in the future. These models can then be used to draw various conclusions about the receptor exposure durations and levels of exposures.

Subsequently, the types of information derived above can be used to evaluate how a person's health might be affected. Typically, a team of scientists with various specialties in environmental sampling and computer analyses, geographic information systems, epidemiology, toxicology, and medicine is assembled to work on this type of investigation. The team uses information from the exposure investigations and other scientific resources to make public health policy decisions, prepare reports, and recommend appropriate public health risk management actions.

Biomonitoring: Utilization of Exposure Biomarkers
Exposure assessment typically has been a rather weak link in the assessment of risk from chemical exposure problems; exposure biomarkers usually will provide some strength to this component of a risk assessment. Indeed, these endpoints provide evidence that exposure has occurred, resulting in absorption by the body; these endpoints also provide the data that may be compared to exposure measurements and analyzed through pharmacokinetic modeling, in order to estimate target tissue dose and risk (Saleh *et al.*, 1994). Furthermore, biomarkers serve as a means of determining aggregate/cumulative risks – by providing a measure of integrated exposure, i.e., exposure that occurs by all routes into the human body or other organism.

Exposure biomarker information seems to be quite important in the evaluation of the impacts from human exposure to a variety of chemicals. Two basic types of biomarkers may generally be defined, namely (Saleh *et al.*, 1994):

(1) Residue analysis (of parent compounds or metabolites) in easily sampled matrices; and
(2) Endpoints that represent interactions between xenobiotic and endogenous components (e.g., enzyme inhibition, protein adducts, receptor complexes, antibody-antigen complexes, and mutation).

Both of the above types of biomarkers are used in biomonitoring studies, in order to assess exposures more fully; biomonitoring consists of the routine analysis of human tissues or excreta for direct or indirect evidence of chemical exposures. For instance, detection of certain compounds (such as pesticides) in the body indicates that an exposure has occurred; that the chemical is bioavailable, having been absorbed; and that a dose to critical body tissues may have been incurred (Saleh *et al.*, 1994). It is noteworthy, however, that several variables do indeed affect the biomarker assessment process. For instance, target tissue dose depends upon the exposure rate as well as the kinetics of the chemical uptake, intake, internal distribution, and storage/elimination. Considering the large variation in susceptibility to toxicant insults of the general population, therefore, neither detection nor determination of concentration of a toxicant in tissues of individuals or the general population has satisfactorily provided a quantitative estimate of risk to human health (Saleh *et al.*, 1994). Indeed, the extent to which a human population is susceptible to a toxic stressor depends not only on the intensity and duration of exposure – but also on the rate of uptake, intake, metabolism, storage, excretion, abundance of target macromolecules at the cellular level, and potential for adaptation to the toxicant (Saleh *et al.*, 1994). Notwithstanding any limitations, the biological monitoring of certain chemical residues and metabolites in the human body is becoming increasingly important in the surveillance of occupationally and environmentally exposed individuals. This is especially true because, biomonitoring data can complement occupational and environmental monitoring data (e.g., personal exposure measurements, ambient and micro-environmental measurements, as well as human activity pattern information) in reducing the uncertainty inherent in exposure or risk assessments.

Environmental and biological monitoring are two key elements in the determination of human exposures to chemicals for risk assessment and risk management decisions. The common methods of the biomonitoring process often involve the chemical analyses of readily sampled matrices (such as urine and blood) for parent compounds and/or metabolites. Also, immuno-chemical methods continue to be developed for screening purposes – and even beyond screening investigations. Although a number of innovative biomarkers for human exposure to chemicals have been reported (such as DNA alterations, protein adducts and changes to enzymatic and immunological systems), measurement of pollutants and their metabolites in blood and urine continues to dominate human biomonitoring efforts (Saleh *et al.*, 1994). To ensure an effective biomonitoring program, it is important to ascertain that, among other things, the most appropriate biological matrix is sampled and that the most appropriate analytes are investigated in the ensuing laboratory analyses using the most appropriate analytical methods.

Physiologically-Based Pharmacokinetic (PBPK) Modeling
The science of pharmacokinetics describes the time course disposition of a xenobiotic, its biotransformed products, and its interactive products within the body. This includes a description of the compound's absorption across the portals of entry, transport and distribution throughout the body, biotransformation by metabolic processes, interactions with biomolecules, and eventual elimination from the body (Saleh *et al.*, 1994). With more emphasis being placed on internal (tissue) dose for quantitating exposure between species, PBPK modeling is finding increasing use in the risk assessment process (Derelanko and Hollinger, 1995).

In a typical PBPK model, each tissue group is described mathematically by a series of differential equations that express the rate of change of a chemical of concern in each compartment. The rate of exchange between compartments is based on species-specific physiological parameters. Also, the number of compartments and their interrelationships will vary depending on the nature of the chemical being modeled.

PBPK models use physiologic and thermodynamic parameters; organ volumes, blood flows, and metabolic rate constants are determined – and these become part of the model. Additional parameters, such as partition coefficients, are thermodynamic – but may also be chemical-specific. In practice, appropriate thermodynamic and biochemical parameters must be determined for each chemical of potential concern.

In general, physiologic models enable a public health risk analyst to quantitatively account for differences in pharmacokinetics that occur between different species, dose levels, and exposure regimens/scenarios. For example, PBPK models have been extensively used to predict the allowable exposure levels in human health risk assessment – based on animal studies through route-to-route, high-to-low dose, and laboratory animal-to-human extrapolations. Indeed, PBPK models can be rather powerful tools for interspecies extrapolations – provided the biological processes are well understood and the pertinent parameter values can be accurately measured. It is noteworthy, however, that no one PBPK model can represent the kinetics of all chemicals.

3.5.4. RISK ASSESSMENT IMPLEMENTATION STRATEGY

A number of techniques are available for conducting risk assessments. Invariably, the methods of approach consist of the several basic procedural elements/components that are further outlined in Chapter 4 of this book. The key issues requiring significant attention in the processes involved will typically involve finding answers to the following questions:

- What chemicals pose the greatest risk?
- What are the concentrations of the chemicals of concern in the exposure media?
- Which exposure routes are the most important?
- Which population groups, if any, face significant risk as a result of the possible exposures?
- What is the range of risks to the affected populations?
- What are the public health implications for any identifiable corrective action and/or risk management alternatives?

As a general guiding principle, risk assessments should be carried out in an iterative fashion, that is appropriately adjusted to incorporate new scientific information and regulatory changes – but with the ultimate goal being to minimize public health and socioeconomic consequences associated with a potentially hazardous situation. Typically, an iterative approach would start with relatively inexpensive screening techniques – and then for hazards suspected of exceeding the *de minimis* risk, further evaluation is conducted by moving on to more complex and resource-intensive levels of data-gathering, model construction, and model application (NRC, 1994).

In general, risk assessment will normally be conducted in an iterative manner that grows in depth with increasing problem complexity. Consider, as an example, a site-specific risk assessment that is used to evaluate/address potential health impacts associated with chemical releases from industrial facilities or hazardous waste sites. A tiered approach is generally recommended in the conduct of such site-specific risk assessments. Usually, this will involve two broad levels of detail – i.e., a 'screening' and a 'comprehensive' evaluation. In the screening evaluation, relatively simple models, conservative assumptions, and default generic parameters are typically used to determine an upper-bound risk estimate associated with a chemical release from the case facility. No detailed/comprehensive evaluation is warranted if the initial estimate is below a pre-established reference or target level (i.e., the *de minimis* risk). On the other hand, if the screening risk estimate is above the *de minimis* risk level, then the more comprehensive/detailed evaluation (that utilizes more sophisticated and realistic data evaluation techniques than were employed in the 'Tier 1' screening) should be carried out. This next step will confirm the existence (or otherwise) of significant risks – which then forms the basis for developing any risk management action plans. The rationale for such a tiered approach is to optimize the use of resources – in that it makes efficient use of time and resources, by applying more advanced and time-consuming techniques to chemicals of potential concern and scenarios only where necessary. Thus is, the comprehensive/detailed risk assessment is performed only when truly warranted. Irrespective of the level of detail, however, a well-defined protocol should always be used to assess the potential risks. Ultimately, a decision on the level of detail (e.g., qualitative, quantitative, or combinations thereof) at which an analysis is carried out will usually be based on the complexity of the situation, as well as the uncertainties associated with the anticipated or predicted risk.

3.6.   Risk Assessment as an Holistic Tool for Environmental and Public Health Management

Risk assessment is a process used to determine the magnitude and probability of actual or potential harm that a hazardous situation poses to human health and the environment. To date, risk assessment has been used in Europe for a relatively constrained set of purposes – chiefly to assess new and existing chemical substances (including pesticides), pharmaceutical products, cosmetics and food additives. There also are proposals and established plans for its application in the occupational health and safety field, as well as for possible use in site remediation decisions in some countries (see, e.g., Cairney, 1995; Ellis and Rees, 1995; HSE, 1989a,b; Smith, 1996). By contrast, risk assessment principles and methodologies have found extensive and a wide variety of applications in the United States for several years. It has typically been used to evaluate many forms of new products (e.g., foods, drugs, cosmetics, pesticides, consumer products); to set environmental standards (e.g., for air and water); to predict the health threat from contaminants in air, water, and soils; to determine when a material is hazardous (i.e., to identify hazardous wastes and toxic industrial chemicals); to set occupational health and safety standards; and to evaluate soil and groundwater remediation efforts (see, e.g., Asante-Duah, 1998; ASTM, 1995; McTernan and Kaplan, 1990; Millner *et al.*, 1992; NRC, 1993, 1995; Shere, 1995; Sittig, 1994; Smith, 1996; Smith *et al.*, 1996; Tsuji and Serl, 1996). For now, risk assessment applications in most of the other parts of the world appear to be limited and sporadic. But that is expected to change before too long, as the world continues to search for cost-effective and credible environmental and public health management tools. In fact, in the wake of the June 1992 UN Conference on Environment and Development in Rio de Janeiro, the global/international community's reliance on risk assessment as an effectual environmental and/or public health management tool is likely to grow well into the future. A growing trend in its use is indeed expected, despite skepticism expressed by some (see, e.g., Shere, 1995) who consider the art and science of risk assessment more as a mythical subject rather than real – and despite the fact that the process may be fraught with several sources of uncertainty.

As an holistic approach to environmental and public health management, risk assessment integrates all relevant environmental and health issues and concerns surrounding a specific problem situation, in order to arrive at risk management decisions that are acceptable to all stakeholders. Among other things, the overall process should generally incorporate information that helps to answer the following pertinent questions:

- Why is the project/study being undertaken?
- How will results and conclusions from the project/study be used?
- What specific processes and methodologies will be utilized?
- What are the uncertainties and limitations surrounding the study?
- What contingency plans exist for resolving newly identified issues?

Also, effective risk communication should be recognized as a very important element of the holistic approach to managing chemical exposure and related environmental hazard problems (Asante-Duah, 1998). Thus, a system for the conveying of risk information derived from a risk assessment should be considered as a very essential integral part of the overall approach.

Finally, to effectively utilize it as a public health management tool, risk assessment should be recognized as a multidisciplinary process that draws on data, information, principles, and expertise from many scientific disciplines – including biology, chemistry, earth sciences, engineering, epidemiology, medicine and health sciences, physics, toxicology, and statistics, among others. Indeed, risk assessment may be viewed as bringing a wide range of subjects and disciplines – from 'archaeology to zoology' – together.

3.7.   Suggested Further Reading

Bate, R. (ed.), 1997. *What Risk? (Science, Politics & Public Health)*, Butterworth-Heinemann, Oxford, UK

Bates, DV, 1994. *Environmental Health Risks and Public Policy*, University of Washington Press, Seattle, Washington

Bromley, DW and K. Segerson (eds.), 1992. *The Social Response to Environmental Risk: Policy Formulation in an Age of Uncertainty*, Kluwer Academic Publishers, Boston, MA

Hamed, MM, 1999. Probabilistic sensitivity analysis of public health risk assessment from contaminated soil, *Journal of Soil Contamination*, 8(3): 285 – 306

Hamed, MM, 2000. Impact of random variables probability distribution on public health risk assessment from contaminated soil, *Journal of Soil Contamination*, 9(2): 99 – 117

Hammitt, JK, 1995. Can more information increase uncertainty? *Chance*, 8(3): 15 – 17

Hammitt, JK and AI Shlyakhter, 1999. The expected value of information and the probability of surprise, *Risk Analysis*, 19(1): 135-152

Hansson, S-O, 1989. Dimensions of risk, *Risk Analysis*, 9(1): 107-112

Hansson, S-O, 1996. Decision making under great uncertainty, *Philosophy of the Social Sciences*, 26(3): 369 – 386

Hansson, S-O, 1996. What is philosophy of risk? *Theoria,* 62: 169 – 186

Joffe, M. and J. Mindell, 2002. A framework for the evidence base to support health impact assessment, *Journal of Epidemiology & Community Health,* 56(2): 132 – 132

Kimmel, CA and DW Gaylor, 1988. Issues in qualitative and quantitative risk analysis for developmental toxicology, *Risk Analysis,* 8: 15 – 20

Pollard, SJ, R. Yearsley, *et al.,* 2002. Current directions in the practice of environmental risk assessment in the United Kingdom, *Environmental Science & Technology,* 36(4): 530 – 538

Richards, D. and WD Rowe, 1999. Decision-making with heterogeneous sources of information, *Risk Analysis,* 19(1): 69 – 81

van Ryzin, J., 1980. Quantitative risk assessment, *Journal of Occupational Medicine,* 22: 321 – 326

Chapter 4

PRINCIPAL ELEMENTS OF
A PUBLIC HEALTH RISK ASSESSMENT
FOR CHEMICAL EXPOSURE PROBLEMS

In planning for public health protection from the likely adverse effects from human exposure to chemicals, the first concern usually relates to whether or not the substance in question possesses potentially hazardous and/or toxic properties. As a corollary, once a 'social chemical' has been determined to present a potential health hazard, then the main concern becomes one of the likelihood for, and the degree of human exposure. Ultimately, risk from human exposure to a chemical of concern is determined to be a function of dose or intake and potency of the substance, viz.:

$$Risk\ from\ chemical\ exposure = [Dose\ of\ chemical] \times [Chemical\ potency] \qquad (4.1)$$

Indeed, both exposure and toxicity information are necessary to fully characterize the potential hazard of a chemical agent – or indeed any other hazardous agent for that matter. This chapter discusses the principal elements and activities necessary for obtaining and integrating the pertinent information that will eventually allow effective public health risk management decisions to be made about chemical exposure problems.

4.1. Characterization of Chemical Exposure Problems

Human exposure to a chemical agent is considered to be an event comprised of the contacting at a boundary between a human body or organ and the chemical-containing medium, at a specific chemical concentration, for a specified time interval. Upon exposure, a receptor generally receives a dose of the chemical – and that may be quite different from the exposed amount; indeed, dose is different from (but occurs as a result of) an exposure (NRC, 1991c). The dose is defined as the amount of the chemical that is absorbed or deposited in the body of an exposed individual over a specified time. A clear understanding of such differences in the exposure parameters is indeed critical to the design of an adequate exposure characterization plan.

The characterization of chemical exposure problems is a process used to establish the presence or absence of chemical hazards, to delineate the nature and degree of the hazards, and to determine possible threats posed by the exposure or hazard situation to human health. The exposure routes (which may consist of inhalation, ingestion, and/or dermal contacts) and duration of exposure (that may be short-term [acute] or long-term [chronic]) will significantly influence the degree of impacts on the affected receptors. The nature and behavior of chemical substances also form a very important basis for evaluating the potential for human exposures to the possible toxic or hazardous constituents of the substance.

While the need for and/or reliance on models and default assumptions is almost always inevitable in most chemical exposure characterization problems, the use of applicable empirical data in exposure assessments is strongly recommended. Information obtained (through monitoring studies) from assessment of direct exposure (e.g., drinking contaminated water) and/or indirect exposure (e.g., accumulation of contaminants via the food chain) should preferably be used. Ideally, the assessment will include monitored levels of the chemical agent in the chemical-containing media, and in human tissues and fluids – and in particular, estimates of the dose at a biologic target tissue(s) where an effect(s) may occur. Such information is necessary to accurately evaluate the potential health risk of exposed populations. Of course, in the absence of complete monitoring information, mathematical exposure assessment models may be employed. These models provide a methodology through which various factors, such as the temporal/spatial distribution of a chemical agent released from a particular source, can be combined to predict levels of human exposures. Still, modeling may not necessarily be viewed as a fully satisfactory substitute for adequate data – but rather as a surrogate to be employed when confronted by compelling needs and inadequate data. Uncertainty associated with these and all methods must be carefully documented and explained to the extent feasible.

4.1.1. FACTORS AFFECTING EXPOSURE CHARACTERIZATION

Several chemical-specific, receptor-specific, and even environmental factors need to be recognized and/or evaluated as an important part of any public health risk management program that is designed to address problems that arise from exposure of the public to various chemical substances. The general types of data and information necessary for the investigation of potential chemical exposure problems relate to the following:

- Identities of the chemicals of concern;
- Concentrations contacted by potential receptors of interest;
- Receptor characteristics;
- Characteristics of the physical and environmental setting that can affect behavior and degree of exposure to the chemicals; and
- Receptor response upon contact with the target chemicals.

In addition, it is necessary to generate information on the chemical intake rates for the specific receptor(s), together with numerous other exposure parameters. Indeed, all parameters that could potentially impact the human health outcomes should be carefully evaluated; this includes the following especially important categories annotated below.

- *Exposure duration and frequency.* A single high-dose exposure to a hazardous agent may result in toxic effects quite different from those following repeated lower dose exposures. Thus, in evaluating the chemical risks, adequate consideration should be given to the duration – namely, acute (usually ≤14 days) *vs.* intermediate (usually 15 – 364 days) *vs.* chronic (usually ≥365 days); the intensity (i.e., dose rate *vs.* total dose); and the frequency (continuous or intermittent) of exposure. These exposure parameters are carefully evaluated, alongside the relevant pharmacokinetic parameters for the constituents of concern.
- *Exposure media and routes.* Exposure to hazardous substances is often a complex phenomenon – entailing exposures via multiple routes and/or media. Thus, all possible exposure media, pathways, and routes should be appropriately investigated and accounted for in the characterization of a chemical exposure situation.
- *Target receptor attributes.* Receptor behavior and activity patterns, such as the amount of time a receptor spends indoors compared with that spent outdoors, and its underlying variability in assessing potential human health effects should be carefully evaluated. Also, it should be recognized that factors such as nutritional status and lifestyle variables (e.g., tobacco smoking, alcohol consumption, and occupation) might all affect the health risks associated with the particular chemical exposure problem under consideration.
- *Potential receptor exposures history.* Chemical exposure effects may occur in populations not only as a result of current exposure to agents but also from past exposures. Thus, past, current, and potential future exposure to hazardous substances should all be carefully evaluated as part of an overall long-term public health risk assessment program.

Indeed, the above listing is by no means complete for the universe of potential exposure possibilities – but certainly represent the critical ones that must certainly be examined rather closely.

Overall, the exposure outcomes depend on the conditions of exposure such as the amount, frequency, duration, and route of exposure (i.e., ingestion, inhalation, and dermal contact). Also, for most environmental chemicals, available health effects information is generally limited to high exposures in studies of humans (e.g., occupational studies of workers) or laboratory animals. Thus, evaluation of potential health effects associated with low levels of exposure generally encountered in the human living and work environments involves inferences based on the understanding of the mechanisms of chemical-induced toxicity. Furthermore, one should be cognizant of the fact that, in general, chemicals frequently affect more than one organ or system in the human body (e.g., liver, kidney, nervous system), and can also produce a variety of health endpoints (e.g., cancer, respiratory allergies, infertility).

4.2.   The Risk Assessment Process

Risk assessment is a scientific process that can be used to identify and characterize chemical exposure-related human health problems. Specific forms of risk assessment generally differ considerably in their levels of detail. Most risk assessments, however, share the same general logic – consisting of four basic elements, namely, hazard assessment, dose-response assessment, exposure assessment, and risk characterization (Figure 4.1).

Hazard assessment describes, qualitatively, the likelihood that a chemical agent can produce adverse health effects under certain environmental exposure conditions. *Dose-response assessment* quantitatively estimates the relationship between the magnitude of exposure and the degree and/or probability of occurrence of a particular health effect. *Exposure assessment* determines the extent of human exposure. *Risk characterization* integrates the findings of the first three components to describe the nature and magnitude of health risk associated with environmental exposure to a chemical substance or a mixture of substances. A discussion of these fundamental elements follows – with more detailed elaboration given in Chapters 5 through 8 of this title, and also elsewhere in the risk analysis literature (e.g., Asante-Duah, 1998; Cohrssen and Covello, 1989; Conway, 1982; Cothern, 1993; Gheorghe and Nicolet-Monnier, 1995; Hallenbeck and Cunningham 1988; Huckle 1991; Kates, 1978; Kolluru *et al.*, 1996; LaGoy, 1994; Lave, 1982; McColl, 1987; McTernan and Kaplan, 1990; Neely, 1994; NRC 1982, 1983, 1994; Paustenbach 1988; Richardson, 1990; Rowe 1977; Suter, 1993; USEPA 1984, 1989; Whyte and Burton, 1980).

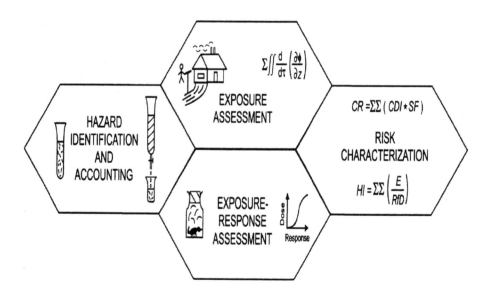

Figure 4.1. Illustrative elements of a risk assessment process

4.2.1. HAZARD IDENTIFICATION AND ACCOUNTING

Hazard identification and accounting involves a qualitative assessment of the presence of, and the degree of hazard that an agent could have on potential receptors. The hazard identification consists of gathering and evaluating data on the types of health effects or diseases that may be produced by a chemical, and the exposure conditions under which public health damage, injury or disease will be produced. It may also involve characterization of the behavior of a chemical within the body and the interactions it undergoes with organs, cells, or even parts of cells. Data of the latter types may be of value in answering the ultimate question of whether the forms of toxic effects determined to be produced by a substance in one population group or in experimental settings are also likely to be produced in humans.

Hazard identification is not a risk assessment *per se.* This process involves simply determining whether it is scientifically correct to infer that toxic effects observed in one setting will occur in other settings – e.g., whether substances found to be carcinogenic or teratogenic in experimental animals are likely to have the same results in humans. In the context of public health risk management for potential chemical exposure problems, this may consist of:

- Identification of chemical exposure sources;
- Compilation of the lists of all chemical stressors present at the locale and impacting target receptors;
- Identification and selection of the specific chemicals of potential concern (that should become the focus of the risk assessment), based on their specific hazardous properties (such as persistence, bioaccumulative properties, toxicity, and general fate and behavior properties); and
- Compilation of summary statistics for the key constituents selected for further investigation and evaluation.

It is noteworthy that, in identifying the chemicals of potential concern, an attempt is generally made to select all chemicals that could possibly represent the major part (usually, $\geq 95\%$) of the risks associated with the relevant exposures.

4.2.2. EXPOSURE-RESPONSE EVALUATION

The *exposure-response evaluation,* (or the *effects assessment*) is the estimation of the relationship between dose or level of exposure to a substance and the incidence and severity of an effect. It considers the types of adverse effects associated with chemical exposures, the relationship between magnitude of exposure and adverse effects, and related uncertainties (such as the weight-of-evidence of a particular chemical's carcinogenicity in humans).

Dose-response assessment involves describing the quantitative relationship between the amount of exposure to a substance and the extent of toxic injury or disease. Data are derived from animal studies or, less frequently, from studies in exposed human populations. There may be many different dose-response relationships for a substance if it produces different toxic effects under different conditions of exposure. It is noteworthy that, even if the substance is known to be toxic, the risks of a substance cannot be ascertained with any degree of confidence unless dose-response relations are quantified.

In the context of chemical exposure problems, this evaluation will generally include a toxicity assessment and/or a dose-response evaluation. The toxicity assessment typically consists of compiling toxicological profiles for the chemicals of potential concern. Dose-response relationships are then used to quantitatively evaluate the toxicity information, and to characterize the relationship between dose of the contaminant administered or received and the incidence of adverse effects on the exposed population. From the quantitative dose-response relationship, appropriate toxicity values can be derived, and this is subsequently used to estimate the incidence of adverse effects occurring in populations at risk for different exposure levels.

4.2.3. EXPOSURE ASSESSMENT AND ANALYSIS

An *exposure assessment* is conducted in order to estimate the magnitude of actual and/or potential receptor exposures to chemicals present in the human environments. The process considers the frequency and duration of the exposures, the nature and size of the populations potentially at risk (i.e., the risk group), and the pathways and routes by which the risk group may be exposed. Indeed, several physical and chemical characteristics of the chemicals of concern will provide an indication of the critical exposure features. These characteristics can also provide information necessary for determining the chemical's distribution, intake, metabolism, residence time, excretion, magnification, and half-life or breakdown to new chemical compounds.

In general, exposure assessments involve describing the nature and size of the population exposed to a substance and the magnitude and duration of their exposure. The evaluation could concern past or current exposures, or exposures anticipated in the future. To complete a typical exposure analysis for a chemical exposure problem, populations potentially at risk are identified, and concentrations of the chemicals of concern are determined in each medium to which potential receptors may be exposed. Finally, using the appropriate case-specific exposure parameter values, the intakes of the chemicals of concern are estimated. The exposure estimates can then be used to determine if any threats exist – based on the prevailing exposure conditions for the particular problem situation.

4.2.4. RISK CHARACTERIZATION AND CONSEQUENCE DETERMINATION

Risk characterization is the process of estimating the probable incidence of adverse impacts to potential receptors under a set of exposure conditions. Typically, the risk characterization summarizes and then integrates outputs of the exposure and toxicity assessments – in order to be able to qualitatively and/or quantitatively define risk levels. The process will usually include an elaboration of uncertainties associated with the risk estimates. Exposures resulting in the greatest risk can be identified in this process – and then mitigative measures can subsequently be selected to address the situation in order of priority, and according to the levels of imminent risks.

In general, risk characterizations involve the integration of the data and analysis of the first three components of the risk assessment process (viz., hazard identification, dose-response assessment, and exposure assessment) – to determine the likelihood that humans will experience any of the various forms of toxicity associated with a substance. (In cases where exposure data are not available, hypothetical risks can be characterized by the integration of hazard identification and dose-response evaluation data alone.) A framework to define the significance of the risk is developed, and all of the assumptions, uncertainties, and scientific judgments from the three preceding steps are also presented.

To the extent feasible, the risk characterization should include the distribution of risk amongst the target populations. Ultimately, an adequate characterization of risks from hazards associated with chemical exposure problems allows risk management and corrective action decisions to be better focused.

4.3. General Considerations in Public Health Risk Assessments

Human health risk assessment for chemical exposure problems may be defined as the characterization of the potential adverse health effects associated with human exposures to chemical hazards. In a typical human health risk assessment process, the extent to which potential receptors have been, or could be exposed to chemical hazards is determined. The extent of exposure is then considered in relation to the type and degree of hazard posed by the chemical(s) – thereby permitting an estimate to be made of the present or future health risks to the populations-at-risk.

Figure 4.2 shows the basic components and steps typically involved in a comprehensive human health risk assessment that is designed for use in environmental and public health risk management programs. Several key aspects of the human health risk assessment methodology are presented in the proceeding chapters of this volume – with additional details provided elsewhere in the literature (e.g., Hoddinott, 1992; Huckle 1991; NRC 1983; Patton, 1993; Paustenbach, 1988; Ricci, 1985; Ricci and Rowe, 1985; USEPA 1984a, 1984b, 1985a, 1986a, 1986b, 1986c, 1986d, 1987a, 1987b, 1989d, 1991, 1992; Van Leeuwen and Hermens).

Invariably, the management of all chemical exposure problems starts with hazard identification, and/or a data collection and data evaluation phase. The data evaluation aspect of a human health risk assessment consists of an identification and analysis of the chemicals associated with a chemical exposure problem that should become the focus of the public health risk management program. In this process, an attempt is generally made to select all chemicals that could represent the major part of the risks associated with case-related exposures; typically, this will consist of all constituents contributing ≥95% of the overall risks. Chemicals are screened based on such parameters as toxicity, carcinogenicity, concentrations of the detected constituents, and the frequency of detection in the sampled matrix.

The exposure assessment phase of the human health risk assessment is used to estimate the rates at which chemicals are absorbed by potential receptors. Since most potential receptors tend to be exposed to chemicals from a variety of sources and/or in different environmental media, an evaluation of the relative contributions of each medium and/or source to total chemical intake could be critical in a multi-pathway exposure analysis. In fact, the accuracy with which such exposures are characterized could be a major determinant of the ultimate validity of the risk assessment.

The quantitative evaluation of toxicological effects consists of a compilation of toxicological profiles (including the intrinsic toxicological properties of the chemicals of concern, which may include their acute, subchronic, chronic, carcinogenic, and/or reproductive effects) and the determination of appropriate toxicity indices (see Chapter 7 and Appendix C).

Finally, the risk characterization consists of estimating the probable incidence of adverse impacts to potential receptors under various exposure conditions. It involves an integration of the toxicity and exposure assessments, resulting in a quantitative estimation of the actual and potential risks and/or hazards due to exposure to each key chemical constituent, and also the possible additive effects of exposure to mixtures of the chemicals of potential concern.

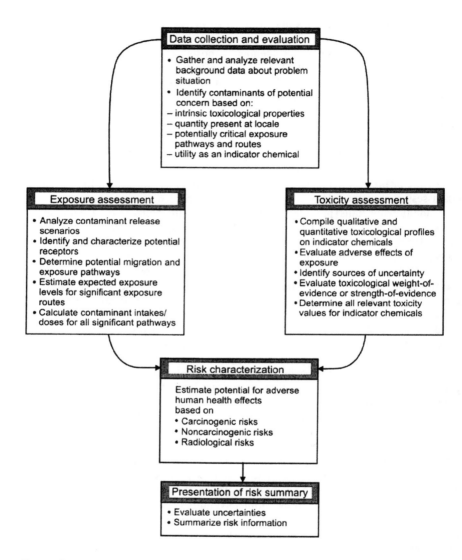

Figure 4.2. A general protocol for the human health risk assessment process: fundamental procedural components of a risk assessment for a chemical exposure problem

4.3.1. DETERMINING EXPOSURE-RELATED HEALTH EFFECTS

Exposure-related health effects of chemical substances introduced into the human living and work environments may be determined within the framework of a public health risk assessment process.

In general, when evaluating the health impact of exposure to hazardous substances, the analyst should consider data from studies of human exposures as well as from the results of experimental animal studies. For health assessment purposes, the use of human data is preferred – because it eliminates uncertainties involved in extrapolating across species. However, human data are often unavailable, particularly for chronic, low-dose exposures. Furthermore, adequate human data are often not available to establish a dose-response relationship. In the absence of adequate human data, therefore, the public health analyst must rely on the results of experimental animal studies. Also, in many chemical exposure situations, exposures must often be characterized as chronic and of low dose. Health effects data and information for such exposures are often lacking. In these types of situations, the health analyst may have to rely on studies that involve shorter exposures and/or higher dose levels. If such studies are used as the basis for a health assessment, the analyst should acknowledge the qualitative and quantitative uncertainties involved in those extrapolations. Finally, it is recommended that estimated chemical exposures be compared to studies or experiments involving comparable routes of exposure – i.e., ingestion, inhalation, and dermal contact. However, in some instances, it may be necessary to use data from studies based on different exposure pathways. Caution should be used when drawing conclusions from such studies because of the uncertainties involved in route-to-route extrapolations – and that is the result of the differences in chemical absorption, distribution, metabolism, and excretion. In addition, a chemical might exert a toxic effect by one route of exposure, but not by another (e.g., chromium is reported to be carcinogenic by inhalation but not by ingestion); such differences should be carefully evaluated.

Indeed, to facilitate the development of responsible public health risk management programs, it is important for the public health analyst to use the best medical and toxicological information available to determine the health effects that may result from exposure to the chemical constituents of concern. Such information can be derived from existing chemical-specific toxicological profiles (e.g., Toxicological Profiles from the ATSDR, and IRIS from the US EPA), standard toxicology textbooks, and scientific journals of environmental toxicology or environmental health. Analysts should also consult on-line databases (such as the Hazardous Substances Data Bank (HSDB) and 'ToxLine') for the most current toxicological and medical information. Furthermore, the analyst should indicate in the health assessment reporting/documentation whether health concerns are for acute, intermediate, or chronic exposures.

4.3.2. EVALUATING FACTORS THAT INFLUENCE ADVERSE HEALTH OUTCOME

To ensure reliable public health policy decisions, the public health analyst should review the various factors that may enhance or mitigate health effects that result from exposure to chemicals present in the human living and work environments. Indeed, among other things, the analyst should also consider other pertinent medical and toxicological information; the health implications for sensitive sub-populations; health implications of past and future exposures; and the effects of corrective/control actions or interventions on human exposure. These particularly important elements are elaborated below.

Public Health Implications of Other Medical and Toxicological Factors
Ideally, the health effects identified by comparing dose estimates with toxicity values during a risk characterization should also be evaluated on the basis of other

toxicological and medical factors that could potentially enhance or mitigate the effects of a chemical exposure. Indeed, as appropriate, several factors should be investigated and their health implications discussed in the health assessment; typical factors that the public health analyst may generally consider in the evaluation of public health outcomes are annotated in Box 4.1. In general, in addition to the medical and toxicological factors identified here, the public health analyst should also consider population-specific factors that may enhance or mitigate health effects associated with exposure to the constituents of concern.

Box 4.1. Typical medical and toxicological factors affecting public health outcomes

- Distribution of chemical within the body (i.e., the fate of the chemical after ingestion, inhalation, or dermal contact)
- Target organs (i.e., physiologic site of major toxicity)
- Toxicokinetics of substance (including possible transfer to cow's milk or nursing mother's milk)
- Enzyme induction (i.e., chemical induction of various enzyme systems may increase or decrease chemical toxicity)
- Cumulative effect of exposures to chemicals that bioaccumulate in the body (e.g., lead, cadmium, organochlorine pesticides)
- Chemical tolerance (i.e., decreased responsiveness to a toxic chemical effect resulting from previous exposure to that chemical or to a structurally related chemical)
- Immediate *versus* delayed effects (i.e., effects observed rapidly after a single exposure *versus* effects that occur after some lapse of time)
- Reversible *versus* irreversible effects (i.e., ability of affected organs to regenerate)
- Local *versus* systemic effects (i.e., whether the effect occurs at the site of first contact, or if the chemical must be absorbed and distributed before the effect is observed)
- Idiosyncratic reactions (i.e., genetically determined abnormal reactivity to a chemical that is qualitatively similar to reactions found in all persons – but may take the form of either extreme sensitivity to low doses or extreme insensitivity to high doses)
- Allergic reactions (i.e., adverse reaction to a chemical resulting from previous sensitization to that chemical or a structurally related one)
- Various other related disease effects (i.e., effect of chemical on previously diseased organ)

Health Implications for Sensitive Sub-populations

Many sub-populations may be identified at a given locale. Each sub-population may have special concerns that must be considered when determining public health implications of a chemical exposure problem. Perhaps the most crucial set of factors that an analyst must weigh are those that influence differential susceptibility to the effects of specific compounds. Age, gender, genetic background, nutritional status, health status, and general lifestyle may each influence the effects of chemical exposures; thus, the analyst should carefully consider the impact that each of these factors may have under a specific chemical exposure scenario for a given population. The key factors are elaborated below.

- *Age of Receptor.* Age-related susceptibility to the toxic effects of chemicals is more widespread than many public health analysts realize. Indeed, at some point in a human lifetime, every person is at an enhanced risk from chemical exposures because of age factors.

  It is generally acknowledged that the very young are a particularly high-risk group that must be protected more stringently from the adverse effects of certain compounds. For example, the US EPA primary drinking water standard for nitrate was set to protect the most susceptible high-risk group – namely, infants in danger of developing methemoglobinemia. Similar age-related sensitivities are reflected in 'allowable' levels set for lead in ambient air and in drinking water, as well as for mercury in aquatic systems. Even so, the very young are not always the age group associated with the most enhanced risk situation. In fact, in some instances, adults are at greater risk of toxicity than infants or children. For example, past studies have shown that the young seem more resistant (than adults) to the adverse effects of renal toxicants such as fluoride and uranyl nitrate. Furthermore, recent acknowledgment that elderly subpopulations may have significantly heightened susceptibility to chemical compounds because of lower functional capacities of various organ systems, reduced capacity to metabolize foreign compounds, and diminished detoxification mechanisms should be recognized.

- *Gender of Receptor.* There is some scientific evidence to support the fact that certain adverse health effects may be mediated through hormonal influences and other factors that are dependent on the sex of the individual receptor. In general, gender-linked differences in toxic susceptibilities have not been extensively investigated. However, as an example, it is well documented that pregnant women are often at significantly greater risk from exposure to beryllium, cadmium, lead, manganese, and organophosphate insecticides than other members of the general population; this is because of the various physiologic modifications associated with the pregnancy. Also, a developing fetus is at greater risk from compounds that exert developmental effects.

- *Biochemical and/or Genetic Susceptibilities.* The presence of subpopulations with certain inherent biochemical and/or genetic susceptibilities should be given careful consideration when evaluating the potential health threats from a chemical exposure problem; this is because a number of studies indicate that genetic predisposition is an important determining factor in numerous diseases. Indeed, studies of some of these 'genetically-determined' diseases have shown an increased susceptibility to the toxic effects of certain chemicals. For example, certain percentages of some ethnic groups are known to suffer from inherited serum alpha-1-antitrypsin deficiency – which predisposes them to alveolar destruction and pulmonary emphysema. Persons with this deficiency are especially sensitive to the effects of certain pollutants. In general, this type of information can be used in conjunction with information on the ethnic makeup of populations in the study area, so as to better evaluate potential toxic effects associated with a chemical exposure problem. In addition, persons who have chronic diseases may also be at increased risk from exposure to certain chemical; for example, individuals with cystic fibrosis are less tolerant of the respiratory and gastrointestinal challenges of some pollutants. Also, persons with hereditary blood disorders, such as sickle-cell anemia, have increased sensitivity to compounds such as benzene, cadmium, and lead – which are suspected 'anemia producers'. Thus, the importance of determining the presence and proximity of facilities such as hospitals or convalescent homes where sensitive subpopulations are likely to be found cannot be overemphasized.

In general, when identifiable groups are known to be at risk from exposure to a chemical source, then it is important to determine from available medical and toxicological literature and databases the nature and magnitude of adverse health effects that could likely result – alongside any confounding factors.

- *Socioeconomic Factors.* Socioeconomic status is not only an important indicator of human susceptibilities to specific pollutants, but such information may also help identify confounding nutritional deficiencies or behaviors that enhance a person's sensitivity to the toxic effects of chemical materials. For instance, studies have shown that dietary deficiencies of vitamins A, C, and E may increase susceptibility to the toxic effects of polychlorinated biphenyls (PCBs) and other chlorinated hydrocarbons, some pesticides, ozone, and various other substances. Other studies have also indicated that deficiencies in trace metals such as iron, magnesium, and zinc exacerbate the toxic potential of fluorides, manganese, and cadmium. In general, populations with sensitivities due to nutritional deficiencies have typically been associated with areas of low socioeconomic status and extreme poverty or in areas with large numbers of indigents. Elderly populations have also been identified as a subgroup at risk of susceptibility because of nutritional deficits.

  Demographic and land-use information can generally be used to help identify the relative socioeconomic status of exposed populations; this information may indeed provide important clues for this step of the health assessment process. In fact, as part of the overall public health risk determination process, the public health analyst must carefully examine demographic information for particular groups on or near the study area or exposure source, and who might be especially sensitive to toxic effects. Any suspected high-risk groups should be explicitly identified in the health assessment report. For instance, locations of daycare centers, schools, playgrounds, recreational areas, hospitals and retirement or convalescent homes on or near a given site should be highlighted as important indications of the presence of sensitive subpopulations. Enumeration of ethnic groups within the population, as well as characterization of socioeconomic status may also indicate sensitive subpopulations near a study area or exposure source. It is noteworthy that, ultimately, information on the number and proximity of people in high-risk subpopulations is vital for developing an optimal public health risk management or mitigation plan.

Overall, subpopulations of special concern should be identified during the public health risk assessment process; those individuals or groups may be at increased risk because of greater sensitivity, compromised health status, concomitant occupational exposures, or indeed a variety of other reasons. Thus, if such individuals or groups do indeed exist, then they should be explicitly identified in the health assessment – and then appropriate recommendations should be made specifically for their protection. Furthermore, other groups that are closely affiliated with high-risk group – for example, families of workers who may be (or have been) exposed through contact with work clothing or other secondary means – should also be carefully evaluated.

Health Implications of Past and Future Exposures
In the attempt to determine the health implications of a chemical exposure problem, and in addressing a population-at-risk's health concerns, the public health analyst should endeavor to include past, current, and potential future exposures in the requisite documentation. In fact, despite the fact that significant exposure may already have occurred, past exposures are difficult to address – especially because they are difficult

to quantify. In general, when evaluating community health concerns about past hazard exposures, the analyst should review all available community-specific health outcome databases, such as morbidity data and disease registries – in order to determine a possible correlation between past and current health outcomes and past exposures. When past exposures have been documented but health studies have not been performed, health effects studies or the review of community health records become very important.

A generally important aspect of the process of determining the public health implications of chemical exposures usually involves establishing a firm difference between what constitutes 'actual' exposures (i.e., expected and/or completed exposures) *vs.* 'potential' exposures (i.e., possible but not necessarily complete exposures). When evaluating future 'actual' and 'potential' exposures, the analyst should also make a determination of the underlying causes for the anticipated exposures (e.g., from the continued use of specific consumer products, etc.) – so that appropriate mitigative measures for such future exposures can be undertaken *a priori*.

Health Implications of Corrective Actions and Interventions
When determining the health implications of a chemical exposure situation, it is important that the analyst takes the effect(s) of remedial actions and other intervention programs into consideration. This is because previous, current, and/or planned remedial or risk management actions can significantly affect conclusions about exposure-related health concerns.

In general, when remedial response measures or other interventions have occurred previously, the analyst should consider the effect that those measures have had on the health of the target population. Similarly, if intervention is already occurring, the analyst should determine what likely effect this might be. Furthermore, the health assessment should be responsive to community health concerns vis-à-vis the remedial actions. In addition, discussions in the health assessment about exposure scenarios should clearly identify and differentiate between those exposure scenarios that are still present *vs.* the exposures that may have occurred in the past – but that have been eliminated or significantly reduced by remedial action or other intervention programs.

4.4. Human Health Risk Assessment in Practice

Quantitative human health risk assessment often becomes an integral part of most environmental and public health risk management programs that are designed to address chemical exposure problems. In the processes involved, four key elements are important in arriving at appropriate risk management solutions – namely, the chemical hazard identification; the chemical toxicity assessment or exposure-response evaluation; the exposure assessment; and the risk characterization. Each of these elements typically will, among other things, help answer the following fundamental questions:

- Chemical hazard identification step – 'what chemicals are present in the human environments of interest?' and 'is the chemical agent likely to have an adverse effect on the potential human receptor?'
- Chemical toxicity assessment or exposure-response evaluation step – 'what is the relationship between human exposure/dose to the chemical of potential concern

and the response, incidence, injury, or disease as a result of the receptor exposure?' That is, 'what harmful effects can be caused by the target chemicals, and at what concentration or dose?'

- Exposure assessment step – 'what individuals, subpopulations, or population groups may be exposed to the chemical of potential concern?' and 'how much exposure is likely to result from various activities of the potential receptor – i.e., what types and levels of exposure are anticipated or observed under various scenarios?'

- Risk characterization step – 'what is the estimated incidence of adverse effect to the exposed individuals or population groups – i.e., what risks are presented by the chemical hazard source?' and 'what is the degree of confidence associated with the estimated risks?'

The basic tasks involved in most human health risk assessments typically will consist of the key components shown in Box 4.2 – and a careful implementation of this framework should provide answers to the above questions. Illustrative examples of the practical application of the process are provided in Chapters 8 and 9. Meanwhile, it cannot be stated enough that there are many uncertainties associated with public health risk assessments. These uncertainties are due in part to the complexity of the exposure-dose-effect relationship, and also the lack of, or incomplete knowledge/information about the physical, chemical, and biological processes within and between human exposure to chemical substances and health effects.

In general, the major sources of uncertainty in public health risk assessments can be attributed to: (i) the use of a wide range of data from many different disciplines (e.g., epidemiology, toxicology, biology, chemistry, statistics, etc.); (ii) the use of many different predictive models and methods in lieu of actual measured data; and (iii) the use of many scientific assumptions and science policy choices (i.e., scientific positions assumed in lieu of scientific data) – in order to bridge the information/knowledge gaps in the risk assessment process. Overall, these diverse elements, along with varying interpretations of the scientific information, can result in divergent results in the risk assessment process – an outcome that often results in some risk assessment controversies. Thus, it is very important to carefully and systematically identify all sources and types of uncertainty and variability – and then present them as an integral part of risk characterization process.

Box 4.2. Illustrative basic outline for a public health risk assessment report

Section Topic	*Basic Subject Matter*
❶ General Overview	♦ Background information on the case problem or locale ♦ The risk assessment process ♦ Purpose and scope of the risk assessment ♦ The risk assessment technique and method of approach ♦ Legal and regulatory issues in the risk assessment ♦ Limits of application for the risk assessment
❷ Data Collection	♦ Chemical exposure sources of potential concern ♦ General case-specific data collection considerations ♦ Assessment of the data quality objectives ♦ Identification of data gathering uncertainties
❸ Data Evaluation	♦ General case-specific data evaluation considerations ♦ Identification, quantification, and categorization of target chemicals ♦ Statistical analyses of relevant chemical data ♦ Screening and selection of the chemicals of potential concern ♦ Identification of uncertainties associated with data evaluation
❹ Exposure Assessment	♦ Characterization of the exposure setting (to include the physical setting and populations potentially at risk) ♦ Identification of the chemical-containing sources/media, exposure pathways, and potentially affected receptors ♦ Determination of the important fate and behavior processes for the chemicals of potential concern ♦ Determination of the likely and significant exposure routes ♦ Development of representative conceptual model(s) for the problem situation ♦ Development of realistic exposure scenarios (to include both current and potential future possibilities) ♦ Estimation/modeling of exposure point concentrations for the chemicals of potential concern ♦ Quantification of exposures (i.e., computation of potential receptor intakes/doses for the applicable exposure scenarios) ♦ Identification of uncertainties associated with exposure parameters
❺ Toxicity Assessment	♦ Compilation of the relevant toxicological profiles of the chemicals of potential concern ♦ Determination of the appropriate and relevant toxicity index parameters ♦ Identification of uncertainties relating to the toxicity information
❻ Risk Characterization	♦ Estimation of the human carcinogenic risks from carcinogens ♦ Estimation of the non-carcinogenic effects for systemic toxicants ♦ Sensitivity analyses of relevant parameters ♦ Identification and evaluation of uncertainties associated with the risk estimates
❼ Risk Summary Discussion	♦ Summarization of risk information ♦ Discussion of all identifiable sources of uncertainties

4.5. Suggested Further Reading

Bate, R. (ed.), 1997. *What Risk? (Science, Politics & Public Health)*, Butterworth-Heinemann, Oxford, UK

Bates, DV, 1994. *Environmental Health Risks and Public Policy*, University of Washington Press, Seattle, Washington

Calow, P. (ed.), 1998. *Handbook of Environmental Risk Assessment and Management*, Blackwell Science, Great Britain

Christakos, G. and DT Hristopulos, 1998. *Spatiotemporal Environmental Health Modelling: A Tractatus Stochasticus*, Kluwer Academic Publishers, Dordrecht, The Netherlands

Crawford-Brown, DJ, 1999. *Risk-Based Environmental Decisions: Methods and Culture*, Kluwer Academic Publishers, Dordrecht, The Netherlands

CSA (Canadian Standards Association), 1991. *Risk Analysis Requirements and Guidelines*, CAN/CSA-Q634-91, Canadian Standards Association (CSA), Rexdale, Ontario, Canada

Dakins, ME, JE Toll, and MJ Small, 1994. Risk-based environmental remediation: decision framework and role of uncertainty, *Environmental Toxicology and Chemistry*, 13(12): 1907 – 1915

Dakins, ME, JE Toll, MJ Small, and KP Brand, 1996. Risk-based environmental remediation: Bayesian Monte Carlo analysis and the expected value of sample information, *Risk Analysis*, 16(1): 67 – 79

Douben, PET, 1998. *Pollution Risk Assessment and Management*, John Wiley & Sons Ltd, Chichester, UK

Health Canada, 1994. *Human Health Risk Assessment for Priority Substances*, Environmental Health Directorate, Canadian Environmental Protection Act, Health Canada, Ottawa, Canada

Linders, JBHJ, 2000. *Modelling of Environmental Chemical Exposure and Risk*, Kluwer Academic Publishers, Dordrecht, The Netherlands

Morris P. and R. Therivel (eds.), 1995. *Methods of Environmental Impact Assessment*, UCL Press, University College London, UK

Neumann, DA and CA Kimmel (eds.), 1999. *Human Variability in Response to Chemical Exposures (Measures, Modeling, and Risk Assessment)*, CRC Press LLC, Boca Raton, FL

Rabl, A., JV Spadaro, and PD McGavran, 1998. Health risks of air pollution from incinerators: a perspective, *Waste Management & Research*, 16(4): 365 – 388

USEPA, 1996. *Radiation Exposure and Risk Assessment Manual*, EPA 402-R-96-016, US Environmental Protection Agency (USEPA), Washington, DC

Chapter 5

CHEMICAL HAZARD DETERMINATION

The first issue in any attempt to conduct a public health risk assessment for chemical exposure problems relates to answering the seemingly straight-forward question: 'does a chemical hazard exist?' Thus, all environmental and public health risk management programs designed for chemical exposure situations usually will start with a hazard identification and accounting; this initial process sets out to determine whether or not the substance in question possesses potentially hazardous and/or toxic properties. This chapter discusses the principal activities involved in the acquisition and manipulation of the pertinent chemical hazard information – that will then allow effective environmental and public health risk management decisions to be made about chemical exposure problems.

5.1. Chemical Hazard Identification: Sources of Chemical Hazards

The chemical hazard identification component of a public health risk assessment involves first establishing the presence of a chemical stressor that could potentially cause adverse human health effects. This process usually includes a review of the major sources of chemical hazards that could potentially contribute to a given chemical exposure and possible risk situation. Indeed, chemical hazards affecting public health risks typically originate from a variety of sources (Box 5.1) – albeit their relative contributions to actual human exposures are not always so obvious. Needless to say, there is a corresponding variability in the range and type of hazards and risks that may be anticipated from different chemical exposure problems.

Oftentimes, qualitative information on potential sources and likely consequences of the chemical hazards is all that is required during this early stage (i.e., the hazard identification phase) of the risk assessment process. To add a greater level of sophistication to the hazard identification process, however, quantitative techniques may be incorporated into this process – to help determine, for instance, the likelihood of an actual exposure situation occurring. The quantitative methods may include a use of mathematical modeling and/or decision analyses techniques to determine chemical fate and behavior attributes following human exposure to a chemical vis-à-vis the likely receptor response upon exposure to the chemical of potential concern. In general, this initial evaluation for a chemical exposure problem should provide insight into the nature and types of chemicals, the populations potentially at risk, and possibly some qualitative ideas about the magnitude of the anticipated risk.

Box 5.1. Examples of major sources of chemical hazards potentially resulting in public health problems

◆ Consumer products (including foods, drinks, cosmetics, medicines, etc.)
◆ Urban air pollution (including automobile exhausts, factory chimney stacks, etc.)
◆ Contaminated drinking water
◆ Industrial manufacturing and processing facilities
◆ Commercial service facilities (such as fuel stations, auto repair shops, dry cleaners, etc.)
◆ Landfills, waste tailings and waste piles
◆ Contaminated lands
◆ Wastewater lagoons
◆ Septic systems
◆ Hazardous materials stockpiles
◆ Hazardous materials storage tanks and containers
◆ Pipelines for hazardous materials
◆ Spills from loading and unloading of hazardous materials
◆ Spillage from hazardous materials transport accidents
◆ Pesticide, herbicide, and fertilizer applications
◆ Contaminated urban runoff
◆ Mining and mine drainage
◆ Waste treatment system and incinerator emissions

5.2. Data Collection and Evaluation Considerations

The process involved in a public health risk assessment for chemical exposure problems will usually include a well-thought out plan for the collection and analysis of a variety of chemical hazard and receptor exposure data. Ideally, and to facilitate this process, project-specific 'work-plans' can be designed to specify the administrative and logistic requirements of the general activities to be undertaken. A typical data collection work-plan that is used to guide the investigation of chemical exposure problems may include, at a minimum, a sampling and analysis plan together with a quality assurance/quality control plan. The general nature and structure for such types of work-plans, as well as further details on the appropriate technical standards for sample collection and sample handling procedures, can be found in the literature elsewhere (e.g., Asante-Duah, 1998; ASTM, 1997b; Boulding, 1994; CCME, 1993; CDHS, 1990; Keith, 1988, 1991; Lave and Upton, 1987; Petts *et al.*, 1997; USEPA, 1989a, 1989b).

In general, all sampling and analysis should be conducted in a manner that maintains sample integrity and encompasses adequate quality assurance and control. Also, specific samples collected should be representative of the target materials that are the source of, and/or 'sink' for, the chemical exposure problem. And regardless of its intended use, it is noteworthy that samples collected for analysis at a remote location are generally kept on ice prior to and during transport/shipment to a certified laboratory for analysis; also, completed chain-of-custody records should accompany the samples to the laboratory.

Indeed, sampling and analysis can become a very important part of the decision-making process involved in the management of chemical exposure problems. Yet, sampling and analysis could also become one of the most expensive and time-consuming aspects of such a public health risk management program. Even of greater concern is the fact that errors in sample collection, sample handling, or laboratory analysis can invalidate the hazard accounting and exposure characterization efforts, and/or add to the overall project costs. All samples that are intended for use in human exposure and risk characterization programs must therefore be collected, handled, and analyzed properly – in accordance with all applicable/relevant methods and protocols. To ultimately produce data of sound integrity and reliability, it is important to give special attention to several issues pertaining to the sampling objective and approach; sample collection methods; chain-of-custody documentation; sample preservation techniques; sample shipment methods; and sample holding times. Asante-Duah (1998) provides a convenient checklist of the issues that should be verified when planning such type of sampling activity.

Overall, highly effective sampling and laboratory procedures are required during the chemical hazard determination process; this is to help minimize uncertainties associated with the data collection and evaluation aspects of the risk assessment. Ultimately, several chemical-specific parameters (such as chemical toxicity or potency, media concentration, ambient levels, frequency of detection, mobility, persistence, bioaccumulative/bioconcentration potential, synergistic or antagonistic effects, potentiation or neutralizing effects, etc.) as well as various receptor information are further used to screen and help select the specific target chemicals that will become the focus of a detailed risk assessment.

5.2.1. DATA COLLECTION AND ANALYSIS STRATEGIES

A variety of data collection and analysis protocols exist in the literature (e.g., Boulding, 1994; Byrnes, 1994; CCME, 1993, 1994; Csuros, 1994; Garrett, 1988; Hadley and Sedman, 1990; Keith, 1992; Millette and Hays, 1994; O'Shay and Hoddinott, 1994; Schulin et al., 1993; Thompson, 1992; USEPA, 1982, 1985a, 1985b, 1992; Wilson et al., 1995) that may be adapted for the investigation of human exposure to chemical constituents found in consumer products and in the human environments. Regardless of the processes involved, however, it is important to recognize the fact that most chemical sampling and analysis procedures offer numerous opportunities for sample contamination from a variety of sources (Keith, 1988). To be able to address and account for possible errors arising from 'foreign' sources, quality control (QC) samples are typically included in the sampling and analytical schemes. The QC samples are analytical 'control' samples that are analyzed in the same manner as the 'field' samples – and these are subsequently used in the assessment of any cross-contamination that may have been introduced into a sample along its life cycle from the field (i.e., point of collection) to the laboratory (i.e., place of analysis).

Invariably, QC samples become an essential component of all carefully executed sampling and analysis program. This is because, firm conclusions cannot be drawn from the investigation unless adequate controls have been included as part of the sampling and analytical protocols (Keith, 1988). To prevent or minimize the inclusion of 'foreign' constituents in the characterization of chemical exposures and/or in a risk assessment, therefore, the concentrations of the chemicals detected in 'control' samples must be compared with concentrations of the same chemicals detected in the 'field' samples. In such an evaluation, the QC samples can indeed become a very important reference datum for the overall evaluation of the chemical sampling data.

In general, carefully designed sampling and analytical plans, as well as good sampling protocols, are necessary to facilitate credible data collection and analysis programs. Sampling protocols are written descriptions of the detailed procedures to be followed in collecting, packaging, labeling, preserving, transporting, storing, and tracking samples. The selection of appropriate analytical methods is also an integral part of the processes involved in the development of sampling plans – since this can strongly affect the acceptability of a sampling protocol. For example, the sensitivity of an analytical method could directly influence the amount of a sample needed in order to be able to measure analytes at pre-specified minimum detection (or quantitation) limits. The analytical method may also affect the selection of storage containers and preservation techniques (Keith 1988; Holmes et al., 1993). In any case, the devices that are used to collect, store, preserve, and transport samples must *not* alter the sample in any manner. In this regard, it is noteworthy that special procedures may be needed to preserve samples during the period between collection and analysis.

Finally, the development and implementation of an overall good quality assurance/quality control (QA/QC) project plan for a sampling and analysis activity is critical to obtaining reliable analytical results. The soundness of the QA/QC program has a particularly direct bearing on the integrity of the sampling as well as the laboratory work. Thus, the general process for developing an adequate QA/QC program, as discussed elsewhere in the literature (e.g., CCME 1994; USEPA 1987a, 1987b, 1992), should be followed religiously. Also, it must be recognized that, the more specific a sampling protocol is, the less chance there will be for errors or erroneous assumptions.

5.2.2. REPORTING OF 'CENSORED' LABORATORY DATA

Oftentimes, in a given set of laboratory samples, certain chemicals will be reliably quantified in some (but not all) of the samples. Data sets may therefore contain observations that are below the instrument or method detection limit, or indeed its corresponding quantitation limit; such data are often referred to as 'censored data' (or 'non-detects' [NDs]). In general, the NDs do not necessarily mean that a chemical is not present at any level – but simply that any amount of such chemical potentially present was probably below the level that could be detected or reliably quantified using a particular analytical method. Thus, this situation may reflect the fact that either the chemical is truly absent at this location or sampled matrix at the time the sample was collected; or that the chemical is indeed present – but at a concentration below the quantitation limits of the analytical method that was employed in the sample analysis.

In fact, all laboratory analytical techniques have detection and quantitation limits below which only 'less than' values may be reported; the reporting of such values provides a degree of quantification for the censored data. In such situations, a decision has to be made as to how to treat such NDs and associated 'proxy' concentrations. The appropriate procedure depends on the general pattern of detection for the chemical in the overall investigation activities (Asante-Duah, 1998; HRI, 1995). In any case, it is customary to assign non-zero values to all sampling data reported as NDs. This is important because, even at or near their detection limits, certain chemical constituents may be of considerable importance in the characterization of a chemical exposure problem. However, uncertainty about the actual values below the detection or quantitation limit can bias or preclude subsequent statistical analyses. Indeed censored data do create significant uncertainties in the data analysis required of the chemical exposure characterization process; such data should therefore be handled in an appropriate manner – for instance, as elaborated in the example methods of approach provided below.

Derivation and Use of 'Proxy' Concentrations

'Proxy' concentrations are usually employed when a chemical is not detected in a specific sampled medium. A variety of approaches are offered in the literature for deriving and using proxy values in environmental data analyses, including the following (Asante-Duah, 1998; HRI, 1995; USEPA, 1989a, 1992):

- *Set the sample concentration to zero.* This assumes that if a chemical was not detected, then it is not present – i.e., the 'residual concentration' is zero. This involves very strong assumptions, and it can rarely be justified that the chemical is not present in the sampled media. Thus, it represents a least conservative (i.e., least health-protective) option.
- *Drop the sample with the non-detect for the particular chemical from further analysis.* This will have the same effect on the data analysis as assigning a concentration that is the average of concentrations found in samples where the chemical was detected.
- *Set the proxy sample concentration to the sample quantitation limit (SQL).* For NDs, setting the sample concentration to a proxy concentration equal to the SQL (which is a quantifiable number used in practice to define the analytical detection limit) makes the fewest assumptions and tends to be conservative, since the SQL represents an upper-bound on the concentration of a ND. This option does indeed offer the most conservative (i.e., most health-protective) approach to chemical hazard accounting and exposure estimation. The approach recognizes that the true distribution of concentrations represented by the NDs is unknown.
- *Set the proxy sample concentration to one-half the SQL.* For NDs, setting the sample concentration to a proxy concentration equal to one-half the SQL assumes that, regardless of the distribution of concentrations above the SQL, the distribution of concentrations below the SQL is symmetrical. (It is noteworthy that, when/if the data are highly skewed, then use of SQL divided by the square-root-of-two (i.e., SQL/$\sqrt{2}$) is recommended, instead of one-half the SQL.)

In the 'worst-case' approach, all NDs are assigned the value of the SQL – which is the lowest level at which a chemical may be accurately and reproducibly quantitated; this approach biases the mean upward. On the other hand, assigning a value of zero to all NDs biases the mean downward. The degree to which the results are biased will depend on the relative number of detects and non-detects in the data set, and also the difference between the reporting limit and the measured values above it. Oftentimes, the common practice seems to utilize the sample-specific quantitation limit for the chemical reported as ND. In fact, the goal in adopting such an approach is to avoid underestimating exposures to potentially sensitive or highly exposed groups such as infants and children, but at the same time attempt to approximate actual 'residual levels' as closely as possible. Ultimately, recognizing that the assumptions in these methods of approach may, in some cases, either overestimate or underestimate exposures, the use of sensitivity analysis to determine the impact of using different assumptions (e.g., ND = 0 vs. ND = SQL/2 vs. ND = SQL/$\sqrt{2}$) is encouraged.

Other methods of approach to the derivation of proxy concentrations may involve the use of 'distributional' methods; unlike the simple substitution methods shown above, distributional methods make use of the data above the reporting limit in order to extrapolate below it (USEPA, 1992). Indeed, even more robust methods may be utilized in such applications for handling censored data sets. In general, selecting the appropriate method requires consideration of the degree of censoring, the goals of the assessment, and the degree of accuracy required.

Notwithstanding the above procedures of deriving and/or using 'proxy' concentrations, re-sampling and further laboratory analysis should always be viewed as the preferred approach to resolving uncertainties that surround ND results obtained from sampled media. Thence, if the initially reported data represent a problem in sample collection or analytical methods rather than a true failure to detect a chemical of potential concern, then the problem could be rectified (e.g., by the use of more sensitive analytical protocols) before critical decisions are made based on the earlier results.

5.3. Statistical Evaluation of Chemical Sampling Data

Once the decision is made to undertake a public health risk assessment, the available chemical exposure data has to be carefully analyzed, in order to arrive at a list of chemicals of potential concern (CoPCs); the CoPCs represent the target chemicals of focus in the risk assessment process. In general, the target chemicals of significant interest to chemical exposure problems may be selected for further detailed evaluation on the basis of several specific important considerations – such as shown in Box 5.2. The use of such selection criteria will allow an analyst to continue with the exposure and risk characterization process only if the chemicals represent potential threats to public health. Next, the proper exposure point concentration (EPC) for the target populations potentially at risk from the CoPCs are determined; an EPC is the concentration of the CoPC in the target material or product at the point of contact with the human receptor.

Box 5.2. Typical important considerations in the screening for chemicals of potential concern for public health risk assessments

---

- ◆ Status as a known human carcinogen *versus* probable or possible carcinogen
- ◆ Status as a known human developmental and reproductive toxin
- ◆ Degree of mobility, persistence, and bioaccumulation
- ◆ Nature of possible transformation products of the chemical
- ◆ Inherent toxicity/potency of chemical
- ◆ Concentration-toxicity score – reflecting concentration levels in combination with degree of toxicity (For exposure to multiple chemicals, the chemical score is represented by a risk factor, calculated as the product of the chemical concentration and toxicity value; the ratio of the risk factor for each chemical to the total risk factor approximates the relative risk for each chemical – giving a basis for inclusion or exclusion as a CoPC)
- ◆ Frequency of detection in target material or product (Chemicals that are infrequently detected may be artifacts in the data due to sampling, analytical, or other problems, and therefore may not be truly associated with the consumer product or target material under investigation)
- ◆ Status and condition as an essential element – i.e., defined as essential human nutrient, and toxic only at elevated doses (For example, Ca or Na generally does not pose a significant risk to public health, but As or Cr may pose a significantly greater risk to human health)

The EPC determination process typically will consist of an appropriate statistical evaluation of the exposure sampling data – especially when large data sets are involved. Statistical procedures used for the evaluation of the chemical exposure data can indeed significantly affect the conclusions of a given exposure characterization and risk assessment program. Consequently, appropriate statistical methods (e.g., in relation to the choice of proper averaging techniques) should be utilized in the evaluation of chemical sampling data. Of special interest, it is noteworthy that, over the years, extensive technical literature has been developed regarding the 'best' probability distribution to utilize in different scientific applications – and such sources should be consulted for appropriate guidance on the statistical tools of choice.

5.3.1. PARAMETRIC VERSUS NONPARAMETRIC STATISTICS

There are several statistical techniques available for analyzing data that are not dependent on the assumption that the data follow any particular statistical distribution. These distribution-free methods are referred to as *nonparametric* statistical tests – and they have fewer and less stringent assumptions. Conversely, several assumptions have to be met before one can use a *parametric* test. Whenever the set of requisite assumptions is met, it is always preferable to use a parametric test – because it is more powerful than the nonparametric test. However, to reduce the number of underlying assumptions required (such as in a hypothesis testing about the presence of specific trends in a data set), nonparametric tests are typically employed.

Nonparametric techniques are generally selected when the sample sizes are small and the statistical assumptions of normality and homogeneity of variance are tenuous. Indeed, nonparametric tests are usually adopted for use in environmental impact assessments because the statistical characteristics of the often messy environmental data make it difficult, or even unwise, to use many of the available parametric methods. It is noteworthy, however, that the nonparametric tests tend to ignore the magnitude of the observations in favor of the relative values or ranks of the data. Consequently, as Hipel (1988) notes, a given nonparametric test with few underlying assumptions that is designed, for instance, to test for the presence of a trend may only provide a 'yes' or 'no' answer as to whether or not a trend may indeed be present in the data. The output from the nonparametric test may not give an indication of the type or magnitude of the trend. To have a more powerful test about what might be occurring, many assumptions must be made – and as more assumptions are formulated, a nonparametric test begins to look more like a parametric test. It is also noteworthy that, the use of parametric statistics requires additional detailed evaluation steps – with the process of choosing an appropriate statistical distribution being an important initial step.

Choice of Statistical Distribution
Of the many statistical distributions available, the Gaussian (or normal) distribution has been widely utilized to describe environmental data; however, there is considerable support for the use of the lognormal distribution in describing such data. Consequently, chemical concentration data for environmental samples have been described by the lognormal distribution, rather than by a normal distribution (Gilbert 1987; Leidel and Busch 1985; Rappaport and Selvin 1987; Saltzman, 1997). Basically, the use of lognormal statistics for the data set X_1, X_2, X_3, ..., X_n requires that the logarithmic transform of these data (i.e., $ln[X_1]$, $ln[X_2]$, $ln[X_3]$, ..., $ln[X_n]$) can be expected to be normally distributed.

In general, the statistical parameters used to describe the different distributions can differ significantly; for instance, the central tendency for the normal distributions is measured by the arithmetic mean, whereas the central tendency for the lognormal distribution is defined by the geometric mean. Ultimately, the use of a normal distribution to describe environmental chemical concentration data, rather than lognormal statistics will often result in significant over-estimation, and may be overly conservative – albeit some investigators may argue otherwise (e.g., Parkhurst, 1998). In fact, Parkhurst (1998) argues that geometric means are biased low and do not represent components of mass balances properly, *whereas* arithmetic means are unbiased, easier to calculate and understand, scientifically more meaningful for concentration data, and more protective of public health. Even so, this same investigator still concedes to the non-universality of this school of thought – and these types of arguments and counter-arguments only go to reinforce the fact that no one particular parameter or distribution may be appropriate for every situation. Consequently, care must be exercised in the choice of statistical methods for the data manipulation exercises carried out during the hazard accounting process.

Goodness-of-Fit Testing
Recognizing that the statistical procedures used in the evaluation of chemical exposure data should generally reflect the character of the underlying distribution of the data set, it is preferable that the appropriateness of any distribution assumed or used for a given data set be checked prior to its application. This verification check can be accomplished by using a variety of goodness-of-fit methods.

Goodness-of-fit tests are formal statistical tests of the hypothesis that a specific set of sampled observations is an independent sample from the assumed distribution. The more common general tests include the Chi-square test and the Kolmogorov-Smirnov test; common goodness-of-fit tests specific for normality and log-normality include the Shapiro-Wilks' test and D'Agostino's test (see, e.g., D'Agostino and Stephens, 1986; Gilbert, 1987; Miller and Freund, 1985; Sachs, 1984). It is noteworthy that, goodness-of-fit tests tend to have notoriously low power – and are generally best for rejecting poor distribution fits, rather than for identifying good fits.

If the data cannot be fitted well enough to a theoretical distribution, then perhaps an empirical distribution function or other statistical methods of approach (such as bootstrapping techniques) should be considered. Another way to assess which probability distribution adequately models the underlying population of a data set is to test the probability of a sample being drawn from a population with a particular probability distribution; one such test is the W-test (Shapiro and Wilk, 1965). The W-test is particularly important in assessing whether a sample is from a population with a normal probability distribution; the W-test can also be used to evaluate if a sample belongs to a population with a lognormal distribution (i.e., after the data has undergone a natural logarithm transformation). It is noteworthy that, the W-test (as developed by Shapiro and Wilk) is limited to a small sample data set size (of 3 to 50 samples). However, a modification of the W-test that allows for its use with larger data sets (up to about 5,000 data points) is also available (e.g., in the formulation subsequently developed by Royston) (Royston, 1995).

5.3.2. STATISTICAL EVALUATION OF 'NON-DETECT' VALUES

During the analysis of environmental sampling data that contains some NDs, a fraction of the SQL is usually assumed (as a proxy or estimated concentration) for non-detectable levels – instead of assuming a value of zero, or neglecting such values. This procedure is typically used, provided there is at least one detected value from the analytical results, and/or if there is reason to believe that the chemical is possibly present in the sample at a concentration below the SQL. The approach conservatively assumes that some level of the chemical could be present (even though a ND has been recorded) and arbitrarily sets that level at the appropriate percentage of the SQL.

In general, the favored approach in the calculation of the applicable statistical values during the evaluation of data containing NDs involves the use of a value of one-half of the SQL. This approach assumes that the samples are equally likely to have any value between the detection limit and zero, and can be described by a normal distribution. However, when the sample values above the ND level are log-normally distributed, it generally may be assumed that the ND values are also log-normally distributed. The best estimate of the ND values for a log-normally distributed data set is the reported SQL divided by the square root of two (i.e., $\dfrac{SQL}{\sqrt{2}} = \dfrac{SQL}{1.414}$) (CDHS, 1990; USEPA, 1989a). Also, in some situations, the SQL value itself may be used if there is strong enough reason to believe that the chemical concentration is closer to this value, rather than to a fraction of the SQL. Where it is apparent that serious biases could result from the use of any of the preceding methods of approach, more sophisticated analytical and evaluation methods may be warranted.

5.3.3. SELECTION OF STATISTICAL AVERAGING TECHNIQUES

Reasonable discretion should generally be exercised in the selection of an averaging technique during the analysis of environmental sampling data. This is because the selection of specific methods of approach to average a set of environmental sampling data can have profound effects on the resulting concentration – especially for data sets from sampling results that are not normally distributed. For example, when dealing with log-normally distributed data, geometric means are often used as a measure of central tendency – in order to ensure that a few very high values do not exert excessive influence on the characterization of the distribution. However, if high concentrations do indeed represent 'hotspots' in a spatial or temporal distribution of the data set, then using the geometric mean could inappropriately discount the contribution of these high chemical concentrations present in the environmental samples. This is particularly significant if, for instance, the spatial pattern indicates that areas of high concentration for a chemical release are in close proximity to compliance boundaries or near exposure locations for sensitive populations (such as children and the elderly).

The geometric mean has indeed been extensively and consistently used as an averaging parameter in the past. Its principal advantage is in minimizing the effects of 'outlier' values (i.e., a few values that are much higher or lower than the general range of sample values). Its corresponding disadvantage is that, discounting these values may be inappropriate when they represent true variations in concentrations from one part of an impacted area or group to another (such as a 'hot-spot' vs. a 'cold-spot' region). As a measure of central tendency, the geometric mean is most appropriate if sample data are lognormally distributed, and without an obvious spatial pattern.

The arithmetic mean – commonly used when referring to an 'average' – is more sensitive to a small number of extreme values or a single 'outlier' compared to the geometric mean. Its corresponding advantage is that true high concentrations will not be inappropriately discounted. When faced with limited sampling data, however, this may not provide a conservative enough estimate of environmental chemical impacts.

In fact, none of the above measures, in themselves, may be appropriate in the face of limited and variable sampling data. Current applications tend to favor the use of an upper confidence limit (UCL) on the average concentration. Even so, if the UCL exceeds the maximum detected value amongst a data pool, then the latter is used as the source term or EPC. It is noteworthy that, in situations where there is a discernible spatial pattern to chemical concentration data, standard approaches to data aggregation and analysis may usually be inadequate, or even inappropriate.

To demonstrate the possible effects of the choice of statistical distributions and/or averaging techniques on the analysis of environmental data, consider a case involving the estimation of the mean, standard deviation, and confidence limits of monthly laboratory analysis data for groundwater from a potential drinking water well. The goal here is to compare the selected statistical parameters based on the assumption that this data is normally distributed *versus* an alternative assumption that the data is lognormally distributed. To accomplish this task, the several statistical manipulations enumerated below are carried out on the 'raw' and log-transformed data for the concentrations of benzene in the groundwater samples shown in Table 5.1.

Table 5.1. Environmental sampling data used to illustrate the effects of statistical averaging techniques on exposure point concentration predictions

Sampling Event	Concentration of Benzene in Drinking Water (μg/L)	
	Original 'raw' data, X	Log-transformed data, $Y = ln(X)$
1	0.049	−3.016
2	0.056	−2.882
3	0.085	−2.465
4	1.200	0.182
5	0.810	−0.211
6	0.056	−2.882
7	0.049	−3.016
8	0.048	−3.037
9	0.062	−2.781
10	0.039	−3.244
11	0.045	−3.101
12	0.056	−2.882

(1) Calculate the following statistical parameters for the 'raw' data: mean, standard deviation, and 95% confidence limits (see standard statistics textbooks for details of procedures involved). The arithmetic mean, standard deviation, and 95 percent confidence limits (95% CL) for a set of n values are defined, respectively, as follows:

$$X_m = \frac{\sum\limits_{i=1}^{n} X_i}{n} \qquad (5.1)$$

$$SD_x = \sqrt{\frac{\sum\limits_{i=1}^{n} (X_i - X_m)^2}{n-1}} \qquad (5.2)$$

$$CL_x = X_m \pm \frac{ts}{\sqrt{n}} \qquad (5.3)$$

where t is the value of the Student t-distribution (refer to standard statistical books) for the desired confidence level (e.g., 95% CL, which is equivalent to a level of significance of $\alpha = 5\%$) and degrees of freedom, $(n-1)$; and s is an estimate of the standard deviation from the mean (X_m). Thus,

$X_m = 0.213$ µg/L
$SD_x = 0.379$ µg/L
$CL_x = 0.213 \pm 0.241$ (i.e., $-0.028 \le CL_x \le 0.454$) and $UCL_x = 0.454$ µg/L

where: X_m = arithmetic mean of 'raw' data; SD_x = standard deviation of 'raw' data; CL_x = 95% confidence interval (95% CI) of 'raw' data; and UCL_x = 95% upper confidence level (95% UCL) of 'raw' data.

Note that, the development of the 95% confidence limits for the untransformed data gives a confidence interval of $0.213 \pm 0.109t = 0.213 \pm 0.241$ (where $t = 2.20$, obtained from the Student t-distribution for $(n-1) = 12-1 = 11$ degrees of freedom) – which indicates a non-zero probability for a negative concentration value.

(2) Calculate the following statistical parameters for the log-transformed data: mean, standard deviation, and 95% confidence limits (see standard statistics textbooks for details of procedures involved). The geometric mean, standard deviation, and 95 percent confidence limits (95% CL) for a set of n values are defined, respectively, as follows:

$$X_{gm} = anti\log\left\{\frac{\sum\limits_{i=1}^{n} \ell n X_i}{n}\right\} \qquad (5.4)$$

$$SD_x = \sqrt{\frac{\sum\limits_{i=1}^{n} (X_i - X_{gm})^2}{n-1}} \qquad (5.5)$$

$$CL_x = X_{gm} \pm \frac{ts}{\sqrt{n}} \qquad (5.6)$$

where t is the value of the Student t-distribution (refer to standard statistical books) for the desired confidence level and degrees of freedom, $(n-1)$; and s is an estimate of the standard deviation of the mean (X_{gm}). Thus,

$Y_{a\text{-}mean} = -2.445$

$SD_y = 1.154$

$CL_y = -2.445 \pm 0.733$ (i.e., a confidence interval from -3.178 to -1.712)

where: $Y_{a\text{-}mean}$ = arithmetic mean of log-transformed data; SD_y = standard deviation of log-transformed data; and CL_y = 95% confidence interval (95% CI) of log-transformed data.

The development of the 95% confidence limits for the log-transformed data gives a confidence interval of $-2.445 \pm 0.333t = -2.445 \pm 0.733$ (where $t = 2.20$, obtained from the student t-distribution for $(n-1) = 12-1 = 11$ degrees of freedom).

Transforming the average of the logarithmic Y values back into arithmetic values yields a geometric mean value of $X_{gm} = e^{-2.445} = 0.087$. Furthermore, transforming the confidence limits of the log-transformed values back into the arithmetic realm yields a 95% confidence interval of 0.042 to 0.180 µg/L, which consist of positive concentration values only. That is,

$X_{gm} = 0.087$ µg/L

$SD_x = 3.171$ µg/L

$0.042 \leq CL_x \leq 0.180$ µg/L

$UCL_x = 0.180$ µg/L

where: X_{gm} = geometric mean for the 'raw' data; SD_x = standard deviation of 'raw' data (assuming lognormal distribution); CL_x = 95% confidence interval (95% CI) for the 'raw' data (assuming lognormal distribution); and UCL_x = 95% upper confidence level (95% UCL) for the 'raw' data (assuming lognormal distribution).

It is obvious that the arithmetic mean, $X_m = 0.213$ µg/L, is substantially larger than the geometric mean of $X_{gm} = 0.087$ µg/L. This may be attributed to the two relatively large sample values in the data set (i.e., sampling events #4 and #5 in Table 5.1) – which tend to strongly bias the arithmetic mean; on the other hand, the logarithmic transform acts to suppress the extreme values. A similar observation can be made for the 95% upper confidence level (UCL) of the normally- and lognormally-distributed data sets. In any event, irrespective of the type of underlying distribution, the 95% UCL is a preferred statistical parameter to use in the evaluation of environmental data, rather than the statistical mean values.

The results from the above example analysis illustrate the potential effects that could result from the choice of one distribution type over another, and also the implications of selecting specific statistical parameters in the evaluation of environmental sampling data. In general, the use of arithmetic or geometric mean values for the estimation of average concentrations would tend to bias the EPC or other related estimates; the 95% UCL offers a better value to use.

5.4. Estimating Chemical Exposure Point Concentrations from Limited Data

In the absence of adequate and/or appropriate field sampling data, a variety of mathematical algorithms and models are often employed to support the determination of chemical exposure concentrations in human exposure media or consumer products. Such types of chemical exposure models are typically designed to serve a variety of purposes, but most importantly the following (Asante-Duah, 1998; Schnoor, 1996):

- To gain better understanding of the fate and behavior of chemicals existing in, or to be introduced into, the human living and work environments.
- To determine the temporal and spatial distributions of chemical exposure concentrations at potential receptor contact sites and/or locations.
- To predict future consequences of exposure under various chemical contacting or loading conditions, exposure scenarios, or risk management action alternatives.

The results from the modeling generally are used to estimate the consequential exposures and risks to potential receptors associated with a given chemical exposure problem.

One of the major benefits associated with the use of mathematical models in public health risk management programs relate to the fact that, environmental concentrations useful for exposure assessment and risk characterization can be estimated for several locations and time-periods of interest. Indeed, since field data are often limited and/or insufficient to facilitate an accurate and complete characterization of chemical exposure problems, models can be particularly useful for studying spatial and temporal variability, together with potential uncertainties. In addition, sensitivity analyses can be conducted by varying specific exposure parameters – and then using models to explore any ramifications reflected by changes in the model outputs.

Ultimately, the effective use of models in public health risk assessment and risk management programs depends greatly on the selection of the models most suitable for this purpose. The type of model selected will be dependent on the overall goal of the assessment, the complexity of the problem, the type of CoPCs, the nature of impacted and threatened media that are being evaluated in the specific investigation, and the type of corrective actions contemplated. A general guidance for the effective selection of models used in chemical exposure characterization and risk management decisions is provided in the literature elsewhere (e.g., Asante-Duah, 1998; CCME 1994; CDHS 1990; Clark, 1996; Cowherd et al., 1985; DOE, 1987; NRC, 1989; Schnoor, 1996; USEPA 1987, 1988a, 1988b; Yong et al., 1992; Zirschy and Harris, 1986). It is noteworthy that, in several typical environmental assessment situations, a 'ballpark' or 'order-of-magnitude' (i.e., a rough approximation) estimate of the chemical behavior and fate is usually all that is required for most analyses – in which case simple analytical models usually will suffice. Some relatively simple example models and equations that are often employed in the estimation of chemical concentrations in air, soil, water, and food products are provided below.

- *Screening Level Estimation of Chemical Volatilization into Shower Air.* A classic scenario that is often encountered in human health risk assessments relates to the volatilization of contaminants from contaminated water into shower air during a bathing/showering activity. A simple/common model that may be used to derive contaminant concentration in air from measured concentration in domestic water consists of a very simple box model of volatilization. In this case, the air concentration is derived from volatile emission rate, by treating the shower as a fixed volume with perfect mixing and no outside air exchange, so that the air concentration increases linearly with time.

In general, the following equation can be used to determine the average air concentration in the bathroom *during* a shower activity (generally for chemicals with a Henry's Law constant of $\geq 2 \times 10^{-7}$ atm-cu m/mol only) (HRI, 1995):

$$C_{sha} = \frac{[C_W \times f \times F_W \times t]}{2 \times [V \times 1000 \ \mu g/mg]} \qquad (5.7)$$

where C_{sha} is the average air concentration in the bathroom during a shower activity; C_W is the concentration of contaminant in the tap water ($\mu g/L$); f is the fraction of contaminant volatilized (unitless); F_W is the water flow rate in the shower (L/hour); t is the duration of shower activity (hours); and V is the bathroom volume (m^3).

The following equation can be used to determine the average air concentration in the bathroom *after* a shower activity (generally for chemicals with a Henry's Law constant of $\geq 2 \times 10^{-7}$ atm-m^3/mol only) (HRI, 1995):

$$C_{sha2} = \frac{[C_W \times f \times F_W \times t]}{[V \times 1000 \ \mu g/mg]} \qquad (5.8)$$

It is noteworthy that, water temperature is a key variable that affects stripping efficiencies and the mass transfer coefficients for the various sources of chemical releases into the shower air.

In the above simplified representations, the models assume that: there is no air exchange in the shower – which assumption tends to overestimate contaminant concentration in bathroom air; there is perfect mixing within the bathroom (i.e., the contaminant concentration is equally dispersed throughout the volume of the bathroom) – which assumption tends to underestimate contaminant concentration in shower air; the emission rate from water is independent of instantaneous air concentration; and the contaminant concentration in the bathroom air is determined by the amount of contaminants emitted into the box (i.e., $[C_W \times f \times F_W \times t]$) divided by the volume of the bathroom (V) (HRI, 1995).

- *Estimation of Household Air Contamination resulting from Volatilization from Domestic Water Supply.* Contaminated water can result in the volatilization of chemicals into residential indoor air – e.g., via shower stalls, bathtubs, washing machines, and dishwashers. Chemical concentrations in household indoor air due to contaminated domestic water may be estimated for volatile chemicals (generally for chemicals with a Henry's Law constant of $\geq 2 \times 10^{-7}$ atm-cu m/mol only), in accordance with the following relationship (HRI, 1995):

$$C_{ha} = \frac{[C_W \times WFH \times f]}{[HV \times ER \times MC \times 1000\ \mu g/mg]} \qquad (5.9)$$

where C_{ha} is the chemical concentration in air (mg/m3); C_W is the concentration of contaminant in the tap water (μg/L); WFH is the water flow through the house (L/day); f is the fraction of contaminant volatilized (unitless); HV is the house volume (m^3/house); ER is the air exchange rate (house/day); and MC is the mixing coefficient (unitless).

It is noteworthy that, water temperature is a key variable that affects stripping efficiencies and the mass transfer coefficients for the various sources of chemical releases into the indoor air.

- *Contaminant Bioconcentration in Meat and Dairy Products.* In many cases, the tendency of certain chemicals to become concentrated in animal tissues relative to their concentrations in the ambient environment can be attributed to the fact that the chemicals are lipophilic (i.e., they are more soluble in fat than in water). Consequently, these chemicals tend to accumulate in the fatty portion of animal tissue. In general, the bioconcentration of chemicals in meat is dependent primarily on the partitioning of chemical compounds to fat deposits (HRI, 1995). Thus,

$$C_x = BCF \times F \times C_W \qquad (5.10)$$

where C_x is the chemical concentration in animal tissue or dairy product; BCF is the chemical-specific bioconcentration factor for tissue fat – indicating the tendency of the chemical to accumulate in fat; F is the fat content of the tissue or dairy product; and C_W is the chemical concentration in water fed to the animal (HRI, 1995; USEPA, 1986).

The concentration of such bioaccumulative chemical in animal tissue (or other animal products for that matter) may be seen to reflect the chemical's inherent BCF.

- *Estimation of Contaminant Concentrations in Fish Tissues/Products.* Fish tissue contaminant concentrations may be predicted from water concentrations using chemical-specific BCFs, which predict the accumulation of contaminants in the lipids of the fish. In this case, the average chemical concentration in fish, based on the concentration in water and a BCF is estimated by the following relationship (HRI, 1995):

$$C_f = C_W \times BCF \times 1000 \qquad (5.11)$$

where C_f is the concentration in fish (μg/kg), C_W is the concentration in water (mg/L), and BCF is the bioconcentration factor.

If fish tissue concentrations are predicted from sediment concentrations, a two-step process is used; first, sediment concentration is used to calculate water concentrations, and then the water concentrations are used to predict fish tissue concentrations. The former is carried out in accordance with the following equation:

$$C_W = \frac{C_{sediment}}{[K_{oc} \times OC \times DN]} \qquad (5.12)$$

where C_W is the concentration of the chemical in water; $C_{sediment}$ is the concentration of the chemical in sediment; K_{oc} is the chemical-specific organic carbon partition coefficient; OC is the organic carbon content of the sediment; DN is the sediment density (relative to water density).

Models can indeed be used for several purposes in the study of chemical exposure and risk characterization problems. In general, the models usually simulate the response of a simplified version of a more complex system. As such, the modeling results are imperfect. Nonetheless, when used in a technically responsible manner, models can provide a very useful basis for making technically sound decisions about a chemical exposure problem. Indeed, models are particularly useful where several alternative scenarios are to be compared. In such cases, all the alternatives are compared on a similar basis; thus, whereas the numerical results of any single alternative may not be exact, the comparative results of showing that one alternative is superior to others will usually be valid.

5.5. Determination of the Level of Chemical Hazard

In order to make an accurate determination of the level of hazard potentially posed by a chemical, it is very important that the appropriate set of exposure data is collected during the hazard identification and accounting processes. It is also very important to use appropriate data evaluation tools in the processes involved. Several of the available statistical methods and procedures finding widespread use in chemical exposure and risk characterization programs can be found in the books on statistics (e.g., Berthouex and Brown, 1994; Cressie, 1994; Fruend and Walpole, 1987; Gibbons, 1994; Gilbert, 1987; Hipel, 1988; Miller and Freund, 1985; Ott, 1995; Sachs, 1984; Sharp, 1979; Wonnacott and Wonnacott, 1972; Zirschy and Harris, 1986). Ultimately, the process/approach used to estimate a potential receptor's EPC will comprise of the following elements:

- Determining the distribution of the chemical exposure/sampling data, and fitting the appropriate distribution to the data set (e.g., normal, lognormal, etc.);
- Developing the basic statistics for the exposure/sampling data – to include calculation of the relevant statistical parameters, such as the upper 95% confidence limit (UCL_{95}); and
- Calculating the EPC – usually defined as the minimum of either the UCL or the maximum exposure/sampling data value – conceptually represented as follows:-
  $EPC = min\,[UCL_{95} \; or \; \text{Max-Value}])$

The derived EPC (that may indeed be significantly different from any field-measured chemical concentrations) represents the 'true' or reasonable exposure level at the potential receptor location of interest – and this value is used in the calculation of the chemical intake/dose for the populations potentially at risk.

5.6. Suggested Further Reading

Allen, D., 1996. *Pollution Prevention for Chemical Processes,* J. Wiley, New York, NY

Aris, R., 1994. *Mathematical Modelling Techniques,* Dover Publications, Inc., New York, NY

Conover, WJ, 1980. *Practical Nonparametric Statistics,* 2nd edition, J. Wiley, New York, NY

Dowdy, D., TE McKone, and DPH Hsieh, 1996. The use of molecular connectivity index for estimating biotransfer factors, *Environmental Science and Technology,* 30: 984 – 989

Draper, NR and H. Smith, 1998. *Applied Regression Analysis, 3rd ed.,* John Wiley & Sons, New York, NY

Driver, J., SR Baker, and D. McCallum, 2001. *Residential Exposure Assessment: A Sourcebook,* Kluwer Academic Publishers, Dordrecht, The Netherlands

Efron, B. and R. Tibshirani, 1993. *An Introduction to the Bootstrap,* Chapman & Hall, New York, NY

Frosig, A., H. Bendixen, and D. Sherson, 2001. Pulmonary deposition of particles in welders: on-site measurements, *Archives of Environmental Health,* 56(6): 513 – 521

Gilpin, A., 1995. *Environmental Impact Assessment (EIA): cutting edge for the twenty-first century,* Cambridge University Press, Cambridge, UK

Isaaks, EH and RM Srivastava, 1989. *Applied Geostatistics,* Oxford University Press, Cambridge, England, UK

Johnson, DL, K. McDade, and D. Griffith, 1996. Seasonal variation in paediatric blood lead levels in Syracuse, NY, USA, *Environmental Geochemistry and Health,* 18: 81 – 88

Levesque, B., P. Ayotte, R. Tardif, *et al.,* 2002. Cancer risk associated with household exposure to chloroform, *Journal of Toxicology and Environmental Health, Part A: Current Issues,* 65(7): 489 – 502

Patil, GP and CR Rao (eds.), 1994. *Handbook of Statistics, Volume 12, Environmental Statistics,* North-Holland, New York

Raloff, J., 1996. Tap water's toxic touch and vapors, *Science News,* 149(6): 84

Schulz, TW and S. Griffin, 1999. Estimating risk assessment exposure point concentrations when data are not Normal or Lognormal, *Risk Analysis,* 19(4): 577 – 584

Chapter 6

EXPOSURE ASSESSMENT: ANALYSIS OF HUMAN INTAKE OF CHEMICALS

Once a 'social' or environmental chemical has been determined to present a potential health hazard, then the main concern shifts to the likelihood for, and degree of, human exposure. The exposure assessment phase of the human health risk assessment is used to estimate the rates at which chemicals are absorbed by potential receptors. Since most potential receptors tend to be exposed to chemicals from a variety of sources and/or in different environmental media, an evaluation of the relative contributions of each medium and/or source to total chemical intake becomes a critical part of most exposure analyses.

Overall, the accuracy with which exposures are characterized can indeed become a major determinant of the ultimate validity of a risk assessment. This chapter discusses the principal exposure evaluation tasks that, upon careful implementation, should allow effective public health risk management decisions to be made about chemical exposure problems.

6.1. Fundamental Concepts and Requirements in the Human Exposure Assessment Process

The characterization of chemical exposure problems is a process used to establish the presence or absence of chemical hazards, to delineate the nature and degree of the hazards, and to determine possible human health threats posed by the exposure or hazard situation. The routes of chemical exposure (which may consist of inhalation, ingestion, and/or dermal contacts), as well as the duration of exposure (that may be short-term [acute] or long-term [chronic]) will significantly influence the level of impacts on the affected receptors. The nature and behavior of the chemical substances of interest also form a very important basis for evaluating the potential for human exposures to its possible toxic or hazardous constituents.

The exposure assessment process is used to estimate the rates at which chemicals are absorbed by the potential human receptor. More specifically, it is generally used to determine the magnitude of actual and/or potential receptor exposures to chemical constituents, the frequency and duration of these exposures, and the pathways via which the target receptor is potentially exposed to the chemicals that they contact from a variety of sources. The exposure assessment also involves describing the nature and size of the population exposed to a substance (i.e., the risk group, which refers to the actual or hypothetical exposed population) and the magnitude and duration of their exposure.

As part of an exposure characterization effort, populations potentially at risk are defined, and concentrations of the chemicals of potential concern (CoPCs) are determined in each medium to which potential receptors may be exposed. Then, using the appropriate case-specific exposure parameters, the intakes of the CoPCs can be estimated. The evaluation could indeed concern past or current exposures, as well as exposures anticipated in the future.

6.1.1. FACTORS AFFECTING HUMAN EXPOSURE TO CHEMICAL HAZARDS

The assessment of human receptor exposure to chemicals requires translating concentrations found in the target consumer product or human environment into quantitative estimates of the amount of chemical that contacts the individual potentially at risk. Contact is expressed by the amount of material per unit body weight (mg/kg-day) that enters the lungs (for an inhalation exposure), enters the gastrointestinal tract (for an ingestion exposure), or crosses the stratum corneum of the skin (for a dermal contact exposure). This quantity is used as a basis for projecting the incidence of health detriment to the human receptor.

To accomplish the human exposure determination task, several important exposure parameters and/or information will typically be acquired (Box 6.1). Also, in terms of chemical exposures, the amount of contacted material that is bioavailable for absorption is very important (see Section 6.1.5).

It is noteworthy that, oftentimes, conservative estimates in the exposure evaluation assume that a potential receptor is always in the same location, exposed to the same ambient concentration, and that there is 100 percent absorption upon exposure. These assumptions hardly represent any real-life situation. In fact, lower exposures will generally be expected under most circumstances – due to the fact that potential receptors will typically be exposed to lower or even near-zero levels of the CoPCs for the period of time spent outside chemical-laden settings.

6.1.2. DEVELOPMENT OF HUMAN EXPOSURE SCENARIOS

The exposure assessment process generally involves several characterization and evaluation efforts – including the following key steps:

- Determination of chemical distributions and behaviors – traced from a 'release' or 'originating' source to the locations for likely human exposure;
- Identification of significant chemical release, migration, and exposure pathways;
- Identification of potential receptors – i.e., the populations potentially at risk;
- Development of conceptual exposure model(s) and exposure scenarios – including a determination of current and future exposure patterns, and the analysis of the environmental fate and persistence of the CoPCs;
- Estimation/modeling of exposure point concentrations for the critical exposure pathways and media; and
- Estimation of chemical intakes for all potential receptors, and for all significant exposure pathways associated with the CoPCs.

Box 6.1. Typical exposure parameters and information necessary for estimating potential receptor exposures

Exposure Route	Relevant Exposure Parameters/Data
• Inhalation	♦ airborne chemical concentrations (e.g., resulting from showering, bathing, and other uses of chemical-based consumer products; or from dust inhalation; etc.) ♦ variation in air concentrations over time ♦ amount of contaminated air breathed ♦ fraction of inhaled chemical absorbed through lungs ♦ breathing rate ♦ exposure duration and frequency ♦ exposure averaging time ♦ average receptor body weight
• Ingestion	♦ concentration of chemical in consumed material (e.g., water, food, drugs/medicines, soils, etc.) ♦ amount of chemical-based material ingested each day (e.g., water ingestion rate; food intake rate; soil ingestion rate; etc.) ♦ fraction of ingested chemical absorbed through wall of gastrointestinal tract ♦ exposure duration and frequency ♦ exposure averaging time ♦ average receptor body weight
• Dermal (Skin) Absorption	♦ concentration of chemical in contacted material (e.g., cosmetics, water, soils, etc.) ♦ amount of daily skin contact (e.g., dermal contact with soil; dermal contact with water; dermal contact with cosmetics; etc.) ♦ fraction of chemical absorbed through skin during contact period ♦ period of time spent in contact with chemical-based material ♦ average contact rate ♦ receptor's contacting body surface area ♦ exposure duration and frequency ♦ exposure averaging time ♦ average receptor body weight

In fact, exposure pathways are one of the most important elements of the exposure assessment process – consisting of the routes that chemical constituents follow to reach potential receptors. Thus, failure to identify and address any significant exposure pathway may seriously detract from the usefulness of any risk assessment, since a complete pathway must be present for receptor exposures to occur. An exposure pathway is considered complete only if all of the following elements are present:

- Chemical hazard source(s)
- Mechanism(s) of chemical contacting by receptors and/or chemical release into the human environment
- Human exposure route(s)
- Receptor intake and/or exposure in the affected media, within the human environment.

Typically, exposure routes are determined by integrating information from an initial environmental characterization with knowledge about potentially exposed populations and their likely behaviors. The significance of the chemical hazard is evaluated on the basis of whether the target chemical could cause significant adverse exposures and impacts.

On the basis of the above, a realistic set of exposure scenarios can be developed for a given chemical exposure/characterization problem. Several specific tasks are usually undertaken to facilitate the development of complete and realistic exposure scenarios; the critical tasks include the following:

- Determine the sources of chemical hazards
- Identify the specific constituents of concern
- Identify the affected environmental or exposure media
- Delineate chemical release and migration pathways
- Identify potential receptors
- Determine potential exposure routes
- Delineate likely and significant chemical contacting rates by receptors, and/or chemical release rates into the human living and work environments
- Construct a representative conceptual exposure model (CEM) for the problem situation.

Indeed, the conceptual model facilitates an assessment of the nature and extent of exposure, as well as helps determine the potential impacts from such exposures. Consequently, in as early a stage as possible during a chemical exposure investigation, all available information should be compiled and analyzed to help develop a representative CEM for the problem situation.

It is noteworthy that the exposure scenario associated with a given hazardous situation may be better defined if the exposure is known to have already occurred. In most cases associated with the investigation of potential chemical exposure problems, however, important decisions may have to be made about exposures that may not yet have occurred – in which case hypothetical exposure scenarios are generally developed to facilitate the problem solution. Ultimately, the type/nature of human exposure scenarios associated with a given exposure situation provides clear direction for the exposure assessment. Also, the exposure scenarios developed for a given chemical exposure problem can be used to support an evaluation of the risks posed by the situation, as well as facilitate the development of appropriate public health risk management decisions.

6.1.3. CHEMICAL INTAKE *VERSUS* DOSE

Intake (also commonly called exposure) is defined as the amount of chemical coming into contact with a receptor's visible exterior body (e.g., skin and openings into the body such as mouth and nostrils), or with the abstract/conceptual exchange boundaries (such as the skin, lungs, or gastrointestinal tract); and *dose* is the amount of chemical absorbed by the body into the bloodstream. In fact, the *internal dose* (i.e., absorbed dose) tends to differ significantly from the (externally) *applied dose* (i.e., exposure or intake); the *internal dose* of a chemical is the amount of a chemical that directly crosses the barrier at the absorption site into the systemic circulation.

The intake value quantifies the amount of a chemical contacted during each exposure event – where 'event' may have different meanings depending on the nature of exposure scenario being considered (e.g., each day's inhalation of an air contaminant constitutes one inhalation exposure event). The quantity of a chemical absorbed into the bloodstream per event – represented by the dose – is calculated by further considering pertinent physiological parameters (such as gastrointestinal absorption rates). Indeed, the internal dose of a chemical is considered rather important for predicting the potential toxic effects of the chemical; this is because, among other things, once in the systemic circulation, the chemical is able to reach all major target organ sites.

In general when the systemic absorption from an intake is unknown, or cannot be estimated by a defensible scientific argument, intake and dose are considered to be the same (i.e., a 100 percent absorption into the bloodstream from contact is assumed). Such an approach provides a conservative estimate of the actual exposures. In any case, intakes and doses are normally calculated during the same step of the exposure assessment; the former multiplied by an absorption factor yields the latter value.

6.1.4. CHRONIC *VERSUS* SUBCHRONIC EXPOSURES

Event-based intake values are converted to final intake values by multiplying the intake per event by the frequency of exposure events, over the timeframe being considered in an exposure assessment. *Chronic daily intake* (CDI), which measures long-term (chronic) exposures, are based on the number of events that are assumed to occur within an assumed lifetime for potential receptors; *subchronic daily intake* (SDI), which represents projected receptor exposures over a short-term period, consider only a portion of a lifetime (USEPA 1989b). The respective intake values are calculated by multiplying the estimated exposure point chemical concentrations by the appropriate receptor exposure and body weight factors.

SDIs are generally used to evaluate subchronic non-carcinogenic effects, whereas CDIs are used to evaluate both carcinogenic risks and chronic non-carcinogenic effects. It is noteworthy that, the short-term exposures can result when a particular activity is performed for a limited number of years or when, for instance, a chemical with a short half-life degrades to negligible concentrations within several months of its presence in a receptor's exposure setting.

6.1.5. INCORPORATING CHEMICAL BIOAVAILABILITY ADJUSTMENTS INTO EXPOSURE CALCULATIONS

The amount of contacted material that is bioavailable for absorption is very important in the evaluation of human exposure to chemicals. Bioavailability can be influenced by external physical/chemical factors such as the form of a chemical in the exposure media, as well as by internal biological factors such as absorption mechanisms within a living organism. For example, the oral bioavailability of a chemical compound is often characterized as a function of two key elements – *bioaccessibility* and *absorption* (Paustenbach *et al.*, 1997). *Bioaccessibility* describes the fraction of the chemical that desorbs from its matrix (e.g., soil, dust, wood, food, drinks, drugs/medicines, etc.) in the gastrointestinal (GI) tract and, therefore, is available for absorption; and *absorption* describes the transfer of a chemical across a biological membrane into the blood circulation.

Bioavailability itself is defined as the fraction of a chemical that is taken up by the body's circulatory system relative to the amount that an organism is exposed to during, for instance, ingestion of the chemical-laden material. Indeed, bioavailability is a rather important concept, because exposure and risk are more closely related to the bioavailable fraction of a chemical than to its total concentration in any given media/matrix – and thus has significant implications for determining any 'safe' levels of chemicals in an exposure medium. It is noteworthy that, bioavailablity can be media-specific; for instance, the bioavailability of metals ingested in a soil matrix is generally believed to be considerably lower than the bioavailability of the same metals ingested in water. The bioavailability of a CoPC may be estimated by multiplying the fraction of the chemical that is bioaccessible and the fraction that is absorbed. Thus, as an example, a bioaccessibility of Pb in soil of 60%, combined with an absorption of Pb in young children of 50% yields a total bioavailability of 30%; and a bioaccessibility of Pb in water of 100%, combined with an absorption of Pb in young children of 50% yields a total bioavailability of 50%.

Bioavailability generally refers to how much of a chemical is 'available' to have an adverse effect on humans or other organisms. Consequently, knowledge of chemical bioavailability can play key roles in risk management decisions. For example, bioavailability adjustments in risk assessment can help establish reduced time and cost necessary for site remediation; bioavailability is inversely related to risk-based cleanup levels – i.e., lower bioavailability results in increased risk-based cleanup levels. Indeed, when risk assessments are adjusted to account for lower case-specific bioavailability, the resulting increase in cleanup levels can, in some cases, reduce remediation costs substantially. This is because, determining the site-specific bioavailability can allow for a revising of the exposure estimates – to more realistically and pragmatically reflect the conditions at a project site.

Overall, bioavailability has a direct relationship to exposure dose and risk. In general, a lower bioavailability means a decrease in exposure dose and risk – and conversely, higher bioavailability implies an increased exposure dose and risk.

6.1.6. CHEMICAL TRANSFORMATION PRODUCTS IN RISK ASSESSMENT: INCORPORATING CHEMICAL DEGRADATION INTO EXPOSURE CALCULATIONS

Many chemicals are transformed to structurally related transformation/daughter products in the environment before they are mineralized (e.g., DDE formed out of DDT) – with each of the resultant transformation products tending to display their own toxicity and persistence. Consequently, it is often imperative to include such transformation products into chemical exposure and risk assessments – albeit this often adds another layer of complexity to the overall exposure and risk assessment process, especially because toxicity data for the daughter products is often lacking.

Various methods of approach may be utilized to incorporate transformation products into exposure and risk assessment of the parent compounds. For instance, Fenner *et al.* (2002) offer some elaborate procedures that integrates the chemical transformation kinetics into the overall assessment – by calculating the environmental exposure to parent compounds and daughter products as they are being formed in the degradation/transformation cascade, and then subsequently developing a corresponding risk quotient.

Indeed, when certain chemical compounds undergo degradation, potentially more toxic daughter products result (such as is the case when trichloroethylene (TCE) biodegrades to produce vinyl chloride). On the other hand, there are situations where the end-products of degradation are less toxic than the parent compounds. Since receptor exposures could be occurring over long time periods, a more valid approach in exposure modeling will be to take chemical degradation (or indeed other transformation processes) into consideration during an exposure assessment. Under such circumstances, if significant degradation is likely to occur, then exposure calculations become much more complicated. In that case, chemical concentrations at exposure or release sources are calculated at frequent and short time intervals, and then summed over the exposure period.

To illustrate the concept of incorporating chemical degradation into exposure assessment, let us assume first-order kinetics for a hypothetical chemical exposure problem. An approximation of the degradation effects for this type of scenario can be obtained by multiplying the chemical concentration data by a degradation factor, *DGF*, defined by:

$$DGF = \frac{(1 - e^{-kt})}{kt} \tag{6.1}$$

where k is a chemical-specific degradation rate constant [days^{-1}] and t is the time period over which exposure occurs [days]. For a first-order decaying substance, k is estimated from the following relationship:

$$T_{1/2} \text{ [days]} = \frac{0.693}{k} \qquad \text{or} \qquad k \text{ [days}^{-1}] = \frac{0.693}{T_{1/2}} \tag{6.2}$$

where $T_{1/2}$ is the chemical half-life, which is the time after which the mass of a given substance will be one-half its initial value.

It is noteworthy that, the degradation factor is usually ignored in most exposure calculations; this is especially justifiable if the degradation product is of potentially equal toxicity, and is present in comparable amounts as the parent compound. In any case, although it cannot always be proven that the daughter products will result in receptor exposures that are at comparable levels to the parent compound, the DGF term is still ignored in most screening-level exposure assessments.

6.2.   Human Exposure Quantification: The Exposure Estimation Model

The analysis of potential human exposures to chemicals in the human environment often involves several complex issues. Invariably, potential receptors may become exposed to a variety of environmental chemicals via several different exposure routes – represented primarily by the inhalation, ingestion, and dermal exposure routes (illustrated by Figure 6.1). These potential human receptor exposures can be evaluated via the calculation of the *average daily dose* (ADD) and/or the *lifetime average daily dose* (LADD).

The carcinogenic effects (and sometimes the chronic non-carcinogenic effects) associated with a chemical exposure problem involve estimating the LADD; for non-carcinogenic effects, the ADD is usually used. The ADD differs from the LADD in that the former is not averaged over a lifetime; rather, it is the average daily dose pertaining to the actual duration of exposure. The *maximum daily dose* (MDD) will typically be used in estimating acute or subchronic exposures.

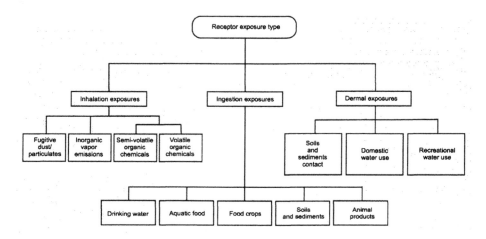

Figure 6.1. Major routes for human exposures to chemicals

In general, human exposures to chemical materials may be conservatively quantified according to the generic equation shown in Box 6.2. The various exposure parameters used in this model may be derived on a case-specific basis, or they may be compiled from regulatory guidance manuals and documents, and related scientific literature (e.g., Binder *et al.*, 1986; Calabrese *et al.*, 1989; CAPCOA, 1990; DTSC 1994; Finley *et al.*, 1994; Hrudey *et al.*, 1996; LaGoy, 1987; Lepow *et al.*, 1974, 1975; OSA 1992; Sedman, 1989; Smith, 1987; Stanek and Calabrese, 1990; Travis and Arms, 1988; USEPA 1987, 1989a, 1989b, 1991, 1992, 1997; Van Wijnen, 1990); these parameters are usually based on information relating to the maximum exposure level that results from specified categories of receptor activity and/or exposures.

As simple illustrative practical example, consider a situation where the average concentration of 1,2-dichlorobenzene in a domestic water supply has been recorded at 1.7 µg/L. Now, it is required to determine the intake for a 70-kg adult who consumes 2 liters of water per day over a 30-year period. The requested chemical intake may be estimated by using the equation shown in Box 6.2. Assuming an exposure frequency of 365 days/year, and also $FI = 1$ and $ABS_f = 1$ for this non-carcinogenic contaminant, the required intake is estimated as follows:

$$EXP = \left[\frac{(C_{medium} \times CR \times CF \times FI \times ABS_f \times EF \times ED)}{(BW \times AT)} \right]$$

Substituting $C_{medium} = 1.7$ µg/L; $CR = 2$ L/day; $CF = 10^{-3}$ mg/µg; $FI = 1$; $ABS_f = 1$; $EF = 365$ d/year; $ED = 30$ years; $BW = 70$ kg; and $AT = (ED \times 365) = (30 \times 365)$ days yields:

$$EXP = \left[\frac{(1.7 \times 2\times10^{-3} \times 1 \times 1 \times 365 \times 30)}{(70 \times 30 \times 365)} \right] \cong 4.86 \times 10^{-5} \, mg \, / \, kg - day$$

The methods by which each specific type of chemical exposure is estimated – including the relevant exposure estimation algorithms/equations for specific major routes of exposure (i.e., inhalation, ingestion, and skin contacting) – are discussed in greater detail below. These algorithms and related ones are elaborated in an even greater detail elsewhere in the literature (e.g., Asante-Duah, 1998; CAPCOA 1990; CDHS 1986; DTSC 1994; McKone, 1989; McKone and Daniels, 1991; NRC, 1991a,b; USEPA 1986c, 1988, 1989a, 1989b, 1991, 1992, 1997). An illustration of the computational steps involved in the calculation of human receptor intakes and doses is also provided below.

Box 6.2. General equation for estimating potential human exposures to chemicals

---

$$EXP = \frac{(C_{medium} \times CR \times CF \times FI \times ABS_f \times EF \times ED)}{(BW \times AT)}$$

where:

EXP	= intake (i.e., the amount of chemical at the exchange boundary), adjusted for absorption (mg/kg-day)
C_{medium}	= average or reasonably maximum exposure concentration of chemical contacted by potential receptor over the exposure period in the medium of concern (e.g., $\mu g/m^3$ [air]; or $\mu g/L$ [water]; or mg/kg [solid materials, such as food and soils])
CR	= contact rate, i.e., the amount of 'chemical-based' medium contacted per unit time or event (e.g., inhalation rate in m^3/day [air]; or ingestion rate in mg/day [food; soil], or L/day [water])
CF	= conversion factor (10^{-6} kg/mg for solid media, or 1.00 for fluid media)
FI	= fraction of intake from 'chemical-based' source (dimensionless)
ABS_f	= bioavailability or absorption factor (%)
EF	= exposure frequency (days/years)
ED	= exposure duration (years)
BW	= body weight, i.e., the average body weight over the exposure period (kg)
AT	= averaging time (period over which exposure is averaged – days)
	= ED x 365 days/year, for non-carcinogenic effects of human exposure
	= LT x 365 days/year = 70 years x 365 days/year, for carcinogenic effects of human exposure (assuming an average lifetime, LT, of 70 years)

6.2.1. POTENTIAL RECEPTOR INHALATION EXPOSURES

Two major types of inhalation exposures are generally considered in the investigation of potential chemical exposure problems (see Figure 6.1) – broadly categorized into the inhalation of airborne fugitive dust/particulates, in which all individuals within approximately 80 km ($\cong$ 50 miles) radius of a chemical release source are potentially impacted; and the inhalation of volatile compounds (i.e., airborne, vapor-phase chemicals). In general, potential inhalation intakes may be estimated based on the length of exposure, the inhalation rate of the exposed individual, the concentration of constituents in the inhaled air, and the amount retained in the lungs; this is conservatively represented by the following relationship:

$$\text{Inhalation Exposure (mg/kg-day)} = \frac{\{GLC \times RR \times CF\}}{BW} \qquad (6.3)$$

where: GLC is the ground-level concentration of constituents of concern ($\mu g/m^3$); RR is the respiration rate of exposed individual (m^3/day); CF is a conversion factor ($= 1mg/1000\mu g = 1.0E-03$ mg/μg); and BW is the body weight of exposed person (kg). Potential receptor inhalation exposures specific to chemical releases associated with wind-borne particulate matter/fugitive dust, and also volatile compounds from airborne vapor-phase emissions are annotated below.

Receptor Inhalation Exposure to Particulates from
Constituents in Fugitive/Airborne Dust
Box 6.3 shows an algorithm that can be used to calculate potential receptor intakes resulting from the inhalation of constituents in wind-borne fugitive dust (CAPCOA 1990; DTSC 1994; USEPA 1988, 1989a, 1989b, 1992, 1997). The constituent concentration in air, C_a, is defined by the ground-level concentration (GLC) – usually represented by the respirable (PM-10) particles – expressed in $\mu g/m^3$. The PM-10 particles consist of particulate matter with physical/aerodynamic diameter of less than 10 microns (i.e., <10 μm) – and it represents the respirable portion of the particulate emissions; this portion is capable of being deposited in thoracic (tracheobronchial and alveolar) portions of the lower respiratory tract. It is noteworthy that, fine particulate matter has also been characterized by $PM_{2.5}$ (i.e., $\leq 2.5\mu$m aerodynamic diameter). Finally, it should be recognized that the total PM exposure of an individual during a given period of time usually consists of exposures to many different particles from various sources whiles the receptor is in different microenvironments. As such, these different human microenvironments should be carefully identified so that the corresponding exposures can be appropriately analyzed/evaluated.

Box 6.3. Equation for estimating inhalation exposure to chemical constituents in fugitive/airborne dust

$$INH_a = \frac{(C_a \times IR \times RR \times ABS_s \times ET \times EF \times ED)}{(BW \times AT)}$$

where:
INH_a	= inhalation intake (mg/kg-day)
C_a	= chemical concentration of airborne particulates (defined by the ground-level concentration [GLC], and represented by the respirable, PM-10 particles) (mg/m^3)
IR	= inhalation rate (m^3/h)
RR	= retention rate of inhaled air (%)
ABS_s	= percent of chemical absorbed into the bloodstream (%)
ET	= exposure time (h/day)
EF	= exposure frequency (days/years)
ED	= exposure duration (years)
BW	= body weight, i.e., the average body weight over the exposure period (kg)
AT	= averaging time (period over which exposure is averaged – days)
	= ED x 365 days/year, for non-carcinogenic effects of human exposure
	= LT x 365 days/year = 70 years x 365 days/year, for carcinogenic effects of human exposure (assuming an average lifetime, LT, of 70 years)

Receptor Inhalation Exposure to Volatile Compounds

Box 6.4 shows an algorithm that can be used to calculate potential receptor intakes resulting from the inhalation of airborne vapor-phase chemicals (CAPCOA 1990; DTSC 1994; USEPA 1988, 1989a, 1989b, 1992, 1997). The vapor-phase contaminant concentration in air is assumed to be in equilibrium with the concentration in the release source.

Box 6.4. Equation for estimating inhalation exposure to vapor-phase chemical constituents

$$INH_{av} = \frac{(C_{av} \times IR \times RR \times ABS_s \times ET \times EF \times ED)}{(BW \times AT)}$$

where:

INH_{av}	= inhalation intake (mg/kg-day)
C_{av}	= chemical concentration in air (mg/m^3) [The vapor-phase contaminant concentration in air is assumed to be in equilibrium with the concentration in the release source.]
IR	= inhalation rate (m^3/h)
RR	= retention rate of inhaled air (%)
ABS_s	= percent of chemical absorbed into the bloodstream (%)
ET	= exposure time (h/day)
EF	= exposure frequency (days/years)
ED	= exposure duration (years)
BW	= body weight, i.e., the average body weight over the exposure period (kg)
AT	= averaging time (period over which exposure is averaged – days)
	= ED × 365 days/year, for non-carcinogenic effects of human exposure
	= LT × 365 days/year = 70 years × 365 days/year, for carcinogenic effects of human exposure (assuming an average lifetime, LT, of 70 years)

It is noteworthy that, showering generally represents an activity that promotes the release of volatile organic chemicals (VOCs) from water – especially due to high turbulence, high surface area, and small droplets of water involved. In fact, contemporary studies have shown that risks from inhalation while showering can be comparable to – if not greater than – risks from drinking contaminated water (Jo *et al.*, 1990a, 1990b; Kuo *et al.*, 1998; McKone, 1987; Richardson *et al.*, 2002; Wilkes *et al.*, 1996). Thus, this exposure scenario represents a particularly important one to evaluate in a human health risk assessment, as appropriate. In this case, the concentration of the contaminants in the shower air is assumed to be in equilibrium with the concentration in the water.

In the case of volatile compounds released whiles bathing, the exposure relationship may be defined by the specific equation shown in Box 6.4A (USEPA, 1988; 1989a,b). Other assumptions used in this model include the following: there is no air exchange in the shower (this assumption tending to overestimate the concentration of contaminants in the air in the bathroom); there is perfect mixing within the bathroom (this assumption tending to underestimate the concentration of contaminants in the air in the shower); and the emission rate from water is independent of instantaneous air concentration.

Box 6.4A. Equation for estimating inhalation exposure to vapor-phase chemical constituents during showering activity

$$INH = [\,C_w \times FV \times \{\frac{ET_1}{(VS \times 2)} + \frac{ET_2}{VB}\}\,] \times \frac{(IR \times RR \times VW \times ABS_s \times EF \times ED)}{(BW \times AT)}$$

$$= [ACB_{sh}\,] \times \frac{(IR \times RR \times VW \times ABS_s \times EF \times ED)}{(BW \times AT)}$$

where:

INH	= inhalation intake whiles showering (mg/kg-day)
C_w	= concentration of contaminant in water – adjusted for water treatment purification factor, T_f, which is the fraction remaining after treatment [i.e., $C_w = C_{w\text{-}source} \times T_f$] (mg/L)
FV	= fraction of contaminant volatilized (unitless)
ET_1	= length of exposure in shower (hrs/day)
ET_2	= length of additional exposure in enclosed bathroom (hrs/day)
VS	= volume of shower stall (m^3)
VB	= volume of bathroom (m^3)
IR	= breathing/inhalation rate (m3/hr)
RR	= retention rate of inhaled air (%)
VW	= volume of water used in shower (L)
	= water flow rate (F_w [L/hr]) x shower duration (hr)
ABS_s	= percent of chemical absorbed into the bloodstream (%)
EF	= exposure frequency (days/year)
ED	= exposure duration (years)
BW	= body-weight (kg)
AT	= averaging time (period over which exposure is averaged – days)
ACB_{sh}	= average air concentration in bathroom during a shower activity

$$= [\,C_w \times FV \times \{\frac{ET_1}{(VS \times 2)} + \frac{ET_2}{VB}\}\,]$$

Note: The concentration of contaminants in water may be adjusted further for environmental degradation, by multiplying by a factor of e^{-kt}, where k (in days^{-1}) is the environmental degradation constant of the chemical and t (in days) is the average time of transit through the water distribution system. This yields a new C_w value to be used for the intakes computation, viz., $C_w{}^* = (C_w)(e^{-kt})$

6.2.2. POTENTIAL RECEPTOR INGESTION EXPOSURES

The major types of ingestion exposures that could affect chemical exposure decisions consist of the oral intake of constituents present in consumer products, food products, waters, and miscellaneous environmental materials (see Figure 6.1). In general, exposure through ingestion is a function of the concentration of the constituents in the material ingested (e.g., soil, water, or food products such as crops, and dairy/beef), the gastrointestinal absorption of the constituent in solid or fluid matrix, and the amount ingested. This can be conservatively estimated by using the following generic types of representative equations:

Water (and other liquids) Ingestion Exposure (mg/kg-day)

$$= \frac{\{CW \times WIR \times GI\}}{BW} \qquad (6.4)$$

$$\text{Soil Ingestion Exposure (mg/kg-day)} = \frac{\{CS \times SIR \times GI\}}{BW} \qquad (6.5)$$

$$\text{Crop Ingestion Exposure (mg/kg-day)} = \frac{\{CS \times RUF \times CIR \times GI\}}{BW} \qquad (6.6)$$

Consumer Products (e.g., dairy and beef) Ingestion Exposure (mg/kg-day)

$$= \frac{\{CD \times FIR \times GI\}}{BW} \qquad (6.7)$$

where: CW is the chemical concentration in water (mg/L); WIR is the water consumption rate (L/day); CS is the chemical concentration in soil (mg/kg); SIR is the soil consumption rate (kg/day); RUF is the root uptake factor; CIR is the crop consumption rate (kg/day); CD is the concentration of chemical in diet (mg/kg) – for grazing animals, the concentration of chemicals in tissue, CT, is $CT = BCF \times F \times CD$, where BCF is the bioconcentration factor (fat basis) for the organism, expressed as [mg/kg fat]/[mg/kg of diet], and F is the fat content of tissues (in [kg fat]/[kg tissue]); FIR is the food (e.g., meat and dairy) consumption (kg/day); GI is the gastrointestinal absorption factor; and BW is the body weight (kg).

The total dose received by the potential receptors from chemical ingestion will, in general, be dependent on the absorption of the chemical across the gastrointestinal (GI) lining. The scientific literature provides some estimates of such absorption factors for various chemical substances. For chemicals without published absorption values and for which absorption factors are not implicitly accounted for in toxicological parameters, absorption may conservatively be assumed to be 100%.

Potential receptor ingestion exposures specific to the oral intake of chemical-impacted waters, the consumption of chemicals in food products, and the incidental ingestion of other contaminated solid matrices (such as soils/sediments) are annotated below.

Receptor Exposure through Ingestion of Constituents in Drinking Water
Exposure to contaminants via the ingestion of contaminated fluids may be estimated using the algorithm shown in Box 6.5 (CAPCOA 1990; DTSC 1994; USEPA 1988, 1989a, 1989b, 1992, 1997). This consists of the applicable relationship for estimating the chemical exposure intake that occurs through the ingestion of drinking water.

As a special situation, receptor exposure through incidental ingestion of constituents in water *during swimming activities* (i.e., the result of the ingestion of contaminated surface water during recreational activities) may be estimated by using the algorithm shown in Box 6.5A.

Box 6.5. Equation for estimating ingestion exposure to constituents in water used for culinary purposes

$$ING_{dw} = \frac{(C_w \times WIR \times FI \times ABS_s \times EF \times ED)}{(BW \times AT)}$$

where:
ING_{dw} = ingestion intake, adjusted for absorption (mg/kg-day)
C_w = chemical concentration in drinking water (mg/L)
WIR = average ingestion rate (liters/day)
FI = fraction ingested from contaminated source (unitless)
ABS_s = bioavailability/gastrointestinal [GI] absorption factor (%)
EF = exposure frequency (days/year)
ED = exposure duration (years)
BW = body-weight (kg)
AT = averaging time (period over which exposure is averaged – days)

Box 6.5A. Equation for estimating incidental ingestion exposure to contaminated surface water during recreational activities

$$ING_r = \frac{(CW \times CR \times ABS_s \times ET \times EF \times ED)}{(BW \times AT)}$$

where:
ING_r = ingestion intake, adjusted for absorption (mg/kg-day)
CW = chemical concentration in water (mg/L)
CR = contact rate (liters/hr)
ABS_s = bioavailability/gastrointestinal [GI] absorption factor (%)
ET = exposure time (hrs/event)
EF = exposure frequency (days/year)
ED = exposure duration (years)
BW = body-weight (kg)
AT = averaging time (period over which exposure is averaged – days)

Receptor Exposure through Ingestion of Constituents in Consumer/Food Products
Typically, exposure from the ingestion of food can occur via the ingestion of plant products, fish, animal products, and mother's milk. A general algorithm for estimating the exposure intake through the ingestion of foods is shown in Box 6.6 – with corresponding relationships defined below for specific types of food products.

- *Ingestion of Plant Products* – Exposure through the ingestion of plant products, ING_p, is a function of the type of plant, gastrointestinal absorption factor, and the fraction of plants ingested that are affected by the chemical constituents of concern. The exposure estimation is performed for each plant type in accordance with the algorithm presented in Box 6.6A (CAPCOA 1990; USEPA 1989a, 1992, 1997).
- *Bioaccumulation and Ingestion of Seafood* – Exposure from the ingestion of chemical constituents in fish (e.g., obtained from contaminated surface water bodies) may be estimated using the algorithm shown in Box 6.6B (USEPA 1987a, 1988, 1989a, 1992, 1997).

Box 6.6. Equation for estimating ingestion exposure to constituents in food products

$$ING_f = \frac{(C_f \times FIR \times CF \times FI \times ABS_s \times EF \times ED)}{(BW \times AT)}$$

where:

ING_f	= ingestion intake, adjusted for absorption (mg/kg-day)
C_f	= chemical concentration in food (mg/kg or mg/L)
FIR	= average food ingestion rate (mg or liters/meal)
CF	= conversion factor (10^{-6} kg/mg for solids and 1.00 for fluids)
FI	= fraction ingested from contaminated source (unitless)
ABS_s	= bioavailability/gastrointestinal [GI] absorption factor (%)
EF	= exposure frequency (meals/year)
ED	= exposure duration (years)
BW	= body weight (kg)
AT	= averaging time (period over which exposure is averaged – days)

Box 6.6A. Equation for estimating ingestion exposure to constituents in plant products

$$ING_p = \frac{(CP_z \times PIR_z \times FI_z \times ABS_s \times EF \times ED)}{(BW \times AT)}$$

where:

ING_p	= exposure intake from ingestion of plant products, adjusted for absorption (mg/kg-day)
CP_z	= chemical concentration in plant type Z (mg/kg)
PIR_z	= average consumption rate for plant type Z (kg/day)
FI_z	= fraction of plant type Z ingested from contaminated source (unitless)
ABS_s	= bioavailability/gastrointestinal [GI] absorption factor (%)
EF	= exposure frequency (days/years)
ED	= exposure duration (years)
BW	= body weight (kg)
AT	= averaging time (period over which exposure is averaged – days)

Box 6.6B. Equation for estimating ingestion exposure to constituents in contaminated seafood

$$ING_{sf} = \frac{(CW \times FIR \times CF \times BCF \times FI \times ABS_s \times EF \times ED)}{(BW \times AT)}$$

where:

ING_{sf}	= total exposure, adjusted for absorption (mg/kg-day)
CW	= chemical concentration in surface water (mg/L)
FIR	= average fish ingestion rate (g/day)
CF	= conversion factor (= 10^{-3} kg/g)
BCF	= chemical-specific bioconcentration factor (L/kg)
FI	= fraction ingested from contaminated source (unitless)
ABS_s	= bioavailability/gastrointestinal [GI] absorption factor (%)
EF	= exposure frequency (days/years)
ED	= exposure duration (years)
BW	= body weight (kg)
AT	= averaging time (period over which exposure is averaged – days)

- *Ingestion of Animal Products* – Exposure resulting from the ingestion of animal products, ING_a, is a function of the type of meat ingested (including animal milk products and eggs), gastrointestinal absorption factor, and the fraction of animal products ingested that are affected by the constituents of concern. The exposure estimation is carried out for each animal product type by using the form of relationship shown in Box 6.6C (CAPCOA 1990; USEPA 1989a, 1992, 1997).

- *Ingestion of Mother's Milk* – Exposure through the ingestion of a mother's milk, ING_m, is a function of the average chemical concentration in the mother's milk, the amount of mother's milk ingested, and gastrointestinal absorption factor – estimated according to the relationship shown in Box 6.6D (CAPCOA 1990; USEPA 1989a, 1992, 1997).

Box 6.6C. Equation for estimating ingestion exposure to constituents in animal products

$$ING_a = \frac{(CAP_z \times APIR_z \times FI_z \times ABS_s \times EF \times ED)}{(BW \times AT)}$$

where:

ING_a	= exposure intake through ingestion of plant products, adjusted for absorption (mg/kg-day)
CAP_z	= chemical concentration in food type Z (mg/kg)
$APIR_z$	= average consumption rate for food type Z (kg/day)
FI_z	= fraction of product type Z ingested from contaminated source (unitless)
ABS_s	= bioavailability/gastrointestinal [GI] absorption factor (%)
EF	= exposure frequency (days/years)
ED	= exposure duration (years)
BW	= body weight (kg)
AT	= averaging time (period over which exposure is averaged – days)

Box 6.6D. Equation for estimating ingestion exposure to chemicals in mother's milk used for breast-feeding

$$ING_m = \frac{(CMM \times IBM \times ABS_s \times EF \times ED)}{(BW \times AT)}$$

where:

ING_m	= exposure intake through ingestion of mother's milk, adjusted for absorption (mg/kg-day)
CMM	= chemical concentration in mother's milk – which is a function of a mother's exposure through all routes and the contaminant body half-life (mg/kg)
IBM	= daily average ingestion rate for breast milk (kg/day)
ABS_s	= bioavailability/gastrointestinal [GI] absorption factor (%)
EF	= exposure frequency (days/years)
ED	= exposure duration (years)
BW	= body weight (kg)
AT	= averaging time (period over which exposure is averaged – days)

Receptor Exposure through Pica and Incidental Ingestion of Soil/Sediment
Exposures resulting from the incidental ingestion of contaminants sorbed onto soils is determined by multiplying the concentration of the constituent in the medium of concern by the amount of soil/material ingested per day and the degree of absorption. The applicable relationship for estimating the resulting exposures is shown in Box 6.7 (CAPCOA 1990; USEPA 1988, 1989a, 1989b, 1992, 1997). In general, it is usually assumed that all ingested soil during receptor exposures comes from a contaminated source, so that the FI term becomes unity.

Box 6.7. Equation for estimating pica and incidental ingestion exposure to contaminated soils/sediments

$$ING_S = \frac{C_S \times SIR \times CF \times FI \times ABS_S \times EF \times ED}{BW \times AT}$$

where:
ING_S = ingestion intake, adjusted for absorption (mg/kg-day)
C_S = chemical concentration in soil (mg/kg)
SIR   = average soil ingestion rate (mg soil/day)
CF   = conversion factor (10^{-6} kg/mg)
FI   = fraction ingested from contaminated source (unitless)
ABS_S = bioavailability/gastrointestinal [GI] absorption factor (%)
EF   = exposure frequency (days/years)
ED   = exposure duration (years)
BW   = body weight (kg)
AT   = averaging time (period over which exposure is averaged – days)

6.2.3. POTENTIAL RECEPTOR DERMAL EXPOSURES

The major types of dermal exposures that could affect chemical exposure decisions consist of dermal contacts with contaminants adsorbed onto or within solid matrices (e.g., cosmetics, soils, etc.), and dermal absorption from contaminated waters and constituents in consumer products such as cosmetics (see Figure 6.1). In general, dermal intake is a function of the chemical concentration in the medium of concern, the body surface area in contact with the medium, the duration of the contact, flux of the medium across the skin surface, and the absorbed fraction – conservatively estimated by the following representative relationships:

$$\text{Dermal Exposure to solid matrix (mg/kg-day)} = \frac{\{SS \times SA \times CS \times UF \times CF\}}{BW} \qquad (6.8)$$

$$\text{Dermal Exposure to water (mg/kg-day)} = \frac{\{WS \times SA \times CW \times UF\}}{BW} \qquad (6.9)$$

where: *SS* is surface dust/materials on skin (mg/cm^2/day); *CS* is chemical concentration in solid matrix (e.g., soil) (mg/kg); *CF* is conversion factor (= 1.00E-06 kg/mg); *WS* is water contacting skin (L/cm^2/day); *CW* is chemical concentration in water (mg/L); *SA* is exposed skin surface area (cm^2); *UF* is uptake factor; and *BW* is body weight (kg).

Potential receptor dermal exposures via dermal contacts with solid matrices containing chemical constituents, and from the dermal absorption of chemicals present in contaminated water media, are annotated below.

Receptor Exposure through Contact/Dermal Absorption from Solid Matrices
The dermal exposures to chemicals in solid materials (e.g., soils and sediments) may be estimated by applying the equation shown in Box 6.8 (CAPCOA 1990; DTSC 1994; USEPA 1988, 1989a, 1989b, 1992, 1997).

Box 6.8. Equation for estimating dermal exposures through contacts with constituents in solid matrices (e.g., contaminated soils)

$$DEX_s = \frac{(C_s \times CF \times SA \times AF \times ABS_s \times SM \times EF \times ED)}{(BW \times AT)}$$

where:
DEX_s	= absorbed dose (mg/kg-day)
C_s	= chemical concentration in solid materials (e.g., contaminated soils) (mg/kg)
CF	= conversion factor (10^{-6} kg/mg)
SA	= skin surface area available for contact, i.e., surface area of exposed skin (cm^2/event)
AF	= solid material to skin adherence factor (e.g., soil loading on skin) (mg/cm^2)
ABS_s	= skin absorption factor for chemicals in solid matrices (e.g., contaminated soils) (%)
SM	= factor for solid materials matrix effects (%)
EF	= exposure frequency (events/year)
ED	= exposure duration (years)
BW	= body weight (kg)
AT	= averaging time (period over which exposure is averaged – days)

Receptor Exposure through Dermal Contact with Waters
and Liquid Consumer Products
Dermal exposures to chemicals in water may occur during domestic use (such as bathing and washing), or through recreational activities (such as swimming or fishing). As a specific example, the dermal intakes of chemicals in ground or surface water and/or from seeps from a contaminated site may be estimated by using the type of equation shown in Box 6.9 (USEPA 1988, 1989a, 1989b, 1992, 1997).

6.3. Establishing 'Exposure Intake Factors' for Use in the Computation of Chemical Intakes and Doses

Several exposure parameters are generally required in order to model the various exposure scenarios typically associated with chemical exposure problems. Oftentimes, default values are obtainable from the scientific literature for some of the requisite parameters used in the estimation of chemical intakes and doses. Table 6.1 shows typical parameters – that represent a generic set of values commonly used in some applications; this is by no means complete, and more detailed information on such parameters can be obtained from various scientific sources (e.g., Calabrese *et al.*, 1989;

CAPCOA 1990; Lepow *et al.*, 1974, 1975; OSA 1992; USEPA 1987b, 1988, 1989a, 1989b, 1991, 1992, 1997).

A spreadsheet for automatically calculating exposure 'intake factors' for varying input parameters that reflect case-specific problem scenarios may be developed (based on the algorithms presented in the preceding sections) to facilitate the computational efforts involved in the exposure assessment (Table 6.2). Some example evaluations for potential receptor groups purportedly exposed through inhalation, soil ingestion (viz., incidental or pica behavior), and dermal contact are discussed in the proceeding sections – simply as an illustration of the computational mechanics involved in the exposure assessment. The same set of units is maintained throughout these illustrative evaluations that follow as were used in the preceding sections.

Box 6.9. Equation for estimating dermal exposures through contacts with contaminated waters

$$DEX_w = \frac{(C_w \times CF \times SA \times PC \times ABS_s \times ET \times EF \times ED)}{(BW \times AT)}$$

where:

DEX_w = absorbed dose from dermal contact with chemicals in water (mg/kg-day)

C_w = chemical concentration in water (mg/L)

CF = volumetric conversion factor for water (1 liter/1000 cm^3)

SA = skin surface area available for contact, i.e., surface area of exposed skin (cm^2)

PC = chemical-specific dermal permeability constant (cm/hr)

ABS_s = skin absorption factor for chemicals in water (%)

ET = exposure time (hr/day)

EF = exposure frequency (days/year)

ED = exposure duration (years)

BW = body weight (kg)

AT = averaging time (period over which exposure is averaged – days).

6.3.1. ILLUSTRATIVE EXAMPLE FOR INHALATION EXPOSURES

The daily inhalation intake of contaminated fugitive dust for various population groups is presented below for both carcinogenic and non-carcinogenic effects. The assumed parameters used in the computational demonstration are provided in Table 6.1, and the electronic spreadsheet automation process shown in Table 6.2.

Table 6.1. An example listing of case-specific exposure parameters

Parameter	Child aged up to 6 years	Child aged 6–12 years	Adult
Physical characteristics			
Average body weight (kg)	16	29	70
Average total skin surface area (cm^2)	6 980	10 470	18 150
Average lifetime (yrs)	70	70	70
Average lifetime exposure period (yrs)	5	6	58
Activity characteristics			
Inhalation rate (m^3/h)	0.25	0.46	0.83
Retention rate of inhaled air (%)	100	100	100
Frequency of fugitive dust inhalation (days/yr)			
‣ off-site residents, schools, and by-passers	365	365	365
‣ off-site workers	-	-	260
Duration of fugitive dust inhalation (outside) (h/day)			
‣ off-site residents, schools, and by-passers	12	12	12
‣ off-site workers	-	-	8
Amount of incidentally ingested soils (mg/day)	200	100	50
Frequency of soil contact (days/yr)			
‣ off-site residents, schools, and by-passers	330	330	330
‣ off-site workers	-	-	260
Duration of soil contact (h/day)			
‣ off-site residents, schools, and by-passers	12	8	8
‣ off-site workers	-	-	8
Skin area contacted by soil (%)	20	20	10
Material characteristics			
Soil-to-skin adherence factor (mg/cm^2)	0.75	0.75	0.75
Soil matrix attenuation factor (%)	15	15	15

Note: The exposure factors represented here are potential maximum exposures (and these produce conservative estimates). Indeed, these could be modified, as appropriate – to reflect the most reasonable exposure patterns anticipated for a project-specific situation; for instance, realistically, soil exposure is reduced from snow cover and rainy days – thus reducing potential exposures for children playing outdoors in a contaminated area. In any case, the sources and/or rationale for the choice of the exposure parameters should be very well supported and adequately documented.

Table 6.2. Example spreadsheet for calculating case-specific 'intake factors' for an exposure assessment

Fugitive Dust Inhalation Pathway

Receptor Group	IR	RR	ET	EF	ED	BW	AT	INH factor
C(1-6)@NCancer	0.25	1	12	365	5	16	1 825	1.88E-01
C(1-6)@Cancer	0.25	1	12	365	5	16	25 550	1.34E-02
C(6-12)@NCancer	0.46	1	12	365	6	29	2 190	1.90E-01
C(6-12)@Cancer	0.46	1	12	365	6	29	25 550	1.63E-02
ResAdult@NCancer	0.83	1	12	365	58	70	21 170	1.42E-01
ResAdult@Cancer	0.83	1	12	365	58	70	25 550	1.18E-01
JobAdult@NCancer	0.83	1	8	260	58	70	21 170	6.76E-02
JobAdult@Cancer	0.83	1	8	260	58	70	25 550	5.60E-02

Soil Ingestion Pathway

Receptor Group	IR	CF	FI	EF	ED	BW	AT	ING factor
C(1-6)@NCancer	200	1.00E-06	1	330	5	16	1 825	1.13E-05
C(1-6)@Cancer	200	1.00E-06	1	330	5	16	25 550	8.07E-07
C(6-12)@NCancer	100	1.00E-06	1	330	6	29	2 190	3.12E-06
C(6-12)@Cancer	100	1.00E-06	1	330	6	29	25 550	2.67E-07
ResAdult@NCancer	50	1.00E-06	1	330	58	70	21 170	6.46E-07
ResAdult@Cancer	50	1.00E-06	1	330	58	70	25 550	5.35E-07
JobAdult@NCancer	50	1.00E-06	1	260	58	70	21 170	5.09E-07
JobAdult@Cancer	50	1.00E-06	1	260	58	70	25 550	4.22E-07

Soil Dermal Contact Pathway

Receptor Group	SA	CF	AF	SM	EF	ED	BW	AT	DEX factor
C(1-6)@NCancer	1 396	1E-06	0.75	0.15	330	5	16	1 825	8.87E-06
C(1-6)@Cancer	1 396	1E-06	0.75	0.15	330	5	16	25 550	6.34E-07
C(6-12)@NCancer	2 094	1E-06	0.75	0.15	330	6	29	2 190	7.34E-06
C(6-12)@Cancer	2 094	1E-06	0.75	0.15	330	6	29	25 550	6.30E-07
ResAdult@NCancer	1 815	1E-06	0.75	0.15	330	58	70	21 170	2.64E-06
ResAdult@Cancer	1 815	1E-06	0.75	0.15	330	58	70	25 550	2.19E-06
JobAdult@NCancer	1 815	1E-06	0.75	0.15	260	58	70	21 170	2.08E-06
JobAdult@Cancer	1 815	1E-06	0.75	0.15	260	58	70	25 550	1.72E-06

Notes:
▸ Notations and units are same as defined in the text.
▸ INH factor = Inhalation factor for calculation of doses and intakes.
▸ ING factor = Soil ingestion factor for calculation of doses and intakes.
▸ DEX factor = Dermal exposure (via skin absorption) factor for calculation of doses and intakes.
▸ C(1-6)@NCancer; C(6-12)@NCancer; ResAdult@NCancer; JobAdult@NCancer = Non-carcinogenic effects for a child aged 1–6 years; child aged 6–12 years; resident adult; and adult worker, respectively.
▸ C(1-6)@Cancer; C(6-12)@Cancer; ResAdult@Cancer; JobAdult@Cancer = Carcinogenic effects for a child aged 1–6 years; child aged 6–12 years; resident adult; and adult worker, respectively.

Estimation of Lifetime Average Daily Dose (LADD) for Carcinogenic Effects
For the fugitive dust inhalation pathway, the LADD (also, the carcinogenic chronic daily intake [CDI]) is estimated for the different population groups (generally pre-selected as representative of the critical receptors in the risk assessment) – and the results are shown below.

- The carcinogenic CDI for children aged up to 6 years is calculated to be:

$$CInh_{(1-6)}$$
$$= \frac{(CA \times IR \times RR \times ABS_s \times ET \times EF \times ED)}{(BW \times AT)}$$
$$= \frac{([CA] \times 0.25 \times 1 \times ABS_s \times 12 \times 365 \times 5)}{(16 \times (70 \times 365))}$$
$$= 1.34 \times 10^{-2} \times ABS_s \times [CA]$$

- The carcinogenic CDI for children aged 6-12 years is calculated to be:

$$CInh_{(6-12)}$$
$$= \frac{(CA \times IR \times RR \times ABS_s \times ET \times EF \times ED)}{(BW \times AT)}$$
$$= \frac{([CA] \times 0.46 \times 1 \times ABS_s \times 12 \times 365 \times 6)}{(29 \times (70 \times 365))}$$
$$= 1.63 \times 10^{-2} \times ABS_s \times [CA]$$

- The carcinogenic CDI for adult residents is calculated to be:

$$CInh_{(adultR)}$$
$$= \frac{(CA \times IR \times RR \times ABS_s \times ET \times EF \times ED)}{(BW \times AT)}$$
$$= \frac{([CA] \times 0.83 \times 1 \times ABS_s \times 12 \times 365 \times 58)}{(70 \times (70 \times 365))}$$
$$= 1.18 \times 10^{-1} \times ABS_s \times [CA]$$

- The carcinogenic CDI for adult workers is calculated to be:

$$CInh_{(adultW)}$$
$$= \frac{(CA \times IR \times RR \times ABS_s \times ET \times EF \times ED)}{(BW \times AT)}$$
$$= \frac{([CA] \times 0.83 \times 1 \times ABS_s \times 8 \times 260 \times 58)}{(70 \times (70 \times 365))}$$
$$= 5.60 \times 10^{-2} \times ABS_s \times [CA]$$

Estimation of Average Daily Dose (ADD) for Non-carcinogenic Effects
For the fugitive dust inhalation pathway, the ADD (also, the non-carcinogenic CDI) is estimated for the different population groups (generally pre-selected as representative of the critical receptors in the risk assessment) – and the results are shown below.

- The non-carcinogenic CDI for children aged up to 6 years is calculated to be:

$$NCInh_{(1-6)}$$
$$= \frac{(CA \times IR \times RR \times ABS_S \times ET \times EF \times ED)}{(BW \times AT)}$$
$$= \frac{([CA] \times 0.25 \times 1 \times ABS_S \times 12 \times 365 \times 5)}{(16 \times (5 \times 365))}$$
$$= 1.88 \times 10^{-1} \times ABS_S \times [CA]$$

- The non-carcinogenic CDI for children aged 6-12 years is calculated to be:

$$NCInh_{(6-12)}$$
$$= \frac{(CA \times IR \times RR \times ABS_S \times ET \times EF \times ED)}{(BW \times AT)}$$
$$= \frac{([CA] \times 0.46 \times 1 \times ABS_S \times 12 \times 365 \times 6)}{(29 \times (6 \times 365))}$$
$$= 1.90 \times 10^{-1} \times ABS_S \times [CA]$$

- The non-carcinogenic CDI for adult residents is calculated to be:

$$NCInh_{(adultR)}$$
$$= \frac{(CA \times IR \times RR \times ABS_S \times ET \times EF \times ED)}{(BW \times AT)}$$
$$= \frac{([CA] \times 0.83 \times 1 \times ABS_S \times 12 \times 365 \times 58)}{(70 \times (58 \times 365))}$$
$$= 1.42 \times 10^{-1} \times ABS_S \times [CA]$$

- The non-carcinogenic CDI for adult workers is calculated to be:

$$NCInh_{(adultW)}$$
$$= \frac{(CA \times IR \times RR \times ABS_S \times ET \times EF \times ED)}{(BW \times AT)}$$
$$= \frac{([CA] \times 0.83 \times 1 \times ABS_S \times 8 \times 260 \times 58)}{(70 \times (58 \times 365))}$$
$$= 6.76 \times 10^{-2} \times ABS_S \times [CA]$$

6.3.2. ILLUSTRATIVE EXAMPLE FOR INGESTION EXPOSURES

The daily ingestion intakes of contaminated soils for various population groups are calculated for both carcinogenic and non-carcinogenic effects. The assumed parameters used in the computational demonstration are provided in Table 6.1, and the electronic spreadsheet automation process shown in Table 6.2.

Estimation of Lifetime Average Daily Dose (LADD) for Carcinogenic Effects
For the soil ingestion pathway, the LADD (also, the carcinogenic CDI) is estimated for the different population groups (generally pre-selected as representative of the critical receptors in the risk assessment) – and the results are shown below.

- The carcinogenic CDI for children aged up to 6 years is calculated to be:

$$
\begin{aligned}
CIng_{(1-6)} \\
&= \frac{(CS \times IR \times CF \times FI \times ABS_S \times EF \times ED)}{(BW \times AT)} \\
&= \frac{([CS] \times 200 \times 1.00E\text{-}06 \times 1 \times ABS_S \times 330 \times 5)}{(16 \times (70 \times 365))} = 8.07E\text{-}07 \times ABS_S \times [CS]
\end{aligned}
$$

- The carcinogenic CDI for children aged 6 to 12 years is calculated to be:

$$
\begin{aligned}
CIng_{(6-12)} \\
&= \frac{(CS \times IR \times CF \times FI \times ABS_S \times EF \times ED)}{(BW \times AT)} \\
&= \frac{([CS] \times 100 \times 1.00E\text{-}06 \times 1 \times ABS_S \times 330 \times 6)}{(29 \times (70 \times 365))} = 2.67E\text{-}07 \times ABS_S \times [CS]
\end{aligned}
$$

- The carcinogenic CDI for adult residents is calculated to be:

$$
\begin{aligned}
CIng_{(adultR)} \\
&= \frac{(CS \times IR \times CF \times FI \times ABS_S \times EF \times ED)}{(BW \times AT)} \\
&= \frac{([CS] \times 50 \times 1.00E\text{-}06 \times 1 \times ABS_S \times 330 \times 58)}{(70 \times (70 \times 365))} = 5.35E\text{-}07 \times ABS_S \times [CS]
\end{aligned}
$$

- The carcinogenic CDI for adult workers is calculated to be:

$$
\begin{aligned}
CIng_{(adultW)} \\
&= \frac{(CS \times IR \times CF \times FI \times ABS_S \times EF \times ED)}{(BW \times AT)} \\
&= \frac{([CS] \times 50 \times 1.00E\text{-}06 \times 1 \times ABS_S \times 260 \times 58)}{(70 \times (70 \times 365))} = 4.22E\text{-}07 \times ABS_S \times [CS]
\end{aligned}
$$

Estimation of Average Daily Dose (ADD) for Non-carcinogenic Effects

For the soil ingestion pathway, the ADD (also, the non-carcinogenic CDI) is estimated for the different population groups (generally pre-selected as representative of the critical receptors in the risk assessment) – and the results are shown below.

- The non-carcinogenic CDI for children aged up to 6 years is calculated to be:

$$NCIng_{(1-6)}$$

$$= \frac{(CS \times IR \times CF \times FI \times ABS_S \times EF \times ED)}{(BW \times AT)}$$

$$= \frac{([CS] \times 200 \times 1.00E\text{-}06 \times 1 \times ABS_S \times 330 \times 5)}{(16 \times (5 \times 365))} = 1.13E\text{-}05 \times ABS_S \times [CS]$$

- The non-carcinogenic CDI for children aged 6 to 12 years is calculated to be:

$$NCIng_{(6-12)}$$

$$= \frac{(CS \times IR \times CF \times FI \times ABS_S \times EF \times ED)}{(BW \times AT)}$$

$$= \frac{([CS] \times 100 \times 1.00E\text{-}06 \times 1 \times ABS_S \times 330 \times 6)}{(29 \times (6 \times 365))} = 3.12E\text{-}06 \times ABS_S \times [CS]$$

- The non-carcinogenic CDI for adult residents is calculated to be:

$$NCIng_{(adultR)}$$

$$= \frac{(CS \times IR \times CF \times FI \times ABS_S \times EF \times ED)}{(BW \times AT)}$$

$$= \frac{([CS] \times 50 \times 1.00E\text{-}06 \times 1 \times ABS_S \times 330 \times 58)}{(70 \times (58 \times 365))} = 6.46E\text{-}07 \times ABS_S \times [CS]$$

- The non-carcinogenic CDI for adult workers is calculated to be:

$$NCIng_{(adultW)}$$

$$= \frac{(CS \times IR \times CF \times FI \times ABS_S \times EF \times ED)}{(BW \times AT)}$$

$$= \frac{([CS] \times 50 \times 1.00E\text{-}06 \times 1 \times ABS_S \times 260 \times 58)}{(70 \times (58 \times 365))} = 5.09E\text{-}07 \times ABS_S \times [CS]$$

6.3.3. ILLUSTRATIVE EXAMPLE FOR DERMAL EXPOSURES

The daily dermal intakes of contaminated soils for various population groups are calculated for both carcinogenic and non-carcinogenic effects. The assumed parameters used in the computational demonstration are provided in Table 6.1, and the electronic spreadsheet automation process shown in Table 6.2.

Estimation of Lifetime Average Daily Dose (LADD) for Carcinogenic Effects
For the soil dermal contact pathway, the LADD (also, the carcinogenic CDI) is estimated for the different population groups (generally pre-selected as representative of the critical receptors in the risk assessment) – and the results are shown below.

- The carcinogenic CDI for children aged up to 6 years is calculated to be:

$CDEX_{(1-6)}$

$$= \frac{(CS \times CF \times SA \times AF \times ABS_S \times SM \times EF \times ED)}{(BW \times AT)}$$

$$= \frac{([CS] \times 1.00E\text{-}06 \times 1396 \times 0.75 \times ABS_S \times 0.15 \times 330 \times 5)}{(16 \times (70 \times 365))}$$

$$= 6.34E\text{-}07 \times ABS_S \times [CS]$$

- The carcinogenic CDI for children aged 6 to 12 years is calculated to be:

$CDEX_{(6-12)}$

$$= \frac{(CS \times CF \times SA \times AF \times ABS_S \times SM \times EF \times ED)}{(BW \times AT)}$$

$$= \frac{([CS] \times 1.00E\text{-}06 \times 2094 \times 0.75 \times ABS_S \times 0.15 \times 330 \times 6)}{(29 \times (70 \times 365))}$$

$$= 6.30E\text{-}07 \times ABS_S \times [CS]$$

- The carcinogenic CDI for adult residents is calculated to be:

$CDEX_{(adultR)}$

$$= \frac{(CS \times CF \times SA \times AF \times ABS_S \times SM \times EF \times ED)}{(BW \times AT)}$$

$$= \frac{([CS] \times 1.00E\text{-}06 \times 1815 \times 0.75 \times ABS_S \times 0.15 \times 330 \times 58)}{(70 \times (70 \times 365))}$$

$$= 2.19E\text{-}06 \times ABS_S \times [CS]$$

- The carcinogenic CDI for adult workers is calculated to be:

$CDEX_{(adultW)}$

$$= \frac{(CS \times CF \times SA \times AF \times ABS_S \times SM \times EF \times ED)}{(BW \times AT)}$$

$$= \frac{([CS] \times 1.00E\text{-}06 \times 1815 \times 0.75 \times ABS_S \times 0.15 \times 260 \times 58)}{(70 \times (70 \times 365))}$$

$$= 1.72E\text{-}06 \times ABS_S \times [CS]$$

Estimation of Average Daily Dose (ADD) for Non-carcinogenic Effects
For the soil dermal contact pathway, the ADD (also, the non-carcinogenic CDI) is estimated for the different population groups (generally pre-selected as representative of the critical receptors in the risk assessment) – and the results are shown below.

- The non-carcinogenic CDI for children aged up to 6 years is calculated as follows:

$$NCDEX_{(1-6)}$$
$$= \frac{(CS \times CF \times SA \times AF \times ABS_S \times SM \times EF \times ED)}{(BW \times AT)}$$
$$= \frac{([CS] \times 1.00E\text{-}06 \times 1396 \times 0.75 \times ABS_S \times 0.15 \times 330 \times 5)}{(16 \times (5 \times 365))}$$
$$= 8.87E\text{-}06 \times ABS_S \times [CS]$$

- The non-carcinogenic CDI for children aged 6 to 12 years is calculated to be:

$$NCDEX_{(6-12)}$$
$$= \frac{(CS \times CF \times SA \times AF \times ABS_S \times SM \times EF \times ED)}{(BW \times AT)}$$
$$= \frac{([CS] \times 1.00E\text{-}06 \times 2094 \times 0.75 \times ABS_S \times 0.15 \times 330 \times 6)}{(29 \times (6 \times 365))}$$
$$= 7.34E\text{-}06 \times ABS_S \times [CS]$$

- The non-carcinogenic CDI for adult residents is calculated to be:

$$NCDEX_{(adultR)}$$
$$= \frac{(CS \times CF \times SA \times AF \times ABS_S \times SM \times EF \times ED)}{(BW \times AT)}$$
$$= \frac{([CS] \times 1.00E\text{-}06 \times 1815 \times 0.75 \times ABS_S \times 0.15 \times 330 \times 58)}{(70 \times (58 \times 365))}$$
$$= 2.64E\text{-}06 \times ABS_S \times [CS]$$

- The non-carcinogenic CDI for adult workers is calculated to be:

$$NCDEX_{(adultW)}$$
$$= \frac{(CS \times CF \times SA \times AF \times ABS_S \times SM \times EF \times ED)}{(BW \times AT)}$$
$$= \frac{([CS] \times 1.00E\text{-}06 \times 1815 \times 0.75 \times ABS_S \times 0.15 \times 260 \times 58)}{(70 \times (58 \times 365))}$$
$$= 2.08E\text{-}06 \times ABS_S \times [CS]$$

6.4.   Receptor Age Adjustments to Human Exposure Factors

Age adjustments are often necessary when human exposures to a chemical occur from childhood through the adult life. Such adjustments are meant to account for the transitioning of a potential receptor from childhood (requiring one set of intake assumptions and exposure parameters) to adulthood (that requires a different set of chemical intake assumptions and exposure parameters). Indeed, in the processes involved in human exposure assessments, it frequently becomes very apparent that contact rates can be quite different for children *vs.* adults. Consequently, carcinogenic risks (that are averaged over a receptor's lifetime) should preferably be calculated by applying the age-adjusted factors shown in Box 6.10 (or similar ones). Further details on the development of age-adjusted factors are provided elsewhere in the literature (e.g., DTSC 1994; OSA 1992; USEPA 1989b, 1997).

The use of age-adjusted factors are especially important in certain specific situations – such as those involving human soil ingestion exposures, which are typically higher during childhood and decrease with age. For instance, because the soil ingestion rate is generally different for children and adults, the carcinogenic risk due to direct ingestion of soil should be calculated using an age-adjusted ingestion factor. This takes into account the differences in daily soil ingestion rates, body weights, exposure fraction, and exposure duration for the two exposure groups – albeit exposure frequency may be assumed to be the same for the two groups. If calculated in this manner, then the estimated exposure/intake factor will result in a more health-protective risk evaluation – compared to, for instance, an 'adult-only' type of assumption. Indeed, in a refined and comprehensive evaluation, it is generally recommended to incorporate age-adjustment factors in the chemical exposure assessment, where appropriate. On the other hand, and for the sake of simplicity, such age adjustments are usually not made part of most screening-level computational processes.

6.5.   Spatial and Temporal Averaging of Chemical Exposure Estimates

Oftentimes, in major public health policy decisions, it becomes necessary to evaluate chemical exposure situations for population groups – rather than for individuals only. In such type of more practical and realistic chemical exposure assessment, it usually is more appropriate and indeed less conservative to estimate chemical exposure to a specific population subgroup over an exposure duration of less than a lifetime, as illustrated by the exposure combination scenarios presented below.

- *Averaging exposure over population age groups – when chemical concentrations are constant in time.* For situations where chemical concentrations are assumed constant over time but for which exposure is to be averaged over population age groups, the chronic daily exposure may be estimated using the following model form (CDHS, 1990; OSA, 1992; USEPA, 1992):

$$CDI = \left\{ \frac{1}{\sum\limits_{a=1}^{NG} AT_a} \right\} \times \left\{ \sum_{a=1}^{NG} [\tfrac{CR}{BW}]_a \times EF_a \times ED_a \right\} \times C_m \qquad (6.10)$$

Box 6.10. Age-adjustment factors to human exposure calculations

---

♦ *Ingestion (mg-yr/kg-d or L-yr/kg-d)*

$$INGf_{adj} = \frac{(MIR_c \times ED_c)}{BW_c} + \frac{(MIR_a \times [ED - ED_c])}{BW_a}$$

$$= \left[\frac{(MIR_c \times ED_c)}{BW_c} \right] + \left[\frac{(MIR_a \times ED_a)}{BW_a} \right]$$

♦ *Dermal contact (mg-yr/kg-d)*

$$DERf_{adj} = \frac{(AF \times SA_c \times ED_c)}{BW_c} + \frac{(AF \times SA_a \times [ED - ED_c])}{BW_a}$$

♦ *Inhalation (m³-yr/kg-d)*

$$INHf_{adj} = \frac{(IRA_c \times ED_c)}{BW_c} + \frac{(IRA_a \times [ED - ED_c])}{BW_a}$$

where:

$INGf_{adj}$	= age-adjusted ingestion factor (mg-yr/kg-d)
$DERf_{adj}$	= age-adjusted dermal contact factor (mg-yr/kg-d)
$INHf_{adj}$	= age-adjusted inhalation factor (m³-yr/kg-d)
MIR_c	= material ingestion rate – child (mg/day or L/day)
MIR_a	= material ingestion rate – adult (mg/day or L/day)
AF	= material adherence factor (mg/cm²)
SA_c	= child's exposed surface area (cm²)
SA_a	= adult's exposed surface area (cm²)
IRA_c	= inhalation rate – child (m³/day)
IRA_a	= inhalation rate – adult (m³/day)
ED	= total exposure duration (years)
ED_c	= exposure duration – child (years)
ED_a	= exposure duration – adult (years)
BW_c	= body weight – child, i.e., the average child body weight over the exposure period (kg)
BW_a	= body weight – adult, i.e., the average adult body weight over the exposure period (kg)

---

where $\left[\frac{CR}{BW}\right]_a$ is the contact rate per unit body weight, averaged over the age group a; EF_a is the exposure frequency of the exposed population in the age group/category a; ED_a is the exposure duration for the exposed population in the age group/category a; AT_a is the averaging time for the age group a; C_m is the concentration in the 'chemical-based' medium contacted; and NG is the number of age groups used to represent the whole population.

- *Averaging exposure over time within a population group – when chemical concentrations vary in time.* For some chemical compounds present in consumer products and/or in the environment, the assumption that concentrations remain constant in time can result in significant overestimation of risks. Consequently, a model that accounts for time-varying concentrations may be utilized in the chemical exposure estimation process.

  Where chemical concentrations in the source medium or material varies with time – such as for cases where there are chemicals volatilizing from a contaminated site, or are being transformed by degradational processes – then exposures or chronic daily intakes for the exposed population may be estimated using the following general model form (CDHS, 1990; OSA, 1992; USEPA, 1992):

$$CDI = \left\{ \frac{[CR_m \times EF \times ED]}{[BW \times AT]} \right\} \times \left\{ \int_{t=0}^{ED} C_m(t)dt \right\} \tag{6.11}$$

  where CR_m is the contact rate in medium m; EF is the exposure frequency of the exposed population; ED is the exposure duration for the exposed population; AT is the averaging time for the population group; and $C_m(t)$ is the time-varying concentration in the 'chemical-based' medium contacted. It should be noted, however, that when one chemical species is transformed such that its concentration decreases in time, then all decay products must also be identified and documented. Exposure to all toxic decay products must then be modeled and accounted for – recognizing also that the concentrations of decay products could actually be increasing with time.

- *Averaging exposure over population age sub-groups – when chemical concentrations vary with time.* In some situations involving time-varying chemical concentrations, it may be decided to estimate the exposure to specific population subgroups over exposure duration of less than a lifetime, and then to use these age subgroups to calculate the lifetime equivalent chemical exposure to an individual drawn at random from the population. Under such circumstances, the following model form can be employed in the chemical exposure estimation (CDHS, 1990; OSA, 1992; USEPA, 1992):

$$CDI = \left\{ \frac{1}{\sum_{a=1}^{NG} AT_a} \right\} \times \left\{ \sum_{a=1}^{NG} [\tfrac{CR}{BW}]_a \times EF_a \times \int_{t=0}^{ED} C_m(t)dt \right\} \tag{6.12}$$

  where $[\frac{CR}{BW}]_a$ is the contact rate per unit body weight, averaged over the age group a; EF_a is the exposure frequency of the exposed population in the age group/category a; ED_a is the exposure duration for the exposed population in the age group/category a; AT_a is the averaging time for the age group a; $C_m(t)$ is the time-varying concentration in the 'chemical-based' medium contacted; and *NG* is the number of age groups used to represent the whole population.

Further details of the evaluation processes involved in the spatial and temporal averaging techniques for chemical exposure problems can be found elsewhere in the literature (e.g., CDHS, 1990; OSA, 1992; USEPA, 1992).

6.6.   Suggested Further Reading

Belzer, RB, 2002. Exposure assessment at a crossroads: the risk of success, *Journal of Exposure Analysis and Environmental Epidemiology,* 12(2): 96 – 103
Bennett, DH, TE McKone, JS Evans, *et al.,* 2002. Defining intake fraction, *Environmental Science & Technology,* 36(9): 206A – 211A
Brooks, SM *et al., 1995. Environmental Medicine,* Mosby, Mosby-Year Book, Inc., St. Louis, Missouri
Clayton, CA, ED Pellizari, and JJ Quackenboss, 2002. National human exposure assessment survey: analysis of exposure pathways and routes for arsenic and lead in EPA Region 5, *Journal of Exposure Analysis and Environmental Epidemiology,* 12(1): 29 – 43
Decaprio, AP, 1997. Biomarkers: coming of age for environmental health and risk assessment, *Environmental Science & Technology,* 37: 1837 – 1848
Gaylor, DW, FF Kadlubar, and FA Beland, 1992. Application of biomarkers to risk assessment, *Environmental Health Perspectives,* 98: 139 – 141
Harris, SA *et al.,* 2002. Development of models to predict dose of pesticides in professional turf applicators, *Journal of Exposure Analysis and Environmental Epidemiology,* 12(2): 130 – 144
Johnson, DL, K. McDade, and D. Griffith, 1996. Seasonal variation in paediatric blood lead levels in Syracuse, NY, USA, *Environmental Geochemistry and Health,* 18: 81 – 88
Lee, BM, SD Yoo, and S. Kim, 2002. A proposed methodology of cancer risk assessment modeling using biomarkers, *Journal of Toxicology and Environmental Health, Part A: Current Issues,* 65(5-6): 341 – 354
Potts, RO and RH Guy, 1992. Predicting skin permeability, *Pharm. Res.,* 9: 663 – 669
Reddy, MB, RH Guy, and AL Bunge, 2000. Does epidermal turnover reduce percutaneous penetration?, *Pharm. Res.,* 17: 1414 – 1419
Roberts, MS and KA Walters (eds.), 1998. *Dermal Absorption and Toxicity Assessment,* Marcel Dekker, Inc., New York, NY
Wallace, L., 1996. Indoor particles: a review, *J. Air & Waste Management Association,* 46: 98 – 126
Wiens, JA and KR Parker, 1995. Analyzing the effects of accidental environmental impacts: approaches and assumptions, *Ecol. Apll.,* 5(4): 1069 – 1083
Williams, DR, JC Paslawski, and GM Richardson, 1996. Development of a screening relationship to describe migration of contaminant vapors into buildings, *J. Soil Contamination,* 5(2): 141 – 156

Chapter 7

EVALUATION OF CHEMICAL TOXICITY

In planning for public health protection from the likely adverse effects of human exposure to chemicals, a primary concern generally relates to whether or not the substance in question possesses potentially hazardous and/or toxic properties. In practice, an evaluation of the toxicological effects typically consists of a compilation of toxicological profiles of the chemicals of concern (including the intrinsic toxicological properties of the chemicals – that may include their acute, subchronic, chronic, carcinogenic, and/or reproductive effects) and a determination of the relevant toxicity indices. This chapter discusses the major underlying concepts, principles, and procedures that are often employed in the evaluation of the hazard effects or toxicity of various chemical constituents found in consumer products and/or in the human environment.

7.1. Fundamental Concepts and Principles in Toxicology

Toxicology, in a broad sense, is the study of poisons and their effects on living organisms. In the context of environmental or public health, toxicology is the study of how specific chemical substances cause injury or undesirable effects to living cells and/or whole organisms. It generally consists of studies conducted to determine several fate and behavior attributes of the chemical, including the following:

- How easily the chemical enters the organism;
- How the chemical behaves in the organism;
- How rapidly the chemical is removed from the organism;
- What cells are affected by the chemical; and
- What cell functions are impaired as a consequence of the chemical exposure.

Toxicity represents the state of being poisonous – and therefore may be said to indicate the state of adverse effects or symptoms being produced by toxicants in an organism. In general, toxicity tends to vary according to both the duration and location of receptor exposure to the toxicant, as well as the receptor-specific responses of the exposed organism (Hughes, 1996).

7.1.1. MECHANISMS OF TOXICITY

Toxicants exert their effects when they interact with cells. Such cellular interaction may occur on the surface of the cell, within the cell, or in the underlying tissues and extracellular (interstitial) space. Chemical characteristics of both the toxicant and cell membrane determine whether any interaction occurs on the surface of the cell or whether the barrier will be effective in keeping the toxicant out of the organism (Hughes, 1996).

In general, after a chemical substance is absorbed following human contact or intake, it travels through the bloodstream. Subsequently, where binding occurs with organs (especially, the liver, the kidneys, and the blood) in the body, toxic effects may result. Depending on the partitioning behavior between the chemical and different biomolecules of the human body (including fat), storage may or may not occur. Chemicals (or their transformation products) that are highly water-soluble and do not have the tendency to partition into fats are rapidly eliminated in the urine. However, organic chemicals characterized by high octanol-water partition coefficients (K_{ow}) (i.e., the hydrophobic compounds) are stored in fat. It is noteworthy that metals may also be stored via binding to fat and other biological molecules of the body – albeit in a different manner from the hydrophobic organic compounds. For example, lead can complex with biological molecules of the central nervous system; cadmium can bind to receptor molecules in the kidney, resulting in renal damage; etc.

Broadly speaking, the important concepts relating to the mechanisms of toxicity for most toxic substances consider the following particularly relevant issues:

- Routes of chemical exposure and absorption;
- Distribution of the toxic chemical through the body;
- The biochemical transformation of the compound;
- Toxicant-receptor interactions;
- Storage of chemical; and
- Excretion of chemical.

Also, it is noteworthy that toxic effects could vary substantially depending on the location of contact and/or absorption of a chemical substance by the human receptor. For instance, whereas asbestos is highly toxic when inhaled, this material does not exhibit any significant degree of toxicity when ingested – possibly attributable to its poor absorption in the gastrointestinal tract.

7.1.2. CATEGORIZATION OF HUMAN TOXIC EFFECTS FROM CHEMICAL EXPOSURES: CARCINOGENICITY *VS.* NON-CARCINOGENICITY

The toxic characteristics of a substance are usually categorized according to the organs or systems they affect (e.g., kidney, liver, nervous system, etc.) or the disease they cause (e.g., birth defects, cancer, etc.). In any case, chemical substances generally fall into one of the two broad categories of 'carcinogens' and 'non-carcinogens' – based, respectively, on their potential to induce cancer and their possession of systemic toxicity effects. Indeed, for the purpose of human health risk assessment, chemicals are usually distinctly categorized into carcinogenic and non-carcinogenic groups.

Chemicals that give rise to toxic endpoints other than cancer and gene mutations are often referred to as 'systemic toxicants' because of their effects on the function of various organ systems; the toxic endpoints are referred to as 'non-cancer' or 'systemic' toxicity. Most chemicals that produce non-cancer toxicity do not cause a similar degree of toxicity in all organs, but usually demonstrate major toxicity to one or two organs; these are referred to as the target organs of toxicity for the chemicals (Klaassen *et al.*, 1986; USEPA, 1989). Also, chemicals that cause cancer and gene mutations commonly evoke other toxic effects (viz., systemic toxicity) as well.

'Threshold' vs. 'Non-threshold' Concepts

Non-carcinogens generally are believed to operate by 'threshold' mechanisms – i.e., the manifestation of systemic effects requires a threshold level of exposure or dose to be exceeded during a continuous exposure episode. Thus, non-cancer or systemic toxicity is generally treated as if there is an identifiable exposure threshold below which there are no observable adverse effects – and which means that, continuous exposure to levels below the threshold will produce no adverse or noticeable health effects.

For many non-carcinogenic effects, protective mechanisms are believed to exist in the mammalian physiological system that must be overcome before the adverse effect of a chemical constituent is manifested. Consequently, a range of exposures exist from zero to some finite value – called the *threshold level* – that can be tolerated by the exposed organism with essentially no likelihood of adverse effects. This characteristic distinguishes systemic endpoints from carcinogenic and mutagenic endpoints, which are often treated as 'non-threshold' processes; that is, the threshold concept and principle is not quite applicable for carcinogens, since it is believed that no thresholds exist for this group. Indeed, carcinogenesis, unlike many non-carcinogenic health effects, is generally thought to be a phenomenon for which risk evaluation based on presumption of a threshold is inappropriate (USEPA, 1989).

In general, it is usually assumed in risk assessments that, any finite exposure to carcinogens could result in a clinical state of disease. This hypothesized mechanism for carcinogenesis is referred to as 'non-threshold' – because there is believed to be essentially no level of exposure to such a chemical that does not pose a finite probability, however small, of generating a carcinogenic response. It is noteworthy, however, that among some professional groups, there is the belief that certain carcinogens require a threshold exposure level to be exceeded to provoke carcinogenic effects (e.g., Wilson, 1996, 1997). In fact, opinion among regulatory scientists seems to have recently been returning to the ancient presumption that at least some cancer-causing substances induce effects through a threshold process (Wilson, 1997). This implies that, for such substances, there exists a finite level of exposure or dose at which no finite response is necessarily indicated.

Mechanisms of Carcinogenicity

An important issue in chemical carcinogenesis relates to the concepts of 'initiators' and 'promoters.' A *promoter* is defined as an agent that results in an increase in cancer induction when it is administered some time after a receptor has been exposed to an *initiator*. Also, a *co-carcinogen* differs from a promoter only in that it is administered at the same time as the initiator. It is believed that initiators, co-carcinogens, and promoters do not usually induce tumors when administered separately. Indeed, it has become apparent that a series of developmental stages is required for carcinogenesis – consisting of the following three processes/steps:

1. *Initiation*, in which genetic damage occurs through a mutation to DNA; this involves a change in the capacity of DNA to function properly.
2. *Promotion*, in which the genetic damage is expressed through the multiplication of cells in which initiation occurred previously.
3. *Progression*, which represents the spreading of cancer through an uncontrolled cell growth.

Many chemical carcinogens are believed to be *complete carcinogens*; these chemicals function as both initiators and promoters (OSTP, 1985). It should be acknowledged, however, that promoters themselves are usually not necessarily carcinogens; these may include dietary fat, alcohols, saccharin, halogenated solvents, and estrogen. Even so, most regulatory agencies do not usually distinguish between initiators and promoters, especially because it is often very difficult to confirm whether a given chemical acts by promotion alone (OSHA, 1980; OSTP, 1985; USEPA, 1984).

Identification of Carcinogens
Both human and animal studies are used in the evaluation of whether chemicals are possible human carcinogens. The strongest evidence for establishing a relationship between exposure to a given chemical and cancer in humans comes from epidemiological studies. These studies of human exposure and cancer must consider the latency period for cancer development, because the exposure to the carcinogen often occurs many years (sometimes 20 to 30 years, or even more) before the first sign of cancer appears. However, the most common method for identifying substances as potential human carcinogens is by long-term animal bioassays. These bioassays provide accurate information about dose and duration of exposure, as well as interactions of the substance with other chemicals or modifiers. In these studies, the chemical, substance, or mixture is administered to one or, usually, two laboratory rodent species over a range of doses and duration of exposure with all experimental conditions carefully chosen to maximize the likelihood of identifying any carcinogenic effects (Huff, 1999).

Indeed, experimental carcinogenesis research is based on the scientific assumption that chemicals causing cancer in animals will have similar effects in humans. It must be acknowledged, however, that it is not possible to predict with complete certainty from animal studies alone which agents, substances, mixtures, and/or exposure circumstances will be carcinogenic in humans. Notwithstanding, all known human carcinogens that have been tested adequately also produce cancers in laboratory animals. In many cases, an agent was found to cause cancer in animals and only subsequently confirmed to cause cancer in humans (Huff, 1993). Even so, it is noteworthy that, laboratory animals' adverse responses to chemicals (of which cancer is only one) do not always strictly correspond to human responses. Yet still, laboratory animals remain the best tool for detecting potential human health hazards of all kinds, including cancer (OTA, 1981; Tomatis *et al.*, 1997).

7.1.3. MANIFESTATIONS OF TOXICITY

Toxic responses, regardless of the organ or system in which they occur, can be of several types (USEPA, 1985b). For some, the *severity* of the injury increases as the dose increases. One of the goals of toxicity studies is to determine the 'no observed effect level' (NOEL) – which is the dose at which no toxic effect is seen in an organism; this parameter becomes an important input in the development of toxicity parameters for use in the risk assessment. For other cases, the severity of an effect may not increase with dose, but the *incidence* of the effects will increase with increasing dose. This type of responses is properly characterized as probabilistic – since increasing the dose increases the probability (i.e., risk) that the abnormality or alteration will develop in the exposed population.

Oftentimes with toxic effects (including cancer), both the severity and the incidence tend to increase as the level of exposure is raised. The increase in severity usually is a result of increased damage at higher doses, whereas the increase in incidence is a result of differences in individual sensitivity. Furthermore, the site at which a substance acts (e.g., kidney, liver, etc.) may change as the dose changes. In general, as the duration of exposure increases, both the NOEL and the doses at which effects appear decreases; in some cases, new effects not apparent upon exposure of short duration may become manifest.

Toxic responses also vary in their degree of *reversibility*. In some cases, an effect will disappear almost immediately following cessation of exposure, whereas at the other extreme, exposures will result in a permanent injury. That is, *reversible* toxic effects are those that can be repaired, usually by a specific tissue's ability to regenerate or mend itself after a chemical exposure, whereas *irreversible* toxic effects are those that cannot be repaired. Most toxic responses tend to fall somewhere between these extremes.

Seriousness is yet another characteristic of a toxic response. Certain types of toxic damage are clearly adverse and are a definite threat to health, whereas other types of effects may not be of obvious health significance.

Finally, it is noteworthy that potential receptor populations (especially humans) tend to be exposed to mixtures of chemicals, as opposed to the single toxic agent scenario often presented in hazard evaluations. Consequently, several outcomes may result from chemical mixtures – including additive, synergistic, and antagonistic effects – that may have to be addressed differently, even if only qualitatively.

7.1.4. DOSE-RESPONSE RELATIONSHIPS

The dose-response relationship is about the most fundamental concept in toxicology. A dose-response relationship exists when there is a consistent mathematical relationship that describes the proportion of test organisms responding to a specific dose of a toxicant/substance for a given exposure period. A number of assumptions usually will need to be considered when attempting to establish a dose-response relationship – most importantly, that the following hold true (Hughes, 1996):

- The observed response is caused by the substance administered to the organism;
- The magnitude of the response is directly related to the magnitude of the dose; and
- It is possible to correctly observe and measure a response.

In general, the relationship between the degree of exposure to a chemical (viz., the dose) and the magnitude of chemical-induced effects (viz., the response) is typically described by a dose-response curve. The typical dose-response curve is sigmoidal – but can also be linear, concave, convex, or bimodal; indeed, the shape of the curve can offer clues to the mechanism of action of the toxin, indicate multiple toxic effects, and identify the existence and extent of sensitive sub-populations (Derelanko and Hollinger, 1995).

Dose-response curves fall into two general categories/groups (Figure 7.1): those in which no response is observed until some minimum (i.e., threshold) dose is reached, and those in which no threshold is apparent – meaning that response is expected for any dose, no matter how small. Indeed, for some chemicals, a very small dose causes no observable effects whereas a higher dose will result in some toxicity, and still higher doses cause even greater toxicity – up to the point of fatality; such chemicals are called *threshold chemicals* (curve B in Figure 7.1). For other chemicals, such as most carcinogens, the threshold concept may not be applicable – in which case no minimum level is required to induce adverse and overt toxicity effects (curve A in Figure 7.1).

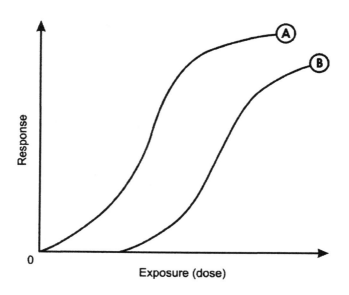

Figure 7.1. Schematic representation of exposure-response relationships:
Illustration of dose-response relationship for (A) = non-threshold chemicals & (B) = threshold chemicals

The most important part of the dose-response curve for a threshold chemical is the dose at which significant effects first begin to show (Figure 7.2). The highest dose that does not produce an observable adverse effect is the 'no-observed-adverse-effect-level' (NOAEL), and the lowest dose that produces an observable adverse effect is the 'lowest-observed-adverse-effect-level' (LOAEL). For non-threshold chemicals, the dose-response curve behaves differently, in that there is no dose that is free of risk.

In general, several important variables (Box 7.1) determine the characteristics of dose-response relationships – and these parameters should be given careful consideration in performing toxicity tests, and when interpreting toxicity data (USEPA, 1985a, 1985b). All these evaluations, however, may be fraught with several uncertainties – best recapped by Rachel Carson's *Silent Spring*, that: "When one is concerned with the mysterious and wonderful functioning of the human body, cause and effect are seldom simple and easily demonstrated relationships. They may be widely separated both in space and time. ... ...The lack of sufficiently delicate methods to detect injury before symptoms appear is one of the great unsolved problems in medicine" (Carson, 1962; 1994).

(a) General schematic of a dose-response curve

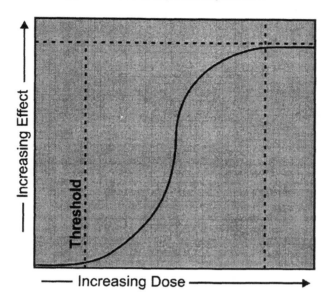

(b) Refined schematic of a dose-response curve – with details added

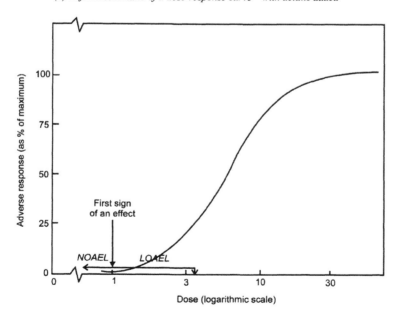

Figure 7.2. Illustrative relationships of a typical dose-response curve for threshold chemicals

Box 7.1. Important parameters/factors to consider in toxicity assessments

♦ *Route of exposure:* The toxicity of some chemicals depends on whether the route of exposure is by inhalation, ingestion, or dermal contact. Also, there may be local responses at the absorption site (viz., lungs, gastrointestinal tract, and skin).

♦ *Duration/frequency of exposure:* The toxicity of many chemicals depends not only on dose (i.e., the amount of chemical contacted or absorbed each day) but also on the length of exposure (i.e., number of days, weeks, or years).

♦ *Test species characteristics:* Differences among species with respect to absorption, excretion or metabolism of chemicals, as well as several other factors (such as genetic susceptibility) should be carefully evaluated in the choice of appropriate animal test species.

♦ *Individual characteristics:* Individual members of a population (especially humans) are not identical, and usually do not respond identically to equal exposures to a chemical. It is therefore important to identify any subgroups that may be more sensitive to a chemical than the general population.

♦ *Toxicological endpoints:* This refers to the nature of toxic effects. The endpoints represent the changes detected in test animals, which become an index of the chemical's toxicity. Some commonly measured endpoints are carcinogenicity, hepatotoxicity (i.e., liver toxicity), mutagenicity, neurotoxicity, renal toxicity, reproductive toxicity, teratogenicity, etc. One of the most important parts of any toxicity study is the selection of the best endpoint to monitor – usually the most sensitive with respect to dose-response changes, the severity of effects, and whether effect is reversible or irreversible.

7.2. Carcinogen Classification Systems

Carcinogenic chemicals are generally classified into several categories, depending on the 'weight-of-evidence' or 'strength-of-evidence' available on the particular chemical's carcinogenicity (Hallenbeck and Cunningham 1988; Huckle, 1991; IARC 1982; USDHS 1989, 1996; USEPA 1986). A chemical's potential for human carcinogenicity is inferred from the available information relevant to the potential carcinogenicity of the chemical, and from judgments as to the quality of the available studies.

In fact, the two carcinogenicity evaluation philosophies (see below) – one based on 'weight-of-evidence' and the other on 'strength-of-evidence' – seem to have found common acceptance and widespread usage. Systems that employ the weight-of-evidence evaluations consider and balance the negative indicators of carcinogenicity with those showing carcinogenic activity (Box 7.2); and schemes using the strength-of-evidence evaluations consider combined strengths of all positive animal tests (vis-à-vis human epidemiology studies and genotoxicity) to rank a chemical without evaluating negative studies, nor considering potency or mechanism (Huckle 1991). Other varying carcinogen classification schemes also exist globally within various regulatory and legislative groups.

In general, carcinogens may be categorized into the following broad identifiable groupings (IARC 1982; Theiss, 1983; USDHS 1989, 1996):

•   *'Known human carcinogens'* – defined as those chemicals for which there exists sufficient evidence of carcinogenicity from studies in humans to indicate a causal relationship between exposure to the agent, substance, or mixture and human cancer.

- *'Reasonably anticipated to be human carcinogens'* – referring to those chemical substances for which there is limited evidence for carcinogenicity in humans and/or sufficient evidence of carcinogenicity in experimental animals. Sufficient evidence in animals is demonstrated by positive carcinogenicity findings in multiple strains and species of animals; in multiple experiments; or to an unusual degree, with regard to incidence, site or type of tumor, or age of onset; or there is less than sufficient evidence of carcinogenicity in humans or laboratory animals.
- *'Sufficient evidence of carcinogenicity'* and *'Limited evidence of carcinogenicity'* – used in the criteria for judging the adequacy of available data for identifying carcinogens; it refers only to the amount and adequacy of the available evidence, and not to the potency of carcinogenic effect on the mechanisms involved.

Conclusions regarding carcinogenicity in humans or experimental animals are based on scientific judgment – with due consideration given to all relevant information. Relevant information includes, but is not limited to, dose-response, route-of-exposure, chemical structure, metabolism, pharmacokinetics, sensitive sub-populations, genetic effects, or other data relating to mechanism of action or factors that may be unique to a given substance. For example, there may be substances for which there is evidence of carcinogenicity in laboratory animals but there are compelling data indicating that the agent acts through mechanisms which do not operate in humans and would therefore not reasonably be anticipated to cause cancer in humans.

In general, evidence of possible carcinogenicity in humans comes primarily from epidemiological studies and long-term animal exposure studies at high doses that have subsequently been extrapolated to humans. Results from these studies are supplemented with information from short-term tests, pharmacokinetic studies, comparative metabolism studies, molecular structure-activity relationships, and indeed other relevant information sources.

Box 7.2. Summary of pertinent/comparative factors affecting the weight-of-evidence for human carcinogens

Factors increasing weight-of-evidence	*Factors decreasing weight-of-evidence*
♦ Evidence of human causality	♦ No evidence or relevant data showing human causality
♦ Evidence of animal effects relevant to humans	♦ No evidence or data on relevance of animal effects to humans
♦ Coherent inferences	♦ Conflicting data
♦ Comparable metabolism and toxicokinetics between species	♦ Metabolism and toxicokinetics between species not comparable
♦ Mode of action comparable across species	♦ Mode of action not comparable across species

7.2.1. WEIGHT-OF-EVIDENCE CLASSIFICATION

A weight-of-evidence approach is used by the US EPA to classify the likelihood that an agent in question is a human carcinogen. This is a classification system for characterizing the extent to which available data indicate that an agent is a human carcinogen (or possesses some other toxic effects such as developmental toxicity). A three-stage procedure is utilized in the process, namely:

- Stage 1 – the evidence is characterized separately for human studies and for animal studies.
- Stage 2 – the human and animal evidence are integrated into a presumptive overall classification.
- Stage 3 – the provisional classification is modified (i.e., adjusted upwards or downwards), based on analysis of the supporting evidence.

The outcome of this process is that chemicals are placed into one of five categories – namely, Groups A-E (Box 7.3) – further discussed below. It is noteworthy that, the guidelines for classification of the weight-of-evidence for human carcinogenicity published by the US EPA (e.g., USEPA 1984, 1986) – consisting of the categorization of the weight-of-evidence into the five groups – are adaptations from those of the International Agency for Research on Cancer (IARC, 1984, 1987, 1988).

Box 7.3. The US EPA weight-of-evidence classification system for potential carcinogens

US EPA Group	*Reference Category*
A	Human carcinogen (i.e., known human carcinogen)
B	Probable human carcinogen:
	B1 indicates limited human evidence
	B2 indicates sufficient evidence in animals and inadequate or no evidence in humans
C	Possible human carcinogen
D	Not classifiable as to human carcinogenicity
E	No Evidence of carcinogenicity in humans (or, Evidence of non-carcinogenicity for humans)

Group A – Human Carcinogen

For this group, there is sufficient evidence from epidemiologic studies to support a causal association between exposure to the agent and human cancer. The following three criteria must be satisfied before a causal association can be inferred between exposure and cancer in humans (Hallenbeck and Cunningham, 1988; USEPA, 1986):

- No identified bias which could explain the association;
- Possibility of confounding factors (i.e., variables other than chemical exposure level which can affect the incidence or degree of the parameter being measured) has been considered and ruled out as explaining the association; and
- Association is unlikely to be due to chance.

Indeed, this group is used only when there is sufficient evidence from epidemiologic studies to support a causal association between exposure to the agents and cancer.

Group B – Probable Human Carcinogen

This group includes agents for which the weight-of-evidence of human carcinogenicity based on epidemiologic studies is 'limited' and also includes agents for which the weight-of-evidence of carcinogenicity based on animal studies is 'sufficient.' The category consists of agents for which the evidence of human carcinogenicity from epidemiologic studies ranges from almost sufficient to inadequate.

This group is divided into two subgroups – reflecting higher (Group B1) and lower (Group B2) degrees of evidence. Usually, category B1 is reserved for agents for which there is limited evidence of carcinogenicity to humans from epidemiologic studies; limited evidence of carcinogenicity indicates that a causal interpretation is credible but that alternative explanations such as chance, bias, or confounding could not be excluded. Inadequate evidence indicates that one of the following two conditions prevailed (Hallenbeck and Cunningham, 1988; USEPA, 1986):

- There were few pertinent data; or
- The available studies, while showing evidence of association, did not exclude chance, bias, or confounding.

When there are inadequate data for humans, it is reasonable to consider agents for which there is sufficient evidence of carcinogenicity in animals as if they presented a carcinogenic risk to humans. Therefore, agents for which there is 'sufficient' evidence from animal studies and for which there is 'inadequate' evidence from human (epidemiological) studies or 'no data' from epidemiologic studies would usually result in a classification of B2 (CDHS 1986; Hallenbeck and Cunningham, 1988; USEPA, 1986).

Group C – Possible Human Carcinogen
This group is used for agents with limited evidence of carcinogenicity in animals in the absence of human data. Limited evidence means that the data suggest a carcinogenic effect, but are limited for the following reasons (Hallenbeck and Cunningham 1988; USEPA, 1986):

- The studies involve a single species, strain, or experiment; or
- The experiments are restricted by inadequate dosage levels, inadequate duration of exposure to the agent, inadequate period of follow-up, poor survival, too few animals, or inadequate reporting; or
- An increase in the incidence of benign tumors only.

Group C classification relies on a wide variety of evidence – including the following (Hallenbeck and Cunningham, 1988; USEPA, 1986): definitive malignant tumor response in a single well conducted experiment that does not meet conditions for 'sufficient' evidence; tumor response of marginal statistical significance in studies having inadequate design or reporting; benign but not malignant tumors, with an agent showing no response in a variety of short-term tests for mutagenicity; and responses of marginal statistical significance in a tissue known to have a high and/or variable background tumor rate.

Group D – Not Classifiable as to Human Carcinogenicity
This group is generally used for agents with inadequate animal evidence of carcinogenicity and also inadequate evidence from human (epidemiological) studies. Inadequate evidence means that, because of major qualitative or quantitative

limitations, the studies cannot be interpreted as showing either the presence or absence of a carcinogenic effect.

Group E – No Evidence of Carcinogenicity in Humans
This group is used to describe agents for indicating evidence of non-carcinogenicity for humans, together with no evidence of carcinogenicity in at least two adequate animal tests in different species, or no evidence in both adequate animal and human (epidemiological) studies. The designation of an agent as being in this group is based on the available evidence and should not be interpreted as a definitive conclusion that the agent will not be a carcinogen under any circumstances.

7.2.2. STRENGTH-OF-EVIDENCE CLASSIFICATION

The International Agency for Research on Cancer (IARC) bases its classification on the strength-of-evidence philosophy. The corresponding IARC classification system, comparable or equivalent to the US EPA description presented above, is shown in Box 7.4 – and further discussed below.

Box 7.4. The IARC strength-of-evidence classification system for potential carcinogens

IARC Group	Category
1	Human carcinogen (i.e., Known human carcinogen)
2	Probable or Possible human carcinogen: 2A indicates limited human evidence (i.e., Probable) 2B indicates sufficient evidence in animals and inadequate or no evidence in humans (i.e., Possible)
3	Not classifiable as to human carcinogenicity
4	No Evidence of carcinogenicity in humans

Group 1 – Known Human Carcinogen
This group is generally used for agents with sufficient evidence from human (epidemiological) studies as to human carcinogenicity. The Group 1 agent is carcinogenic to humans.

Group 2 – Probable or Possible Human Carcinogens
This category includes agents for which, at one extreme, the degree of evidence of carcinogenicity in humans is almost sufficient, as well as agents for which, at the other extreme, there are no human data but for which there is experimental evidence of carcinogenicity. Agents are assigned to either 2A (probably carcinogenic) or 2B (possibly carcinogenic) on the basis of epidemiological, experimental and other relevant data.

Group 2A – Probable Human Carcinogen. This group is generally used for agents for which there is sufficient animal evidence, evidence of human carcinogenicity, or at least limited evidence from human (epidemiological) studies. These are probably carcinogenic to humans, with (usually) at least limited human evidence.

This category is used when there is limited evidence of carcinogenicity in humans and sufficient evidence of carcinogenicity in experimental animals. Exceptionally, an agent may be classified into this category solely on the basis of limited evidence of carcinogenicity in humans or of sufficient evidence of carcinogenicity in experimental animals strengthened by supporting evidence from other relevant data.

Group 2B − Possible Human Carcinogen. This group is generally used for agents for which there is sufficient animal evidence but inadequate evidence from human (epidemiological) studies, or where there is limited evidence from human (epidemiological) studies in the absence of sufficient animal evidence. These are probably carcinogenic to humans, but (usually) have no human evidence.

This category is generally used for agents that indicate limited evidence in humans, in the absence of sufficient evidence in experimental animals. It may also be used when there is inadequate evidence of carcinogenicity in humans, or when human data are nonexistent but there is sufficient evidence of carcinogenicity in experimental animals. In some instances, an agent for which there is inadequate evidence or no data in humans but limited evidence of carcinogenicity in experimental animals together with supporting evidence from other relevant data may be placed in this group.

Group 3 − Not Classifiable
This group is generally used for agents for which there are inadequate animal evidence and inadequate evidence from human (epidemiological) studies − but where there is sufficient evidence of carcinogenicity in experimental animals. The Group 3 agent is not classifiable as to its carcinogenicity to humans − and agents are placed in this category when they do not fall into any other group.

Group 4 − Non-carcinogenic to Humans
This group is generally used for agents for which there is evidence to support a lack of carcinogenicity. The Group 4 agent is probably not carcinogenic to humans − and this category is used for agents for which there is evidence suggesting lack of carcinogenicity in humans together with evidence suggesting lack of carcinogenicity in experimental animals.

In some circumstances, agents for which there is inadequate evidence of (or no data on) carcinogenicity in humans but evidence suggesting lack of carcinogenicity in experimental animals, consistently and strongly supported by a broad range of other relevant data, may be placed into this group.

7.3. Evaluation of Chemical Toxicity

Typically, the identification of toxic substances begins with the retrieval of a variety of pertinent information that is available on the suspected agent (Box 7.5) (CDHS, 1986; Smith, 1992). Toxicity tests may reveal that a substance produces a wide variety of adverse effects on different organs or systems of the human body, or that the range of effects is narrow. Also, some effects may occur only at the higher doses used, and only the most sensitive indicators of a substance's toxicity may be manifest at the lower doses (USEPA, 1985b).

Box 7.5. Typical information requirements for the identification of chemical toxicity

- Physical and chemical properties
- Routes of exposure
- Metabolic and pharmacokinetic properties
- Structure-activity relationships
- Toxicological effects
- Short-term tests
- Long-term animal tests
- Human epidemiologic studies
- Clinical data

General methods of chemical toxicity assessment that are commonly used for determining the hazardous nature of substances include the following (Lave, 1982; NRC, 1991; Talbot and Craun, 1995):

- Case clusters
- Structural toxicology (structure-activity studies)
- Laboratory study of simple test systems
- Long-term animal bioassays
- Human (epidemiologic) studies

Case clusters are based on the identification of an abnormal pattern of disease. This procedure tends to be more powerful in identifying hazards especially when the resulting condition is extremely rare; the method is not very powerful in situations when the health condition is quite common in the general population. Since the population at risk is essentially never known in detail, the case cluster method necessarily yields no conclusive evidence, only rather vague suspicions. *Structural toxicology* involves searching for similarities in chemical structure that might identify toxicological categories, such as carcinogens. The structure-activity studies seek to evaluate toxicity based on the substance's chemical structure; for instance, the close association between mutagens and carcinogens lead to a general presumption that mutagenic substances are also carcinogenic. *Animal bioassays* are laboratory experimentations, generally with rodents. In these types of studies, statistical models are used to extrapolate from animal bioassays to humans. *Epidemiology* is a more scientific, systematic form of case cluster analysis – with an attempt to control for confounding factors in the experimental design or statistical analysis. It examines the occurrence of disease in human populations and tries to determine the causes.

A comprehensive toxicity assessment with respect to chemical exposure problems is generally accomplished in two steps: hazard effects assessment and dose-response assessment. These steps are briefly discussed below, and in greater detail elsewhere in the literature (e.g., Casarett and Doull, 1975; Klaassen *et al.*, 1986, 1996; Lave, 1982; NRC, 1991; Talbot and Craun, 1995; USEPA 1989).

7.3.1. HAZARD EFFECTS ASSESSMENT

Hazard effects assessment is the process used to determine whether exposure to an agent can cause an increase in the incidence of an adverse health effect (e.g., cancer, birth defects, etc.); it involves a characterization of the nature and strength of the evidence of causation. The process involves gathering and evaluating data on the types of health injury or disease that may be produced by a chemical, and on the conditions of exposure under which injury or disease is produced. Hazard assessment may also involve characterizing the behavior of a chemical within the receptor's body and the interactions it undergoes with organs, cells, or even parts of cells. Data of the latter types may be of value in answering the ultimate question of whether the forms of toxicity known to be produced by a substance in one population group or in experimental settings, are also likely to be produced in humans.

The overall purpose of a hazard assessment is to review and evaluate data pertinent to answering questions relating to two key issues – namely, whether an agent may pose a hazard to potential receptors; and under what circumstances an identified hazard may be manifested. Toxicity studies are generally conducted to identify the nature of health damage produced by a substance and the range of doses over which damage is produced (Box 7.6) (USEPA, 1985b). The assessment of the toxicity of a chemical substance involves identification of the adverse effects that the chemical causes and systematic study of how these effects depend upon dose, route and duration of exposure, and test organisms. This information is typically derived from studies falling into one of the following general protocols/categories (Cohrssen and Covello, 1989; Derelanko and Hollinger, 1995; Moeller, 1997; USEPA, 1985a, 1985b):

Box 7.6. Summary reasons for conducting toxicity studies

- ◆ To identify the specific organs or systems of the body that may be damaged by a substance
- ◆ To identify specific abnormalities or diseases (such as cancer, birth defects, nervous disorders, or behavioral problems) that a substance may produce
- ◆ To establish the conditions of exposure and dose that give rise to specific forms of damage or disease
- ◆ To identify the specific nature and course of the injury or disease produced by a substance
- ◆ To identify the biological processes that underlie the production of observable damage or disease

---

- *Laboratory animal studies,* which evaluate the toxicity of a chemical with special reference and/or ultimate goal to predicting the toxicity in humans. Testing protocols in animals are designed to identify the principal adverse effects of a chemical as a function of dose, route of exposure, species and sex of test animals, and duration of exposure.

- *Clinical case studies in humans,* in which there are case-by-case investigations of the symptoms and diseases in humans who are exposed to a toxic substance at doses high enough to call for medical attention or intervention. Exposures may be accidental (e.g., a farmer applying pesticide without proper protection) or, in rare cases, intentional (e.g., suicide or homicide cases). Tragically, this sort of direct toxicological observation is especially valuable in characterizing toxic responses of clinical significance in humans – far better than extrapolations from laboratory animals to humans.

- *Epidemiologic studies,* which seek to determine whether a correlation exists between chemical exposure and frequency of disease or health problems in large groups of human populations. It involves the examination of persons who have been inadvertently exposed to one or more chemical agents. Indeed, despite the many problems inherent to epidemiological studies (especially the various biases, confounding factors, and inadequate quantitation of exposure), these studies offer a major advantage over those conducted with animals – namely, the direct observation of effects in humans. The major advantages of epidemiological studies are that they are based on large numbers of humans and exposure levels are usually sub-clinical. Thus, the data are directly relevant, with no need to extrapolate from animal data or to make projections from a small number of humans exposed to a high dose of the chemical, as for clinical studies.

It is noteworthy that, since much of the uncertainty associated with risk assessment arises from the extrapolation of animal data to humans, quality epidemiological studies can indeed significantly reduce or eliminate such uncertainty. Usually, however, the availability of quality epidemiological studies is limited, and both human and animal data are used together in the risk assessment process.

7.3.2. DOSE-RESPONSE ASSESSMENT AND QUANTIFICATION

Dose-response assessment is the process of quantitatively evaluating toxicity information and characterizing the relationship between the dose of the chemical administered or received (i.e., exposure to an agent) and the incidence of adverse health effects in the exposed populations. The process consists of estimating the potency of the specific compounds by the use of dose-response relationships. In the case of carcinogens, for example, this involves estimating the probability that an individual exposed to a given amount of chemical will contract cancer due to that exposure; potency estimates may be given as 'unit risk factor' (expressed in $\mu g/m^3$) or as 'potency slopes' (in units of $[mg/kg\text{-}day]^{-1}$). Data are generally derived from animal studies or, less frequently, from studies in exposed human populations.

The dose-response assessment first addresses the relationship of dose to the degree of response observed in an experiment or a human study. When chemical exposures are outside the range of observations, extrapolations are necessary – in order to be able to estimate or characterize the dose relationship. The extrapolations will typically be made from high to low doses, from animal to human responses, and/or from a specific route of exposure to a different one. The details of the extrapolation mechanics are beyond the scope of this discussion, but are elaborated elsewhere in the literature (e.g., Brown, 1978; CDHS 1986; Crump, 1981; Crump and Howe, 1984; Gaylor and Kodell, 1980; Gaylor and Shapiro, 1979; Hogan, 1983; Krewski and Van Ryzin, 1981).

In general, even if a substance is known to be toxic, the risks associated with the substance cannot be ascertained with any degree of confidence unless dose-response relationships are quantified. The dose-response relationships are typically used to determine what dose of a particular chemical causes specific levels of toxic effects to potential receptors. In fact, there may be many different dose-response relationships for any given substance if it produces different toxic effects under different conditions of exposure. In any case, the response of a given toxicant depends on the mechanism of its action; for the simplest scenario, the response, R, is directly proportional to its concentration, $[C]$, so that:

$$R = k \cdot [C] \tag{7.1}$$

where k is a rate constant. This would be the case for a chemical that metabolizes rapidly. Even so, the response and the value of the rate constant would tend to differ for different risk groups of individuals and for unique exposures. If, for instance, the toxicant accumulates in the body, the response is better defined as follows:

$$R = k \cdot [C] \cdot t^n \tag{7.2}$$

where t is the time and n is a constant. For cumulative exposures, the response would generally increase with time. Thus, the cumulative effect may show as linear until a threshold is reached, after which secondary effects begin to affect and enhance the responses. Also, the cumulative effect may be related to what is referred to as the 'body burden' (BB). The body burden is determined by the relative rates of absorption (ABS), storage (STR), elimination (ELM), and biotransformation (BTF) – in accordance with the following relationship (Meyer 1983):

$$BB = ABS + STR - ELM - BTF \tag{7.3}$$

Each of the factors involved in the quantification of the body burden is dependent on a number of biological and physiochemical factors.

In fact, the response of an individual to a given dose cannot be truly quantitatively predicted since it depends on many extraneous factors, such as general health and diet of individual receptors. Nonetheless, from the quantitative dose-response relationship, toxicity values can be derived and used to estimate the incidence of adverse effects occurring in potential receptors at different exposure levels.

It is noteworthy that, the fundamental principles underlying the dose-response assessment for carcinogenic chemicals remain arguable – especially in relation to the tenet that there is some degree of carcinogenic risk associated with every potential carcinogen, no matter how small the dose. The speculation and/or belief that chemically induced cancer is a non-threshold process/phenomenon may be false, but represents a conservative default policy necessary to ensure adequate protection of human health – albeit this shortcoming should be kept in perspective in consequential policy decisions about the assessment.

The Nature of Dose-Response Extrapolation Models
Three major classes of mathematical extrapolation models are often used for relating dose and response in the sub-experimental dose range, namely:

- Tolerance Distribution models – including Probit, Logit, and Weibull;
- Mechanistic models – including One-hit, Multi-hit, and Multi-stage; and
- Time-to-Occurrence models – including Lognormal and Weibull.

Indeed, other independent models – such as linear, quadratic, and linear-cum-quadratic – may also be employed for this purpose. The details of these are beyond the scope of this discussion, but are elaborated elsewhere in the literature (e.g., Brown, 1978; CDHS 1986; Crump, 1981; Crump and Howe, 1984; Gaylor and Kodell, 1980; Gaylor and Shapiro, 1979; Hogan, 1983; Krewski and Van Ryzin, 1981, Tan, 1991). In any case, the primary models used to extrapolate from non-threshold effects associated with carcinogenic responses that are observed at high doses to responses at low doses include the following (Derelanko and Hollinger, 1995; Jolley and Wang, 1993; Tan, 1991):

- *Linearized multistage (LMS) model* – which assumes that there are multiple stages to cancer; it fits curve to the experimental data, and is linear from the upper confidence level to zero. That is, it is based on the assumption that the induction of irreversible self-replicating toxic effects is the result of a number of random biological events, the occurrence of each being in strict linear proportion to the dose rate.
- *One-hit model* – which assumes there is a single stage for cancer and that one molecular or radiation interaction induces malignant change; it is a very conservative model, and corresponds to the simplest mechanistic model of carcinogenesis. This model is based on the concept that a response will occur after the target has been 'hit' by a single biologically effective unit of dose.
- *Multi-hit model* – which assumes several interactions are needed before a cell becomes transformed; it is the least conservative model. This model is based on an extension of the one-hit model, assuming that more than one 'hit' is required to induce a response.
- *Probit model* – which assumes probit (lognormal) distribution for tolerance of exposed population; it is appropriate for acute toxicity, but questionable for cancer.
- *Physiologically-based pharmacokinetic (PB-PK) models* – which incorporate pharmacokinetic and mechanistic data into the extrapolation; they possess data-rich requirements with great promise for extensive utilization in the future, as more biological data becomes available. This type of model quantifies the relationship between the exposure to a carcinogen and the dose of the biologically active component of the chemical – and then incorporates the kinetics of metabolic processes that may change a chemical's toxicity.

It is noteworthy that, most of the techniques used to compensate for assessment uncertainties (such as the use of large safety factors, conservative assumptions, and extrapolation models) are designed to err on the side of safety. For these reasons, many regulatory agencies tend to use the so-called linearized multistage (LMS) model for the sake of conservatism. In fact, several models have been proposed for the quantitative extrapolations of carcinogenic effects to low dose levels; however, among these models, the LMS model – that conservatively assumes linearity at low doses – seems to be favored by regulatory agencies such as the US EPA (USEPA, 1986). Of course, alternative models that do not assume a linear relationship, and that are generally less conservative also exist; in any case, more recent guidelines seem to be favoring/incorporating the 'state-of-the-art' in carcinogenic risk assessment models. In general, however, the choice of model is determined by its consistency with the current understanding of the mechanisms of carcinogenesis.

It must be acknowledged here that, there often is no sound basis (in a biological sense) for choosing one model over another. When applied to the same data, the various models can produce a wide range of risk estimates. The frequently recommended model – i.e., the LMS model – produces among the highest estimates of risk, and thus seems to provide a greater margin of protection for human health. However, this model does not provide a 'best estimate' or point estimate of risk, but rather an upper-bound probability that the actual risk will be less than the predicted risk 95 percent of the time. Indeed, using the LMS to extrapolate from high-dose to low-dose effects can lead to erroneous conclusions about risk for many animal carcinogens; in this light, the use of other appropriate extrapolation models is encouraged. In any case, given that no single model will apply for all chemicals, it is important to identify risk on a case-by-case basis. In fact, Huckle (1991) suggests a presentation of the best estimate of risk (or range, with an added margin of safety) from two or three appropriate models, or a single value based on 'weight-of-evidence' – rather than simply using the LMS model. Exceptions may occur, however, for cases of chemicals that have not been sufficiently studied.

7.4. Determination of Toxicological Parameters for Human Health Risk Assessments

In the processes involved in the assessment of human health risks that result from chemical exposures, it often becomes necessary to compare receptor chemical intakes with doses shown to cause adverse effects in humans or experimental animals. The dose at which no effects are observed in human populations or experimental animals is referred to as the 'no-observed-effect-level' (NOEL). Where data identifying a NOEL are lacking, a 'lowest-observed-effect-level' (LOEL) may be used as the basis for determining safe threshold doses.

For acute effects, short-term exposures/doses shown to produce no adverse effects are involved; this is called the 'no-observed-adverse-effect-level' (NOAEL). A NOAEL is an experimentally determined dose at which there has been no statistically or biologically significant indication of the toxic effect of concern. In cases where a NOAEL has not been demonstrated experimentally, the term 'lowest-observed-adverse-effect level' (LOAEL) is used.

In general, for chemicals possessing carcinogenic potentials, the LADD is compared with the NOEL identified in long-term bioassay experimental tests; for chemicals with acute effects, the MDD is compared with the NOEL observed in short-term animal studies. An elaboration on the derivation of the relevant toxicological parameters commonly used in human health risk assessments follows below, with further in-depth discussions to be found in the literature elsewhere (e.g., Dourson and Stara 1983; USEPA 1985b, 1986a, 1989a, 1989c, 1989d).

7.4.1. TOXICITY PARAMETERS FOR NON-CARCINOGENIC EFFECTS

Traditionally, risk decisions on systemic toxicity are made using the concept of 'acceptable daily intake' (ADI), or by using the so-called reference dose (RfD). The ADI is the amount of a chemical (in mg/kg body weight/day) to which a receptor can be exposed to on a daily basis over an extended period of time – usually a lifetime – without suffering a deleterious effect. The RfD is defined as the maximum amount of a chemical (in mg/kg body weight/day) that the human body can absorb without experiencing chronic health effects. For exposure of humans to the non-carcinogenic effects of environmental chemicals, the ADI or RfD is used as a measure of exposure that is considered to be without adverse effects. Although often used interchangeably, RfDs are based on a more rigorously defined methodology, and is therefore preferred over ADIs.

The RfD provides an estimate of the continuous daily exposure of a non-carcinogenic substance for the general human population (including sensitive subgroups) which appears to be without an appreciable risk of deleterious effects. RfDs are established as thresholds of exposure to toxic substances below which there should be no adverse health impact. These thresholds are established on a substance-specific basis for oral and inhalation exposures, taking into account evidence from both human epidemiologic and laboratory toxicologic studies. Also, subchronic RfD is typically used to refer to cases involving only a portion of the lifetime, whereas chronic RfD is associated with lifetime exposures.

The reference concentration (RfC) of a chemical – like the RfD – represents an estimate of the exposure that can occur on a daily basis over a prolonged period, with a reasonable anticipation that no adverse effect will occur from that exposure. In contrast to RfDs, however, RfCs are expressed in units of concentration in an environmental medium (e.g., mg/m^3 or $\mu g/L$). RfCs generally pre-suppose continuous exposure, with an average inhalation rate and body weight; it may therefore be inappropriate to use them in 'non-standard' exposure scenarios.

To derive a RfD or RfC for a non-cancer critical effect, the common practice is to apply standard 'uncertainty factors' (UFs) to the NOAEL, LOAEL, or a benchmark dose/concentration, $BMCL_x$ (US EPA, 1995c). ($BMCL_x$ is defined as the lower 95% confidence limit of the dose that will result in a level of $x\%$ response; e.g., $BMCL_{10}$ is the lower 95% confidence limit of a dose for a 10% increase in a particular response.) The UFs are used to account for the extrapolation uncertainties (e.g., inter-individual variation, interspecies differences, exposure duration, etc.) and database adequacy. A modifying factor (MF) is also used to account for the confidence in the critical study(s) used in the derivation of the RfD or RfC. Replacements for default UFs are used when chemical-specific data are available to modify these standard values; this is known as the 'data-derived' approach. Moreover, the use of pharmacokinetic or dosimetry models can obviate the need for an UF to account for differences in toxicokinetics across species. It is noteworthy that, a number of related factors can indeed result in significant uncertainties in the RfD or RfC values. Among these is the selection of different observed effects as a critical effect – which may vary within and across available studies. Also significant is the choice of different data sets for the identification of the NOAEL, LOAEL, or benchmark dose analysis; the use of different values for the various UFs; and additional judgments that impact the MF.

In general, both the RfD and RfC represent estimates of the exposure that can occur on a daily basis over a prolonged period, with a reasonable expectation that no adverse effect will occur from that exposure. In assessing the chronic and subchronic effects of non-carcinogens and also non-carcinogenic effects associated with carcinogens, the experimental dose value (e.g., NOEL) is usually divided by a safety (or uncertainty) factor to yield the RfD – as illustrated below.

When no toxicological information exists for a chemical, concepts of structure-activity relationships may have to be employed to derive acceptable intake levels by influence and analogy analysis vis-à-vis closely related or similar compounds. In such cases, some reasonable degree of conservatism is suggested in any judgment call to be made.

Derivation of RfDs and RfCs
The RfD is a 'benchmark' dose operationally derived from the NOAEL by consistent application of general 'order-of-magnitude' uncertainty factors (UFs) (also called 'safety factors') that reflect various types of data sets used to estimate RfDs. In addition, a modifying factor (MF) is sometimes used that is based on professional judgment of the

entire database of the specific chemical. More generally stated, RfDs (and ADIs) are calculated by dividing a NOEL (i.e., the highest level at which a chemical causes no observable changes in the species under investigation), a NOAEL (i.e., the highest level at which a chemical causes no observable adverse effect in the species being tested), or a LOAEL (i.e., that dose rate of chemical at which there are statistically or biologically significant increases in frequency or severity of adverse effects between the exposed and appropriate control groups) derived from human or animal toxicity studies by one or more uncertainty and modifying factors.

Typically, RfDs are calculated using a single exposure level together with uncertainty factors that account for specific deficiencies in the toxicological database. Both the exposure level and uncertainty factors are selected and evaluated in the context of the available chemical-specific literature. After all the toxicological, epidemiologic, and supporting data have been reviewed and evaluated, a key study is selected that reflects optimal data on the critical effect. Dose-response data points for all reported effects are examined as a component of this review. USEPA (1989b) discusses specific issues of particular significance in this endeavor – including the types of response levels (ranked in order of increasing severity of toxic effects as NOEL, NOAEL, LOAEL, and FEL [the Frank effect level, defined as overt or gross adverse effects]) that are considered in deriving RfDs. Ultimately, the RfD (or ADI) can be determined from the NOAEL (or LOAEL) for the critical toxic effect by consistent application of UFs and a MF, in accordance with the following relationship:

$$\text{Human dose (e.g., } ADI \text{ or } RfD) = \frac{\text{Experimental dose (e.g., NOAEL)}}{(\text{UF x MF})} \tag{7.4}$$

or, specifically:

$$RfD = \frac{\text{NOAEL}}{(\text{UF x MF})} \tag{7.5}$$

or, more generally:

$$RfD = \frac{[NOAEL \text{ or } LOAEL]}{\left[\sum_{i=1}^{n} UF_i \times MF\right]} \tag{7.6}$$

The derivation of a RfC is a parallel process that is appropriately based on a 'no-observed-adverse-effect-concentration' (NOAEC) or 'lowest-observed-adverse-effect-concentration' (LOAEC). Alternatively, a RfC may be derived from a RfD, taking into account the exposure conditions of the study used to derive the RfD.

Determination of the Uncertainty and Modifying Factors. Overall, the uncertainty factor used in deriving the RfD reflects scientific judgment regarding the various types of data used to estimate RfD values. It is used to offset the uncertainties associated with extrapolation of data, etc. Generally, the UF consists of multiples of 10 (although values less than 10 could also be used), each factor representing a specific area of uncertainty inherent in the extrapolations from the available data. For example, a factor of 10 may be introduced to account for the possible differences in responsiveness between humans and animals in prolonged exposure studies. Indeed, for interspecies extrapolation of toxic effects seen in experimental animals to what might occur in exposed humans, an UF of up to 10-fold is generally recommended. This is usually viewed as consisting of two parts: one that accounts for metabolic or pharmacokinetic differences between the species; and another that addresses pharmacodynamic differences (i.e., differences between the response of human and animal tissues to the chemical exposure). A second factor of 10 may be used to account for variation in susceptibility among individuals in the human population. Indeed, exposed humans are known to vary considerably in their response to toxic chemical and drug exposures due to age, disease states, and genetic makeup – particularly in genetic polymorphisms for enzymes (isozymes) for detoxifying chemicals. The resultant UF of 100 has been judged to be appropriate for many chemicals. For other chemicals, with databases that are less complete (for example, those for which only the results of subchronic studies are available), an additional factor of 10 (leading to a UF of 1,000) might be judged to be more appropriate. For certain other chemicals, and based on well-characterized responses in sensitive humans (as in regards to the effect of fluoride on human teeth), an UF as small as 1 might be selected (Dourson and Stara 1983). Finally an additional 10-fold UF may be used to account for possible carcinogenicity. Box 7.7 provides the general guidelines for the process of selecting uncertainty and modifying factors for the derivation of RfDs (Dourson and Stara 1983; USEPA 1986b, 1989a, 1989b).

Box 7.7. General guidelines for selecting uncertainty and modifying factors in the derivation of RfDs

✦ *Standard Uncertainty Factors (UFs):*
♦ Use a 10-fold factor when extrapolating from valid experimental results in studies using prolonged exposure to average healthy humans. This factor is intended to account for the variation in sensitivity among the members of the human population, due to heterogeneity in human populations, and is referenced as "10H". Thus, if NOAEL is based on human data, a safety factor of 10 is usually applied to the NOAEL dose to account for variations in sensitivities between individual humans.
♦ Use an additional 10-fold factor when extrapolating from valid results of long-term studies on experimental animals when results of studies of human exposure are not available or are inadequate. This factor is intended to account for the uncertainty involved in extrapolating from animal data to humans and is referenced as "10A". Thus, if NOAEL is based on animal data, the NOAEL dose is divided by an additional safety factor of 10, to account for differences between animals and humans.
♦ Use an additional 10-fold factor when extrapolating from less than chronic results on experimental animals when there are no useful long-term human data. This factor is intended to account for the uncertainty involved in extrapolating from less than chronic (i.e., subchronic or acute) NOAELs to chronic NOAELs and is referenced as "10S".
♦ Use an additional 10-fold factor when driving an RfD from a LOAEL, instead of a NOAEL. This factor is intended to account for the uncertainty involved in extrapolating from LOAELs to NOAELs and is referenced as "10L".
♦ Use an additional up to 10-fold factor when extrapolating from valid results in experimental animals when the data are 'incomplete.' This factor is intended to account for the inability of any single animal study to adequately address all possible adverse outcomes in humans, and is referenced as "10D."

✦ *Modifying Factor (MF):*
♦ Use professional judgment to determine the MF, which is an additional uncertainty factor that is greater than zero and less than or equal to 10. The magnitude of the MF depends upon the qualitative professional assessment of scientific uncertainties of the study and data base not explicitly treated above – e.g., the completeness of the overall data base and the number of species tested. The default value for the MF is 1.

In general, the choice of the UF and MF values reflect the uncertainty associated with the estimation of a RfD from different human or animal toxicity databases. For instance, if sufficient data from chronic duration exposure studies are available on the threshold region of a chemical's critical toxic effect in a known sensitive human population, then the UF used to estimate the RfD may be set at unity (1). That is, these data are judged to be sufficiently predictive of a population sub-threshold dose, so that additional UFs are not needed (USEPA, 1989b).

Illustrative Examples of the RfD Derivation Process. Some hypothetical example situations involving the determination of RfDs based on information on NOAEL, and then also on LOAEL, are provided below.

- *Determination of the RfD for a Hypothetical Example Using the NOAEL.* Consider the case of a study made on 250 animals (e.g., rats) that is of subchronic duration, yielding a NOAEL dosage of 5 mg/kg/day. Then,

$$UF = 10H \times 10A \times 10S = 1,000$$

In addition, there is a subjective adjustment (represented by the MF), based on the high number of animals (250) per dose group:

$$MF = 0.75$$

These factors then give UF x MF = 750, so that

$$RfD = \frac{NOAEL}{(UF \times MF)} = \frac{5}{750} = 0.007 \text{ (mg/kg/day)}$$

- *Determination of the RfD for a Hypothetical Example Using the LOAEL.* If the NOAEL is not available, and if 25 mg/kg/day had been the lowest dose from the test that showed adverse effects, then

$$UF = 10H \times 10A \times 10S \times 10L = 10,000$$

Using again the subjective adjustment of MF = 0.75, one obtains

$$RfD = \frac{LOAEL}{(UF \times MF)} = \frac{25}{7500} = 0.003 \text{ (mg/kg/day)}$$

The Benchmark Dose (BMD) Approach

The RfD or RfC for humans is often derived from animal experiments – and the NOAEL (which represent the highest experimental dose for which no adverse health effects have been documented) has often been a starting point for its calculation. However, using the NOAEL in determining RfDs and RfCs has long been recognized as having significant limitations – especially because it is limited to one of the doses in the study and is dependent on study design; it does not account for variability in the estimate of the dose-response; it does not account for the slope of the dose-response curve; and it cannot be applied when there is no NOAEL, except through the application of an uncertainty factor (Crump, 1984; Kimmel and Gaylor, 1988; USEPA, 1995). As an alternative to the use of NOAEL and LOAEL in the determination of the

RfD or RfC in the non-cancer risk evaluation, other methodologies are becoming increasingly popular – and more so with the so-called 'benchmark dose' (BMD) approach. An important goal of the BMD approach is to define a starting point of departure for the computation of a reference value (viz., RfD or RfC), or a slope factor, that is more independent of study design.

The use of BMD methods involve fitting mathematical models to dose-response data and using the different results to select a BMD that is associated with a predetermined benchmark response – such as a 10% increase in the incidence of a particular lesion, or a 10% decrease in body weight changes. The BMD has been defined as a lower confidence limit on the effective dose associated with some defined level of effect – e.g., a 5% or 10% increase in response. That is, it is the confidence limit on the dose that elicits adverse responses in a fraction (often 5% or 10%) of the experimental animals; confidence limits characterize uncertainty in the dose that affects a specified fraction of the animals. A dose-response relationship is fitted to the bioassay data points, and a confidence limit for that relationship is determined. The BMD is the dose yielding the desired response rate (e.g., 5%, or 10%) based on the curve representing the confidence limit. Unlike the NOAEL, the BMD makes use of all the bioassay data and is not constrained to dose values administered in the experiment. On the other hand, there is no scientific basis for selecting a particular response rate for the BMD (e.g., 5% *vs.* 10%), and neither does the BMD easily address 'continuous' responses (e.g., as reflected by changes in body weight) – since it depends on identifying the dose at which a certain fraction of animals is 'affected' by the toxicant.

Inter-Conversions of Non-carcinogenic Toxicity Parameters
Usually, the RfD for inhalation exposure is reported both as a concentration in air (mg/m^3) and as a corresponding inhaled dose (mg/kg-day). When determining the toxicity value for inhalation pathways, the inhalation RfC $[mg/m^3]$ should be used when available. The RfC can also be converted to equivalent RfD values (in units of dose [mg/kg-day]) by multiplying the RfC term by an average human inhalation rate of $20 \ m^3/day$ (for adults) and dividing it by an average adult body weight of 70kg. That is,

$$RfD_i \ [mg/kg\text{-}day] = \frac{RfC \ [mg/m^3] \times 20 \ m^3/day}{70 \ kg} = 0.286 \ RfC \qquad (7.7)$$

RfD values associated with oral exposures (and reported in mg/kg-day) can also be converted to a corresponding concentration in drinking water (usually called the 'drinking water equivalent level', DWEL), as follows:

$$DWEL \ [mg/L \ in \ water] = \frac{oral \ RfD \ (mg/kg\text{-}day) \times body \ weight \ (kg)}{ingestion \ rate \ (L/day)}$$

$$= \frac{RfD_0 \ (mg/kg\text{-}day) \times 70 \ (kg)}{2 \ (L/day)} = 35 \ RfD_0 \qquad (7.8)$$

This derivation assumes 2 L/day of water consumption by a 70-kg adult.

Making Risk Management Decisions

In the risk characterization process, a comparison is generally made between the RfD and the estimated exposure dose (EED) (or the so-called 'regulatory dose' [RgD]); the EED should include all sources and routes of exposure involved. If the EED is less than the RfD, the need for regulatory concern may be small.

Another alternative measure that is also considered useful to risk managers is the 'margin of exposure' (MOE), which is the magnitude by which the NOAEL of the critical toxic effect exceeds the EED – where both are expressed in the same units. The MOE is defined as follows:

$$MOE = \frac{NOAEL}{EED} \qquad (7.9)$$

As an example of its utility of the MOE, suppose the EED for humans exposed to a chemical substance (with a RfD of 0.005 mg/kg-day) under a proposed use pattern is 0.02 mg/kg-day (i.e., the EED is greater than the RfD), then:

$$NOAEL = RfD \times (UF \times MF) = 0.005 \times 1,000 = 5 \text{ mg/kg-day}$$

and

$$MOE = \frac{NOAEL}{EED} = \frac{5 \text{ (mg/kg/day)}}{0.02 \text{ (mg/kg/day)}} = 250$$

Because the EED exceeds the RfD (and the MOE is less than the [UF x MF] of 1,000), the risk manager will need to look carefully at the data set as well as the assumptions for both the RfD and the exposure estimates. The MOE may indeed be used as a surrogate for risk; as the MOE becomes larger, the risk becomes smaller.

7.4.2. TOXICITY PARAMETERS FOR CARCINOGENIC EFFECTS

Under a no-threshold assumption for carcinogenic effects, exposure to any level of a carcinogen is considered to have a finite risk of inducing cancer. An estimate of the resulting excess cancer per unit dose (called the unit cancer risk, or the cancer slope factor) – that represents the cancer potency factor (CPF) – is generally used to develop risk decisions for chemical exposure problems.

In general, the CPF is used in human health risk assessments to estimate an upper-bound lifetime probability of an individual developing cancer as a result of exposure to a given level of a potential carcinogen. This represents a slope factor derived from a mathematical function (e.g., the LMS model) used to extrapolate the probability of incidence of cancer from a bioassay in animals using high doses to that expected to be observed at the low doses, likely to be found in chronic human exposures.

It is noteworthy that cancer dose-response assessment generally involves many scientific judgments regarding the following: the selection of different data sets (e.g., benign and malignant tumors, or their precursor responses) for extrapolation; the choice of low dose extrapolation approach based on the interpretation and assessment of the mode of action for the selected tumorigenic response(s); the choice of extrapolation models and methods to account for differences in dose across species; and the selection of the point of departure for low dose extrapolation. Indeed, many judgments usually have to be made in the many steps of the assessment process in the face of data variability. Also, different science policy choices and default procedures and methods are used to bridge data and knowledge gaps. Consequently, it is generally recognized that significant uncertainty exists in the cancer risk estimates.

Two specific toxicity parameters for expressing carcinogenic hazards based on the dose-response function find common application in human health risk assessments, namely:

- Cancer slope factor ([C]SF) – that expresses the slope of the dose-response function in dose-related units (i.e., $[mg/kg\text{-}day]^{-1}$); and
- Unit risk factor (URF) – that expresses the slope in concentration-based units (i.e., $[\mu g/m^3]^{-1}$).

Typically, the [C]SFs are used when evaluating risks from oral or dermal exposures, whereas the URFs are used to evaluate risks from inhalation exposures.

The SF, also called cancer potency factor or potency slope, is a measure of the carcinogenic toxicity or potency of a chemical. It is the plausible upper-bound estimate of the probability of a response per unit intake of a chemical over a lifetime – represented by the cancer risk (proportion affected) per unit of dose (i.e., risk per mg/kg/day).

In evaluating risks from chemicals found in certain human environmental settings, dose-response measures may also be expressed as risk per concentration unit – yielding the URF (also called unit cancer risk, UCR or unit risk, UR) values. These measures may include the unit risk factor for air (viz., an inhalation URF), and the unit risk for drinking water (viz., an oral URF). The continuous lifetime exposure concentration units for air and drinking water are usually expressed in micrograms per cubic meter ($\mu g/m^3$) and micrograms per liter ($\mu g/L$), respectively.

When no toxicological information exist for a chemical, structural similarity factors, etc. can be used to estimate cancer potency units for the chemicals lacking such values, but that are suspected carcinogens. For instance, in the case of a missing URF, this concept may be used to derive a surrogate parameter for the chemical with unknown URF – for example, by estimating the geometric mean of a number of similar compounds whose URFs are known, and then using this as the surrogate value. Also, in recent times, PB-PK modeling has become a preferred approach to both dose estimation and interspecies scaling of inhalation exposures, where data are available to support such efforts.

Derivation of SFs and URFs

The determination of carcinogenic toxicity parameters often involves the use of a variety of mathematical extrapolation models. In fact, scientific investigators have developed numerous models to extrapolate and estimate low-dose carcinogenic risks to humans from the high-dose carcinogenic effects usually observed in experimental animal studies. Such models yield an estimate of the upper limit in lifetime risk per unit of dose (or the unit cancer risk).

In the estimation of carcinogenic potency, the type of extrapolation employed for a given chemical depends on the existence of data to support linearity or nonlinearity, or indeed a biologically based or case-specific model (USEPA, 1996f). The more popular approach amongst major regulatory agencies (such as the US EPA) involves a use of the LMS model – especially because of the conservative attributes of this model.

The Linearized Multistage Model. Mathematically, the multistage model may be expressed as follows:

$$P(d) = 1 - \exp\left[-(q_0 + q_1 d + q_2 d^2 + \ldots\ldots + q_k d^k)\right] \qquad (7.10)$$

where $P(d)$ is the lifetime probability of developing a tumor at a given dose, d, of carcinogen; q_0 is a constant that accounts for the background incidence of cancer (i.e., occurring in the absence of the carcinogen under consideration); and q_1, q_2, ... q_k are coefficients that allow the data to be expressed to various powers of the dose of carcinogen, in order to obtain the best fit of the model to the data. To determine the extra risk above the background rate at dose, d, the above equation takes the form:

$$P_e(d) = 1 - \exp\left[-(q_1 + q_2 d^2 + \ldots\ldots + q_k d^k)\right] \qquad (7.11)$$

At low doses, the extra risk is approximated by:

$$P_e(d) = q_1 d \qquad (7.12)$$

The linearized multistage model uses animal tumor incidence data to compute maximum likelihood estimates (MLE) and upper 95% confidence limits (UCL$_{95}$) of risk associated with a particular dose. The true risk is very unlikely to be greater than the UCL, may be lower than the UCL, and could be even as low as zero. In fact, the linearized multistage model yields upper bound estimates of risks that are a linear function of dose at low doses and are frequently used as a basis for regulation.

Overall, the linearized multistage model is known to make several conservative assumptions that results in highly conservative risk estimates – and thus yields over-estimates of actual URFs for carcinogens; in fact, the actual risks may be substantially lower than that predicted by the upper bounds of this model (Paustenbach, 1988). Even so, such approach is generally preferred since it allows analysts to err on the side of safety – and therefore offer better protection of public health.

Inter-conversion of Carcinogenic Toxicity Parameters
The URF estimates the upper-bound probability of a 'typical' or 'average' person contracting cancer when continuously exposed to one microgram per cubic meter ($1\,\mu g/m^3$) of the chemical over an average (70-year) lifetime. Potency estimates are also given in terms of the potency slope factor (SF), which is the probability of contracting cancer as a result of exposure to a given lifetime dose (in units of mg/kg-day).

The SF can be converted to URF (also, unit risk, UR or unit cancer risk, UCR), by adopting several assumptions. The most critical requirement is that the endpoint of concern must be a systematic tumor – in order that the potential target organs will experience the same blood concentration of the active carcinogen regardless of the method of administration. This implies an assumption of equivalent absorption by the various routes of administration. The basis for these conversions is the assumption that, at low doses, the dose-response curve is linear, so that:

$$P(d) = SF \times [DOSE] \qquad (7.13)$$

where P(d) is the response (probability) as a function of dose; SF is the cancer potency slope factor ($[mg/kg\text{-}day]^{-1}$); and [DOSE] is the amount of chemical intake (mg/kg-day). The inter-conversions between URF and SF are given below.

- *Inter-conversions of the Inhalation Potency Factor.* Risks associated with a unit chemical concentration in air may be estimated in accordance with following mathematical relationship:

Air unit risk

= risk per $\mu g/m^3$ (air)

= [slope factor (risk per mg/kg/day)] $\times$ [$\dfrac{1}{\text{body weight (kg)}}$]

$\qquad \times$ [inhalation rate (m^3/day) $\times$ 10^{-3} (mg/μg)

Thus, the inhalation potency can be converted to an inhalation URF by applying the following conversion factor:

$[(kg\text{-}day)/mg] \times [1/70 \text{ kg}] \times [20 \text{ m}^3/day] \times [1 \text{ mg}/1{,}000 \text{ }\mu g] = 2.86 \times 10^{-4}$

Consequently, the lifetime excess cancer risk from inhaling $1\mu g/m^3$ concentration for a full lifetime is estimated as:

$$URF_i \,(\mu g/m^3)^{-1} \;=\; (2.86 \times 10^{-4}) \times SF_i \qquad\qquad (7.14)$$

Conversely, the SF_i can be derived from the URF_i as follows:

$$SF_i = (3.5 \times 10^3) \times URF_i \qquad\qquad (7.15)$$

The assumptions used in the above derivations involve a 70-kg body weight, and an average inhalation rate of $20 m^3$/day.

- *Inter-conversions of the Oral Potency Factor.* Risks associated with a unit chemical concentration in water may be estimated in accordance with following mathematical relationship:

Water unit risk

= risk per μg/L (water)

= [slope factor (risk per mg/kg/day)] $\times$ [$\dfrac{1}{\text{body weight (kg)}}$]

$\qquad \times$ ingestion rate (L/day) $\times$ 10^{-3} (mg/μg)

Thus, the ingestion potency can be converted to an oral URF value by applying the following conversion factor:

$[(kg\text{-}day)/mg] \times [1/70kg] \times [2L/day] \times [1mg/1{,}000\mu g] = 2.86 \times 10^{-5}$

Consequently, the lifetime excess cancer risk from ingesting $1\,\mu g/L$ concentration for a full lifetime is:

$$URF_O\,(\mu g/L)^{-1} = (2.86 \times 10^{-5}) \times SF_O \qquad (7.16)$$

Or, alternatively the potency, SF_O, can be derived from the unit risk as follows:

$$SF_O = (3.5 \times 10^4) \times URF_O \qquad (7.17)$$

The assumptions used in the above derivations involve a 70-kg body weight, and an average water ingestion rate of 2 L/day.

7.4.3. THE USE OF SURROGATE TOXICITY PARAMETERS

Risk characterizations generally should consider every likely exposure route in the evaluation process. However, toxicity data may not always be available for each route of concern, in which case the use of surrogate values – that may include extrapolation from data for another exposure route – may be required. Extrapolations may be possible for some cases where there is reliable information on the degree of absorption of materials by both routes of exposure – assuming the substance is not locally more active by one route. In any case, this type of extrapolation can become useful approximations to employ – at least for preliminary risk assessments.

In general, toxicity parameters used in risk assessments are dependent on the route of exposure. However, as an example, oral RfDs and SFs have been used for both ingestion and dermal exposures to some chemicals that affect receptors through a systemic action – albeit this will be inappropriate if the chemical affects the receptor contacts through direct local action at the point of application. In fact, in several (but certainly not all) situations, it is quite appropriate to use oral SFs and RfDs as surrogate values to estimate systemic toxicity as a result of dermal absorption of a chemical (DTSC 1994; USEPA 1989b, 1992). It is noteworthy, however, that direct use of the oral SF or oral RfD does not account for differences in absorption and metabolism between the oral and dermal routes – leading to increasingly uncertain outcomes. Also, in the evaluation of the inhalation pathways, when an inhalation SF or RfD is not available for a compound, the oral SF or RfD may be used in its place – especially for screening-level analyses. Similarly, inhalation SFs and RfDs may be used as surrogates for both ingestion and dermal exposures for those chemicals lacking oral toxicity values – also adding to the uncertainties in the evaluations.

In addition to the uncertainties caused by route differences, further uncertainty is introduced by the fact that the oral dose-response relationships are based on potential (i.e., administered) dose, whereas dermal dose estimates are absorbed doses. Ideally, these differences in route and dose type should be resolved via pharmacokinetic modeling. Alternatively, if estimates of the gastrointestinal absorption fraction are available for the compound of interest in the appropriate vehicle, then the oral dose-response factor, unadjusted for absorption, can be converted to an absorbed dose basis as follows:

$$RfD_{absorbed} = RfD_{administered} \times ABS_{GI} \qquad (7.18)$$

$$SF_{absorbed} = \frac{SF_{administered}}{ABS_{GI}} \qquad (7.19)$$

On the average, absorption fractions corresponding to approximately 10% and 1% are typically applied to organic and inorganic chemicals, respectively.

For the most part, direct toxic effects on the skin have not been adequately evaluated for several chemicals encountered in the human work and living environments. This means that, it may be inappropriate to use the oral slope factor to evaluate the risks associated with a dermal exposure to carcinogens such as benz(a)pyrene, that are believed to cause skin cancer through a direct action at the point of application. Indeed, depending on the chemical involved, the use of an oral SF or oral RfD for the dermal route is likely to result in either an over- or under-estimation of the risk or hazard. Consequently, the use of the oral toxicity value as a surrogate for a dermal value will tend to increase the uncertainty in the estimation of risks and hazards. However, this approach is not generally expected to significantly under-estimate the risk or hazard relative to the other routes of exposure that are evaluated in the risk assessment (DTSC 1994; USEPA, 1992).

In principle, it is generally not possible to extrapolate between exposure routes for some substances that produce localized effects dependent upon the route of exposure. For example, a toxicity value based on localized lung tumors that result only from inhalation exposure to a substance would not be appropriate for estimating risks associated with dermal exposure to the substance. Thus, it may be appropriate to extrapolate dermal toxicity values *only* from values derived for oral exposure. In fact, it is recommended that oral toxicity reference values *not* be extrapolated casually from inhalation toxicity values, although some such extrapolations may be performed on a case-by-case basis (USEPA, 1989b).

Of course, other methods of approach that are quite different from the above may be used to generate surrogate toxicity values. For instance, in some situations, toxicity values to be used in characterizing risks are available only for certain chemicals within a chemical class – and this may require a different evaluation approach. In such cases, rather than simply eliminating those chemicals without toxicity values from a quantitative evaluation, it usually is prudent to group data for such class of chemicals well-defined categories (e.g., according to structure-activity relationships or other similarities) for consideration in the risk assessment. Such grouping should not be based solely on toxicity class or carcinogenic classifications. It must be acknowledged that significant uncertainties will likely result by using this type of approach as well. Hence, if and when this type of grouping is carried out, the rationale should be explicitly stated and adequately documented in the risk assessment summary – emphasizing the fact that the action may have produced over- or under-estimates of the true risk.

Overall, the introduction of additional uncertainties in an approach that relies on surrogate toxicity parameters cannot be over-emphasized – and such uncertainties should be properly documented as part of the overall risk evaluation process.

Route-to-route Extrapolation of Toxicological Parameters

For systemic effects away from the site of entry, an inhalation toxicity parameter, TP_{inh} [mg/m^3], may be converted to an oral value, TP_{oral} [mg/kg-d], or vice versa, by using the following type of relationship (van Leeuwen and Hermens, 1995):

$$TP_{inh} \times IR \times t \times BAF_{inh} = TP_{oral} \times BAF_{oral} \times BW \qquad (7.20)$$

where IR is the inhalation rate [m^3/h]; t is the time [h]; BAF_r is the bioavailability for route r, for which default values should be used if no data exists [e.g., use 1 for oral exposure, 0.75 for inhalation exposure, and 0 (in the case of very low or very high lipophilicity or high molecular weight) or 1 (in the case of intermediate lipophilicity and low molecular weight) for dermal exposure]; and BW is the body weight [kg].

A dermal toxicity parameter for systemic effects, TP_{derm} [mg/kg-d] can also be derived from the TP_{oral} [mg/kg-d] or the TP_{inh} [mg/m^3] values as follows (van Leeuwen and Hermens, 1995):

$$TP_{derm} = TP_{oral} \times \frac{BAF_{oral}}{BAF_{derm}} \qquad (7.21)$$

$$TP_{derm} = \frac{TP_{inh} \times IR \times t}{BW} \times \frac{BAF_{inh}}{BAF_{derm}} \qquad (7.22)$$

It is noteworthy that, route-to-route extrapolation introduces additional uncertainty into the overall risk assessment process; such uncertainty can be reduced by utilizing physiologically-based pharmacokinetic (PB-PK) models. Indeed, PB-PK models are particularly useful for predicting disposition differences due to exposure route differences, if sufficient pharmacokinetic data is available (van Leeuwen and Hermens, 1995).

Examples of the Route-to-route Extrapolation Process. As an illustrative example, some of the more common processes involved in route-to-route extrapolation exercises are provided below for both the carcinogenic and non-carcinogenic effects of chemical constituents.

- *Non-carcinogenic Effects.* Oftentimes, reference concentrations (RfCs) for inhalation exposures are extrapolated from oral reference doses (RfDs) for adults by using the following relationship:

$$\text{Extrapolated RfC [mg/m}^3] = RfD_{oral} \text{ [mg/kg-day]} \times \frac{70 \text{ [kg]}}{20 \text{ [m}^3/\text{day]}} \qquad (7.23)$$

It should be noted here, however, that for this simplistic approximation, dosimetric adjustments have not been made to account for respiratory tract deposition efficiency and distribution; physical, biological, and chemical factors; and other aspects of exposure (e.g., discontinuous exposures) that affect chemical uptake and clearance (USEPA, 1996). Consequently, this simple extrapolation method relies on the implicit assumption that the route of administration is irrelevant to the dose delivered to a target organ – an assumption not supported by the principles of dosimetry or pharmacokinetics.

- *Carcinogenic Effects.* For carcinogens, unit risk factors (URFs) for inhalation exposures may be extrapolated from oral carcinogenic slope factors (SFs) for adults by using the following relationship (assuming a 100% absorption via inhalation):

Extrapolated URF $[(\mu g/m^3)^{-1}]$

$$= SF_{oral} [(mg/kg\text{-}day)^{-1}] \times \frac{20 \ [m^3/day]}{70 \ [kg]}$$

$$\times [\text{inhalation absorption rate, } 100\%] \times 10^{-3} \ [mg/\mu g] \qquad (7.24)$$

Using the extrapolated URF, risk-specific air concentrations can be calculated as a lifetime average exposure concentration, as follows:

$$\text{Extrapolated air concentration } [\mu g/m^3] = \frac{\text{Target risk [e.g. } 10^{-6}]}{\text{URF } [(\mu g/m^3)^{-1}]} \qquad (7.25)$$

In general, inhalation values should *not* be extrapolated from oral values – if at all avoidable. Even so, situations arise when it becomes necessary to rely on such approximations to make effective environmental and public health risk management decisions.

Toxicity Equivalence Factors
A toxicity equivalence factor (TEF) procedure is one used to derive quantitative dose-response estimates for substances that are members of a certain category or class of agents. In fact, the TEF approach has been extensively used for the hazard assessment of different classes of toxic chemical mixtures. The assumptions implicit in the utilization of the TEF approach include the following significant ones (NATO/CCMS, 1988; Safe, 1998):

- The individual compounds all act through the same biologic or toxic pathway;
- The effects of individual chemicals in a mixture are essentially additive at sub-maximal levels of exposure;
- The dose-response curves for different congeners should be parallel; and
- The organotropic manifestations of all congeners must be identical over the relevant range of doses.

TEFs are based on shared characteristics that can be used to order the class members by carcinogenic potency when cancer bioassay data are inadequate for this purpose. The ordering is by reference to the characteristics and potency of a well-studied member or members of the class. Other class members are then indexed to the reference agent(s) by using one or more shared characteristics to generate their TEFs. Examples of shared characteristics that may be used include receptor-binding characteristics; results of biological activity assays related to carcinogenicity; or structure-activity relationships.

The TEFs are usually indexed at increments of a factor of 10. Very good data, however, may permit a smaller increment to be used.

To date, adequate data to support the use of TEFs has been found in only a limited class of compounds – most prominently, the dioxins (USEPA, 1989c). Dioxin and dioxin-like compounds are structurally related groups of chemicals from the family of halogenated aromatic hydrocarbons. Depending on the number of chlorine-substituted positions, there are several congeners in each group. Apparently, the most toxic and the most studied congener is TCDD. Thus, TEFs have been developed to compare the relative toxicity of individual dioxin-like compounds to that of TCDD; the TEFs are numerical factors that express the toxicity of an individual PCDD or PCDF relative to the toxicity of TCDD, the highly toxic and best studied among the 210 congeners. This comparison is based on the assumption that dioxin and dioxin-like compounds act through the same mechanism of action. The TEF for TCDD is set at one (1), whereas TEF values for all other dioxin-like compounds are less than one. Ultimately, the toxicity equivalent (TEQ) of TCDD (or the toxicity equivalency concentration) is calculated by multiplying the exposure level of a particular dioxin-like compound by its TEF. The TEQs are used to assess the risk of exposure to a mixture of dioxin-like compounds. A TEQ is defined by the product of the concentration, C_i, of an individual 'dioxin-like compound' in a complex environmental mixture (i.e., concentration of the i-th congener) and the corresponding TCDD toxicity equivalency factor (TEF_i) for that compound (relative to 2,3,7,8-Tetrachlorodibenzo-p-dioxin, i.e., 2,3,7,8-TCDD). The total TEQs is the sum of the TEQs for each of the congeners in a given mixture, viz.:

$$\text{Total TEQs} = \sum_{i=1}^{n} (C_i \times TEF_i) \tag{7.26}$$

TEFs are based on congener-specific data and the assumption that the toxicity of dioxin-like chemicals is additive. The TEF scheme compares the relative toxicity of individual dioxin-like compounds to that of TCDD, which is the most toxic halogenated aromatic hydrocarbon. Ultimately, the total TCDD equivalents (TEQ) is then used in conjunction with a cancer potency or reference exposure level to estimate cancer risk or non-cancer hazard index, respectively.

It is noteworthy that, even though they are distinctly different compounds, some PCBs also exhibit dioxin-like toxicity. Thus, a similar approach that utilizes the TEF concept for PCB congeners can also be found in the literature. When PCB congener concentrations are available, the usual PCB slope-factor approach can be supplemented by analysis of dioxin TEQs – in order to evaluate the dioxin-like toxicity. Risk from the dioxin-like congeners is evaluated using TEFs – and this is be added to risks from the rest of the mixture.

A basic premise of the TEF methodology is the presence of a common biologic end-point, or in the case of multiple end-points, a common mechanism of action. A second assumption is the additivity of effects. In fact, these assumptions are inherent in all TEF-schemes – and thus, the accuracy of all TEF-schemes will be affected by situations where such assumptions are not applicable. It is also noteworthy that, for more complex mixtures containing compounds that act through multiple pathways to give both similar and different toxic responses, the TEF/TEQ approach may *not* be appropriate (Safe, 1998).

International TEFs (ITEFs). TEF schemes for PCDDs and PCDFs have been developed or adopted by many governmental institutions throughout the industrialized world. Of special interest, the International TEFs (ITEFs) scheme was the result of an international conference, convened for the purpose of reaching a consensus for a uniform TEF scheme based on available whole animal non-cancer and cancer data, short term exposures, and *in vitro* data (NATO/CCMS, 1988a,b).

To incorporate the results from many studies, the ITEF scheme used the assumption that all effects are initiated by the interaction of a PCDD or PCDF congener with a specific receptor protein, the *Ah* receptor. The rationale for this assumption comes from analyses in which biologic endpoints were found to be highly correlated with known *Ah* receptor associated effects. The ITEF scheme focuses on those congeners that are preferentially absorbed and accumulated in mammalian tissue over a long period of time and exhibit a similar spectrum of toxicities as 2,3,7,8 TCDD – namely, PCDDs/PCDFs in which positions 2-,3-,7-, and 8- are substituted by chlorine.

The choice of an ITEF for each 2,3,7,8-PCDD/PCDF congener has been based on a synthesis of data from cancer studies, long-term toxicity studies, subchronic effects, acute toxicity studies, and receptor binding and enzyme induction. Greater weight has been given to results from long-term studies, but information of short-term studies has also been considered. In the absence of long-term studies, data from short-term whole animal and/or *in vitro* studies have been used. A specific formula was not applied to the various data – but rather the final individual ITEF was based on the professional judgment of the aggregate data available for the individual congener.

Relative Potency Factors
Another variation of the TEF concepts described above is often encountered within other families of compounds. For example, although several polycyclic aromatic hydrocarbons (PAHs) have been classified as probable human carcinogens (viz., USEPA Group B or IARC Group 2A), cancer slope factors for this family is often associated with only benzo[*a*]pyrene (BaP). Consequently, quantitative risk estimates for PAH mixtures have often assumed that all carcinogenic PAHs are equipotent to BaP. A preferred approach, however, involves using an estimated order of potential potency relative to BaP. This allows a potency-weighted total concentration to be calculated, as follows:

$$PEC = \sum_{i=1}^{n} (RP_i \times C_i) \tag{7.27}$$

where *PEC* is the potency equivalency concentration; RP_i is the relative potency for the *i*-th PAH; and C_i is the concentration of the *i*-th PAH.

Indeed, PAHs seem almost ubiquitous in most environmental settings – and therefore, it is important to be able to accurately assess the risks from exposure to this family of compounds. However, there is *not* an adequate scientific data on which to base a cancer SF for all PAHs, other than BaP. Consequently, the risk for PAHs has often been based on assessing all the PAHs found in an environmental sampling analysis by using the SF for BaP. Since BaP is one of the two most potent PAHs (with dibenz[*a,h*]anthracene being the other), assessing all PAHs as if they were BaP greatly overestimates risks. The use of the relative potencies for PAHs, therefore, offers great improvements in ensuring that risks are better represented. Yet still, it must be cautioned that the applications of TEFs for PAHs usually require a more detailed knowledge of the complete composition of these mixtures as well as the TEFs of all active components. Thus, the approach may be most useful for defined PAH mixtures containing only parent hydrocarbons (Safe, 1998).

7.5. Mechanisms of Action and the Determination of Human Health Hazard Effects

Chemical hazard effects evaluation is generally conducted as part of a risk assessment in order to, qualitatively and/or quantitatively, determine the potential for adverse effects from receptor exposures to chemical stressor(s). For most chemical exposure problems, this usually will comprise of intricate toxicological evaluations, the ultimate goal of which is to derive reliable estimates of the amount of chemical exposure that may be considered 'tolerable' (or 'acceptable' or 'reasonably safe') for humans. The relevant toxicity parameters that are generated during this process usually will be dependent on the mechanism and/or mode of action for the particular toxicant, on the receptor of interest.

Mechanism of action is defined as the complete sequence of biological events that must occur to produce an adverse effect. In cases where only partial information is available, the term *mode of action* is used to describe only major (but not all) biological events that are judged to be sufficient to inform about the shape of the dose-response curve beyond the range of observation. For effects that involve the alteration of genetic material (e.g., most cancers, heritable mutations), there are theoretical reasons to believe that such a mode of action would not show a threshold or dose below which there are no effects. On the other hand, a threshold is widely accepted for most other health effects, based on considerations of compensatory homeostasis and adaptive mechanisms. The threshold concept presumes that there exists a range of chemical exposures (from zero up to some finite value) that can be tolerated by an individual without adverse effects. Accordingly, different approaches have traditionally been used to evaluate the potential carcinogenic effects and also health effects other than cancer (namely, 'non-cancer' effects).

Typically, public health risk assessments for chemical exposure problems rely heavily on 'archived' toxicity indices/information developed for specific chemicals. A summary listing of such toxicological parameters – represented by the cancer SF (for carcinogenic effects) and the RfD (for non-cancer effects) – is provided in Table C.1 (Appendix C) for some representative chemicals commonly found in consumer products and/or in the human living and work environments. A more complete and up-to-date listing may be obtained from a variety of toxicological databases – such as the Integrated Risk Information System (IRIS) database (developed and maintained by the US EPA); the International Register of Potentially Toxic Chemicals (IRPTC) database (from UNEP); the International Toxicity Estimates for Risks (ITER) database (from TERA); etc. (see Appendix B). Where toxicity information does not exist at all, a decision may be made to estimate toxicological data from that of similar compounds (i.e., with respect to molecular weight and structural similarities; etc.).

7.6. Suggested Further Reading

ATSDR, 1997. *Technical Support Document for ATSDR Interim Policy Guideline: Dioxin and Dioxin-like Compounds in Soil*, Agency for Toxic Substances and Disease Registry (ATSDR), US Department of Health and Human Services, Atlanta, GA

Crosby, DG, 1998. *Environmental Toxicology and Chemistry*, Oxford University Press, New York, NY

Klaassen, CD, 2002. Xenobiotic transporters: another protective mechanism for chemicals, *International Journal of Toxicology*, 21(1): 7 – 12

NRC (National Research Council), 2002. *Scientific Frontiers in Developmental Toxicology and Risk Assessment*, National Academy Press, Washington, DC

Ottoboni, MA, 1997. *The Dose Makes the Poison (A Plain Language Guide to Toxicology)*, 2nd edition, Van Nostrand Reinhold, ITP, New York

van Leeuwen, FXR, 1997. Derivation of toxic equivalency factors (TEFs) for dioxin-like compounds in humans and wildlife, *Organohalogen Componds*, 34: 237 – 269

Chapter 8

CHEMICAL RISK CHARACTERIZATION
AND UNCERTAINTY ANALYSES

Fundamentally, risk characterization consists of estimating the probable incidence of adverse impacts to potential receptors, under the various exposure conditions associated with a chemical hazard situation. It involves an integration of the hazard effects and exposure assessments – in order to arrive at an estimate of the health risk to the exposed population. In fact, all information derived from each step of a chemical exposure/risk assessment is integrated and utilized during the risk characterization – to project the degree and severity of adverse health effects in the populations potentially at risk.

Risk characterization is indeed the final step in the risk assessment process, and it becomes the first input into risk management programs. Thus, risk characterization serves as a bridge between risk assessment and risk management – and is therefore a key factor in the ultimate decision-making process that is developed to address chemical exposure problems. Through probabilistic modeling and analyses, uncertainties associated with the risk evaluation process can be assessed properly and their effects on a given decision accounted for systematically. In this manner the risks associated with given decisions may be delineated and then appropriate corrective measures taken accordingly. This chapter elaborates the mechanics of the risk characterization process, together with an elaboration of the uncertainties that surround the overall process.

8.1. Fundamental Issues and Considerations Affecting the Risk Characterization Process

The risk characterization process consists of an integration of the toxicity and exposure assessments – resulting in a quantitative estimation of the actual and potential risks and/or hazards associated with a chemical exposure problem. Broadly stated, risk from human exposure to chemicals is a function of dose or intake and potency, viz.:

Risk from chemical exposure = [*Dose of chemical*] x [*Chemical potency*]   (8.1)

During risk characterization, chemical-specific toxicity information is compared against both field measured and estimated chemical exposure levels (and in some cases, those levels predicted through fate and behavior modeling) in order to determine whether concentrations associated with a chemical exposure problem are of significant concern. The process also considers the possible additive effects of exposure to mixtures of the chemicals of potential concern.

Depending on the nature of populations potentially at risk from a chemical exposure problem, different types of risk measures or parameters may be employed in the risk characterization process. Typically, the cancer risk estimates and hazard quotient/hazard index estimates are the measures of choice used to define potential risks to human health (see Sections 8.2 and 8.3) – and that is required to support effectual public health risk management programs. Consequently, the health risks to potentially exposed populations resulting from chemical exposures are characterized through a calculation of non-carcinogenic hazard quotients and indices and/or carcinogenic risks (CAPCOA, 1990; CDHS, 1986; USEPA, 1986a, 1989a).

In general, risks associated with the inhalation and non-inhalation pathways may be estimated in accordance with the following generic relationships:

Risk for inhalation pathways

$$= \text{Ground-level concentration (GLC)} \ [\mu g/m^3] \times \text{Unit risk} \ [(\mu g/m^3)^{-1}] \qquad (8.2)$$

Risk for non-inhalation pathways

$$= \text{Dose} \ [mg/kg\text{-}day] \times \text{Potency slope} \ [(mg/kd\text{-}day)^{-1}] \qquad (8.3)$$

These measures can then be compared with benchmark criteria/standards in order to arrive at risk decisions about a chemical exposure problem.

It is noteworthy that, any risk assessment should clearly delineate the strengths and weaknesses of the data, the assumptions made, the uncertainties in the methodology, and the rationale used in reaching the conclusions (e.g., similar or different routes of exposure, and metabolic differences between humans and test animals). Furthermore, the hazard and risk assessment of human exposure to chemicals must also take into account scenarios where chemical interactions may significantly influence toxic outcomes. Chemical interactions are indeed very important determinants in evaluating the potential hazards and risks of exposure to chemical mixtures (Safe, 1998).

Ultimately, an effective risk characterization should fully, openly, and clearly characterize risks and disclose the scientific analyses, uncertainties, assumptions, and science policies that underlie decisions throughout the risk assessment and risk management processes. Also, a health risk assessment/characterization is only as good as its component parts – i.e., the hazard characterization, the dose-response analysis, and the exposure assessment. Confidence in the results of a risk assessment is thus a function of the confidence in the results of the analysis of these key elements, and indeed their corresponding ingredients. Overall, several important issues usually will have very significant bearing on the processes involved in completing risk characterization tasks designed to support effective public health risk management programs; the particularly important topics are discussed below.

8.1.1. CORRECTIONS FOR 'NON-STANDARD' POPULATION GROUPS

In the risk estimation process, the exposure information is combined with dose-response information. In the processes involved, care must be taken to ensure that the assumptions about population parameters in the dose-response analysis are consistent with the population parameters used in the exposure analysis; procedures for insuring such consistency is provided elsewhere in the literature elsewhere (e.g., USEPA, 1997; West et al., 1997). In general, when the population of interest is different in comparison with the 'standard' population assumed in the dose-response assessments, then the dose-response parameter may need to be adjusted. Furthermore, when the population of interest is different from the population from which the often-used default exposure factors were derived, then the exposure factor may need to be adjusted. A good example of a non-standard sub-population would be a sedentary hospital population with lower than $20m^3$/day air intake rates (as is often assumed for most 'standard' population groups). Also, an example of such a sub-population relates to mean body weight (that is different from the assumed standard of 70kg); for instance, females usually may be assumed to have an average body weight of 60kg, and also children's body weights will be dependent on their age.

As an example of the requisite procedures for modifying standard parameters for non-standard populations, consider a recommended value for the average consumption of tap water by adults in a population group to be 1.4 liters per day. Assume the drinking water unit risk for chemical X is 8.3 x 10^{-6} per μg/L, and that this was calculated from the slope factor assuming the standard intake, I_{ws}, of 2 liters per day. Then, for the population group drinking 1.4 liters of tap water per day, the corrected drinking water unit risk should be (USEPA, 1997):

$$[8.3 \times 10^{-6}] \times \left[\frac{1.4}{2}\right] = 5.8 \times 10^{-6} \text{ per μg/L}$$

Subsequently, the risk to the average individual is then estimated by multiplying this value by the average concentration (in units of μg/L).

Next, if the body weight, W^p, of the population of interest differs from the body weight, W^s, of the population from which the standard exposure values were derived, then a modeling adjustment may have to be made in estimating the intake of food, water, and air in this population (USEPA, 1997; West et al., 1997). Another example using the procedures provided by the US EPA (USEPA, 1997) involves estimating the risk specifically for women drinking the water contaminated with chemical X. If it is assumed that the women have an average body weight of 60kg, the correction factor for the drinking water unit risk (disregarding the correction discussed above with respect to consumption rate) is:

$$\left[\frac{70}{60}\right]^{2/3} = 1.11$$

Thus, the corrected water unit risk for chemical X is:

$$[8.3 \times 10^{-6}] \times [1.11] = 9.2 \times 10^{-6} \text{ per μg/L}$$

As indicated previously, the risk to the average individual is subsequently estimated by multiplying this value by the water concentration.

8.1.2. ADJUSTMENTS FOR CHEMICAL ABSORPTION:
ADMINISTERED *VS.* ABSORBED DOSE

Absorption adjustments may be necessary to ensure that the exposure estimate and the toxicity value being compared during the risk characterization are both expressed as absorbed doses, or both expressed as administered doses (i.e., intakes). Adjustments may also be necessary for different vehicles of exposure (e.g., water, food, or soil) – although, in most cases, the unadjusted toxicity value will provide a reasonable or conservative estimate of risk. Furthermore, adjustments may be needed for different absorption efficiencies, depending on the medium of exposure. In particular, correction for fractional absorption is generally appropriate when interaction with environmental media or other chemicals may alter absorption from that expected for the pure compound. Correction may also be necessary when assessment of exposure is via a different route of contact than what was utilized in the experimental studies used to establish the toxicity parameters (i.e., the SFs and RfDs from Chapter 7).

In general, only limited toxicity reference values exist for dermal exposure; consequently, oral values are frequently used to assess risks from dermal exposures (USEPA, 1989d). On the other hand, most RfDs and some carcinogenic SFs usually are expressed as the amount of substance administered per unit time and unit body weight, *whereas* exposure estimates for the dermal route of exposure are eventually expressed as absorbed doses. Thus, for dermal exposures, it may become particularly important to adjust an oral toxicity value from an administered to an absorbed dose – generally carried out as indicated below (USEPA, 1989d).

- *Adjustment of an administered dose to an absorbed dose RfD.* The administered dose RfD (RfD_{adm}) of a chemical with oral absorption efficiency, ABS, in the species on which the RfD is based, may be adjusted to an absorbed dose RfD (RfD_{abs}). This is achieved by simply multiplying the unadjusted RfD by the absorption efficiency percent – i.e., RfD_{abs} = RfD_{adm} x ABS, which can be compared with the amount estimated to be absorbed dermally.

- *Adjustment of an administered dose to an absorbed dose SF.* The administered dose SF (SF_{adm}) of a chemical with oral absorption efficiency, ABS, in the species on which the SF is based, may be adjusted to an absorbed dose SF (SF_{abs}). This is achieved by simply dividing the unadjusted SF by the absorption efficiency percent – i.e., $SF_{abs} = \dfrac{SF_{adm}}{ABS}$, which can be used to estimate the cancer risk associated with the estimated absorbed dose for the dermal route of exposure.

- *Adjustment of an exposure estimate to an absorbed dose.* If the toxicity value is expressed as an absorbed rather than an administered dose, then it may become necessary to convert the exposure estimate from an intake into an absorbed dose for comparison. The unadjusted exposure estimate or intake (CDI_{adm}) of a chemical with absorption efficiency, ABS, may be converted to an 'adjusted exposure' or absorbed dose (CDI_{abs}). This is achieved by simply multiplying the unadjusted CDI by the absorption efficiency percent – i.e., CDI_{abs} = CDI_{adm} x

ABS, which can be used in comparisons with the RfD or SF that is based on an absorbed, not administered dose.

It is noteworthy that, for evaluations of the dermal exposure pathway, if the oral toxicity value is already expressed as an absorbed dose, it is not necessary to adjust the toxicity value. Also, exposure estimates should not be adjusted for absorption efficiency if the toxicity values are based on administered dose. Furthermore, in the absence of reliable information, 100% absorption is usually used for most chemicals; for metals, an approximately 10% absorption may be considered a reasonable upper-bound for other than the inhalation exposure route.

Absorption efficiency adjustment procedures are discussed elsewhere in the literature (e.g., USEPA 1989d, 1992). In general, absorption factors should not be used to modify exposure estimates in those cases where absorption is inherently factored into the toxicity/risk parameters used for the risk characterization. Thus, 'correction' for fractional absorption is appropriate only for those values derived from experimental studies based on absorbed dose. That is, absorbed dose should be used in risk characterization only if the applicable toxicity parameter (e.g., SF or RfD) has been adjusted for absorption; otherwise, intake (unadjusted for absorption) are used for the calculation of risk levels.

8.1.3. AGGREGATE EFFECTS OF CHEMICAL MIXTURES AND MULTIPLE EXPOSURES

Oftentimes in the study of human exposures to chemical hazards, there is the need to undertake aggregate and cumulative exposure and risk assessments. Indeed, it is quite important to consider both aggregate and cumulative exposures in the making effectual risk assessment and risk management decisions – as well in the process of setting chemical tolerance or safe levels for human exposures. In general, aggregate exposures may occur across different pathways and media that contribute to one or more routes of an individual receptor's exposure – which then becomes the basis for determining cumulative risks. Cumulative risk refers to effects from chemicals that have a common mode of toxicological action – and thus have aggregate exposure considerations as part of the assessment process (Clayton, et al., 2002).

While some chemical hazard situations involve significant exposure to only a single compound, most instances of chemical exposure problems can involve concurrent or sequential exposures to a mixture of compounds that may induce similar or dissimilar effects over exposure periods ranging from short-term to a lifetime (USEPA, 1984a, 1986b). Evaluating mixtures of chemicals is indeed one of the areas of risk assessment with obvious uncertainties; this is especially so, because several types of interactions in chemical mixtures are possible, including the following:

- Additive – i.e., the effects of the mixture equals that of adding the effects of the individual constituents
- Synergistic – i.e., the effects of the mixture is greater than obtained by adding the effects of the individual constituents
- Antagonistic – i.e., the effects of the mixture is less than obtained by adding the effects of the individual constituents

Of particular concern are those mixtures where the effects are synergistic. Unfortunately, the toxicology of complex mixtures is not very well understood – complicating the problem involved in the assessment of the potential for these compounds to cause various health effects. Nonetheless, there is the need to assess the cumulative health risks for the chemical mixtures, despite potential large uncertainties that may exist. The risk assessment process must, therefore, address the multiple endpoints or effects, and also the uncertainties in the dose-response functions for each effect.

In fact, potential receptors are typically exposed not to isolated chemical sources, but rather to a complex, dilute mixture of many origins. Considering how many chemicals are present in the wide array of consumer products, and in the human environments, there are virtually infinite number of combinations that could constitute potential synergisms and antagonisms. In the absence of any concrete evidence of what the interactive effects might be, however, an additive method that simply sums individual chemical effects on a target organ is usually employed in the evaluation of chemical mixtures.

Finally, in combining multi-chemical risk estimates for multiple chemical sources, it should be noted here that, if two sources do not affect the same individual or subpopulation, then the sources' individual risk estimates (and/or hazard indices) do not affect each other – and, therefore, these risks should not be combined. Thus, one should not automatically sum risks from all sources evaluated for a chemical exposure problem – unless if it has been determined/established that such aggregation is appropriate.

Carcinogenic Chemical Effects
The common method of approach in the assessment of chemical mixtures assumes additivity of effects for carcinogens when evaluating multiple carcinogens – albeit alternative procedures that are more realistic and/or less conservative have been proposed for certain situations by some investigators (e.g., Bogen, 1994; Chen *et al.*, 1990; Gaylor and Chen, 1996; Kodell and Chen, 1994; Slob, 1994). In any case, prior to the summation of aggregate risks, estimated cancer risks should preferably be segregated by weight-of-evidence (or strength-of-evidence) category for the chemicals of concern – the goal being to provide a clear understanding of the risk contribution of each category of carcinogen.

Systemic (Non-cancer) Chemical Effects
For multiple chemical exposures to non-carcinogens and the non-carcinogenic effects of carcinogens, constituents should be grouped by the same mode of toxicological action (i.e., those that induce the same physiologic endpoint – such as liver or kidney toxicity). Cumulative non-carcinogenic risk is evaluated through the use of a hazard index that is generated for each health or physiologic 'endpoint'.

Physiologic/toxicological endpoints that will normally be considered with respect to chronic toxicity include: cardiovascular systems (CVS); central nervous system (CNS); gastrointestinal (GI) system; immune system; reproductive system (including teratogenic and developmental effects); kidney (i.e., renal); liver (i.e., hepatic); and the respiratory system. Listings of chemicals with their associated non-carcinogenic toxic effects on specific target organ/system can be found in such databases as IRIS (Integrated Risk Information System), as well as in the literature elsewhere (e.g., Cohrssen and Covello, 1989a; USEPA, 1996).

In fact, in a strict sense, constituents should not be grouped together unless they have the same toxicological/physiologic endpoint. Thus, in a well-defined risk characterization exercise, it becomes necessary to segregate chemicals by organ-specific toxicity – since strict additivity without consideration for target-organ toxicities could over-estimate potential hazards (USEPA 1986b, 1989d). Consequently, the 'true' hazard index is preferably calculated only after putting chemicals into groups with same physiologic endpoints.

8.1.4. FUNDAMENTAL CONSIDERATIONS IN THE HEALTH ASSESSMENT OF CARCINOGENS

Cancer risk assessment by necessity involves a number of assumptions – most of which reflect scientific and policy judgments. In general, in the absence of data to the contrary, a substance that has been shown to cause cancer in animals is presumed to pose a potential carcinogenic risk to humans. However, as more knowledge on particular agents and the oncogenic process in general becomes available, the position on these issues becomes subject to change. A number of fundamental but critical issues affecting the health risk assessment of carcinogens are enumerated below.

Qualitative Issues
Several qualitative issues affect the health assessment of carcinogens – most importantly, the issues identified below (IARC, 1987; NTP, 1991;USEPA, 1986).

- *'Weight of Evidence'*. A weight-of-evidence or a strength-of-evidence approach may be adopted in evaluating all the relevant case data. The types of evidence that may be used for qualitatively identifying carcinogens include case studies, epidemiological studies, long-term animal bioassays, short-term tests, and structure-activity relationships.

  Specific factors that are typically evaluated in determining if a substance poses a carcinogenic risk to humans include, but are not limited to, the quality of the toxicity studies (namely, relating to the choice of appropriate control groups; sufficient number of animals; administration route; dose selection; tumor types; etc.), and the relevance of animal data to humans. A narrative statement may be used to incorporate the weight/strength-of-evidence conclusions – i.e., in lieu of alphanumeric designations alone being used to convey qualitative conclusions regarding the chemical carcinogenicity.

- *Mechanistic Inference and Species Concordance.* Carcinogenesis is generally viewed as a multistage process – proceeding from initiation, through promotion, and progression. Carcinogens may work through mechanisms that directly or indirectly affect the genome. Currently, it is usually assumed that many or most carcinogens are characterized by the absence of a threshold in eliciting a tumorigenic response. However, the presence or absence of a threshold for one step in the multistage process of carcinogenesis does not necessarily imply the presence or absence of a threshold for other steps or the entire process. For example, carcinogenic effects of some agents may result from non-physiological responses to the agents, such as extensive organ damage. Under such circumstances, the relevance of the animal data to humans should be evaluated on a case-by-case basis

– with a view towards extending its assessment effort beyond the dominant paradigm of carcinogenesis (i.e., initiation, promotion, and progression).

- *Exposure Route Specificity.* In the analysis of potential carcinogenic risk of chemical agents to humans, it is important to address the issue of exposure route specificity. For some agents, exposure results in adverse health effects via one route only. For example, while chronic oral exposures to an agent may not result in cancer in animals and/or humans, the same agent may be carcinogenic via inhalation in the same species. Accordingly, the potential health risk of toxic substances should be evaluated by taking into account the relevant route(s) of exposure. In the absence of data to the contrary, an agent that is carcinogenic via one route may be considered to be a potential carcinogen via alternate routes as well.

- *Role of Epidemiological Data.* Epidemiological studies provide direct information on the carcinogenic risk of chemical agents to humans. For this reason, in evaluating the potential human cancer risks, a higher weight may be assigned to well-designed and well-executed epidemiological studies than to animal studies of comparable quality. Notwithstanding, the observational nature of such studies, as well as the use of indirect measures of exposure, sometimes constrains interpretation of the data.

    In general, descriptive epidemiological studies may be useful in generating/refining hypotheses that suggest further in-depth studies. These studies also provide limited information on causal relationships. Alternatively, analytical epidemiological investigations, such as case-control or cohort studies, can provide the basis for testing causal associations – and are an invaluable resource in public health decisions. The causal association of toxic chemical exposure and cancer is greatly enhanced when studies show: relationships without significant bias, a temporal sequence of exposure and response, consistency with other studies, strength of association, a dose- response relationship, and biologic plausibility.

    It is noteworthy that, although an agent may not have been shown to be a carcinogen in a well-designed epidemiological study, a potential association between exposure to the agent and human cancer cannot be completely ruled out. The potential for an association will remain – especially if relevant animal data suggest that a carcinogenic effect exist. This premise would also apply in the case of health effects other than cancer.

- *Sensitive and Susceptible Populations.* Certain populations may be at a higher risk of developing cancer due to several factors – including exposure to unusually high levels of carcinogens, genetic predisposition, age, and other host factors (such as physiological and nutritional status). Thus, it is quite important to carefully identify these susceptible populations and independently address the associated public health concerns for the particularly sensitive group(s).

- *Structure-Activity Relationships.* Information on the physical, chemical, and toxicological characteristics as well as the environmental fate of many hazardous substances exist within the scientific communities. Thus, some correlation can be made between the structures of some hazardous substances and the properties they exhibit. Indeed, the use of structure-activity relationships to derive preliminary estimates of both the environmental and toxicological characteristics of hazardous substances for which little or no information is available could become very crucial in some risk characterization programs. However, a great deal of scientific judgment may be required in interpreting these results, since these methods may need to be refined and validated. Also, conclusions derived by such approaches may be inadequate as surrogates for human or other bioassay data.

- *Chemical Interactions.* Health evaluations are often complicated by the fact that multiple hazardous substances may be of concern at specific locales. Given the paucity of empirical data and the complexity of this issue, it is often assumed that, in the absence of information regarding the interaction of these substances, their effects are additive. In any case, such assessments should also be accompanied by qualitative weight-of-evidence type of statement on the possibilities for interactive effects, whether they are potentiation, additivity, antagonism, and/or synergism. Ideally, these conclusions are based on insights regarding the mechanism of action of individual components – as relates to the potential for interaction among components of the mixture.

Quantitative Issues

Several quantitative issues affect the health assessment of carcinogens – most importantly, the issues identified below (IARC, 1987; NTP, 1991;USEPA, 1986).

- *Dose Scaling.* Conversion of exposure levels derived from experimental animal studies to humans is an equivocal process because of recognized differences among species – e.g., life span and body size, as well as pharmacokinetic and genetic factors, among others. Although a number of default scaling factors have been proposed in various scientific works, no single scaling approach may be considered as being universally appropriate. Indeed, the use of any default approach to scaling is at best a crude approximation, and all factors responsible for interspecies differences must be considered in dose/exposure conversions among species when selecting extrapolation methods. Thus, empirically derived data relevant to dose scaling are preferred – and this should be used preferentially, whenever available. It is noteworthy that, extrapolation may not be necessary if epidemiological data are used to assess potential carcinogenic risk; however, differences in individual sensitivity must still be taken into account.

- *Pharmacokinetics and Pharmacodynamics.* It is quite important to carry out health assessments in populations that have been exposed to carcinogens in the past or that are currently exposed to these agents. In assessing the potential carcinogenic risks of chemical agents, information on the 'delivered' target dose rather than the exposure dose may help in developing a more accurate assessment of the possible carcinogenicity of an agent. Thus, the development and use of physiologically-based pharmacokinetic models for estimating the magnitude and time course of exposure to agents at target sites in animal models may be an important exercise. Once data from the animal models have been appropriately validated, they can then be used to estimate corresponding target tissue doses in humans. Furthermore, it should be recognized that the estimation of lifetime cancer risks is further complicated when available data are derived from less-than-lifetime exposures, and that pharmacokinetic insights may be of great help in addressing this issue.

- *Mechanistic Considerations and Modeling.* Health assessment for potential carcinogens must take into consideration dose-response relationships from all available relevant studies. In chronic bioassays, animals are often exposed to levels of the chemical agent that are, for practical reasons, far higher than levels to which humans are likely to be exposed in the environment. Therefore, mathematical models are used to extrapolate from high to low dose. The selection of models depends on the known or presumed mechanism of action of the agent, and on science policy considerations. In the absence of sufficient information to choose among several equally plausible models, preference should perhaps be given to the more conservative (i.e., protective) models.

  The multistage model is widely used for low-dose extrapolation for genotoxic agents. It is based on the premise that a developing tumor proceeds through several different stages before it is clinically detectable. In the low-dose region, this multistage model is frequently linear, and it is assumed that a threshold, below which effects are not anticipated, does not exist.

  It must be recognized that no single mathematical model is appropriate in all cases, and that the incorporation of new information on mechanism and pharmacokinetics, among other factors, will increase the model's usefulness and facilitate the selection of the most appropriate mathematical model. It must be acknowledged, however, that existing mathematical models for low-dose extrapolation may not quite be appropriate for nongenotoxic agents. Indeed, more information on biological mechanism is needed to determine if there are threshold exposure levels for nongenotoxic agents. For these reasons, where feasible, the presentation of a range of plausible potency estimates should be used to convey quantitative conclusions.

- *Individual vs. Population Risk – The Role of Molecular Epidemiology.* Recent advances in biomolecular technology have resulted in the development of highly sensitive methods for measuring biomarkers of exposure, effects, and susceptibility (Shields and Harris, 1991; Johnson and Jones, 1992). Biomarkers have the potential to serve as bridges between experimental and epidemiological studies of carcinogens, insofar as they reflect biochemical or molecular changes associated with exposure to carcinogens.

  Biomarkers, such as DNA adducts, may be used as indices of the biologically effective doses – reflecting the amount of the potential carcinogen or its metabolite that has interacted with a cellular macromolecule at the target site. Furthermore, markers of early biologic effect, such as activated oncogenes and their protein products, and/or loss of suppressor gene activity, may indicate the occurrence of possibly irreversible toxic effects at the target site. Also, genetic markers may suggest the presence of heritable predispositions or the effects of other host factors, such as lifestyle or prior disease. Thus, molecular epidemiology – that combines experimental models, molecular biology, and epidemiology – provides an opportunity to estimate individual cancer risk, and to better define the health implications of chemical exposure problems for members of exposed populations

(NRC, 1991). It should be noted, however, that extensive work is needed before biomarkers can be used as prognostic indicators.

8.2. Carcinogenic Risk Effects:
Estimation of Carcinogenic Risks to Human Health

For potential carcinogens, risk is defined by the incremental probability of an individual developing cancer over a lifetime as a result of exposure to a carcinogen. The risk of developing cancer can be estimated by combining information about the carcinogenic potency of a chemical and exposure to the substance. Specifically, carcinogenic risks are estimated by multiplying the route-specific cancer slope factor (which is the upper 95% confidence limit of the probability of a carcinogenic response per unit intake over a lifetime of exposure) by the estimated intakes; this yields the excess or incremental individual lifetime cancer risk.

The carcinogenic effects of the constituents associated with potential chemical exposure problems are typically calculated using the linear low-dose and one-hit models, represented by the following relationships (USEPA 1989d):

$$\text{Linear low-dose model, } CR = CDI \times SF \qquad (8.4)$$

$$\text{One-hit model, } CR = 1 - \exp(-CDI \times SF) \qquad (8.5)$$

where CR is the probability of an individual developing cancer (dimensionless); CDI is the chronic daily intake for long-term exposure (i.e., averaged over receptor lifetime) (mg/kg-day); and SF is the cancer slope factor ($[\text{mg/kg-day}]^{-1}$). The linear low-dose model is based on the so-called 'linearized multistage' (LMS) model – which assumes that there are multiple stages for cancer; the 'one-hit' model assumes that there is a single stage for cancer, and that one molecular or radiation interaction induces malignant change – making it very conservative. In reality, and for all practical purposes, the linear low-dose cancer risk model is valid only at low risk levels (i.e., estimated risks <0.01). For situations where chemical intakes may be high (i.e., potential risks >0.01), the one-hit model represents the more appropriate algorithm to use.

As a simple illustrative example calculation of human health carcinogenic risk, consider a situation where PCBs from abandoned electrical transformers have leaked into a groundwater reservoir that serves as a community water supply source. Environmental sampling and analysis conducted in a routine testing of the public water supply system showed an average PCB concentration of 2 µg/L. Thence, the pertinent question is: 'what is the individual lifetime cancer risk associated with drinking water exposure from this source?' Assuming the only exposure route of concern is from water ingestion, and using a cancer oral SF of 7E-02 (obtained from Table C.1 in Appendix C), then the cancer risk attributable to this exposure scenario is estimated as follows:

$$\begin{aligned}
\text{Cancer risk} &= SF_o \times CDI_o \\
&= SF_o \times C_w \times 0.0149 \\
&= (7 \times 10^{-2}) \times (2 \text{ µg/L} \times 10^{-3} \text{ mg/µg}) \times 0.0149 = 2.1 \times 10^{-6}
\end{aligned}$$

Similar evaluations can indeed be carried out for the various media and exposure routes.

As noted above in Section 8.1.3, the method of approach for assessing the cumulative health risks from chemical mixtures generally assumes additivity of effects for carcinogens when evaluating chemical mixtures or multiple carcinogens. Thus, for multiple carcinogenic chemicals and multiple exposure routes/pathways, the aggregate cancer risk for all exposure routes and all chemicals of concern associated with a potential chemical exposure problem can be estimated using the algorithms shown in Boxes 8.1A and 8.1B. The combination of risks across exposure routes is based on the assumption that the same receptors would consistently experience the reasonable maximum exposure via the multiple routes. Hence, if specific routes do not affect the same individual or receptor group, risks should not be combined under those circumstances.

As a rule-of-thumb, incremental risks of between 10^{-4} and 10^{-7} are generally perceived as being reasonable and adequate for the protection of human health – with 10^{-6} often used as the 'point-of-departure'. In reality, however, populations may be exposed to the same constituents from sources unknown or unrelated to a specific study. Consequently, it is preferable that the estimated carcinogenic risk is well below the 10^{-6} benchmark level – in order to allow for a reasonable margin of protectiveness for populations potentially at risk. Indeed, if a calculated cancer risk exceeds the 10^{-6} benchmark, then the health-based criterion for the chemical mixture has been exceeded and the need for corrective measures and/or risk management actions must be given serious consideration.

8.2.1. POPULATION EXCESS CANCER BURDEN

The two important parameters or measures often used for describing carcinogenic effects are the individual cancer risk and the estimated number of cancer cases (i.e., the cancer burden). The individual cancer risk from simultaneous exposure to several carcinogens is assumed to be the sum of the individual cancer risks from each individual chemical. The risk experienced by the individual receiving the greatest exposure is referred to as the 'maximum individual risk'.

To assess the population cancer burden associated with a chemical exposure problem, the number of cancer cases due to an exposure source within a given community can be estimated by multiplying the individual risk experienced by a group of people by the number of people in that group. Thus, if 10 million people (as an example) experience an estimated cancer risk of 10^{-6} over their lifetimes, it would be estimated that 10 (i.e., 10 million x 10^{-6}) additional cancer cases could occur for this group. The number of cancer incidents in each receptor area can be added to estimate the number of cancer incidents over an entire region. Hence, the excess cancer burden, B_{gi}, is given by:

$$B_{gi} = \sum (R_{gi} \times P_g) \qquad (8.6)$$

where: B_{gi} is the population excess cancer burden for i^{th} chemical for exposed group, G; R_{gi} is the excess lifetime cancer risk for i^{th} chemical for the exposed population group, G; P_g is the number of persons in exposed population group, G. Assuming cancer burden from each carcinogen is additive, the total population group excess cancer burden is given by:

Box 8.1A. The linear low-dose model for the estimation of low-level carcinogenic risks

$$\text{Total Cancer Risk, } TCR_{lo\text{-}risk} = \sum_{j=1}^{p} \sum_{i=1}^{n} (CDI_{ij} \times SF_{ij})$$

and

$$\text{Aggregate/Cumulative Total Cancer Risk, } ATCR_{lo\text{-}risk} = \sum_{k=1}^{s} \left\{ \sum_{j=1}^{p} \sum_{i=1}^{n} (CDI_{ij} \times SF_{ij}) \right\}$$

where:
TCR = probability of an individual developing cancer (dimensionless)
CDI_{ij} = chronic daily intake for the ith chemical and j^{th} route (mg/kg-day)
SF_{ij} = slope factor for the ith chemical and j^{th} route ([mg/kg-day]$^{-1}$)
n = total number of carcinogens
p = total number of pathways or exposure routes
s = total number for multiple sources of exposures to receptor
(e.g., dietary, drinking water, occupational, residential, recreational, etc.)

Box 8.1B. The one-hit model for the estimation of high-level carcinogenic risks

$$\text{Total Cancer Risk, } TCR_{hi\text{-}risk} = \sum_{j=1}^{p} \sum_{i=1}^{n} [1 - \exp(-CDI_{ij} \times SF_{ij})]$$

and

$$\text{Aggregate/Cumulative Total Cancer Risk, } ATCR_{hi\text{-}risk}$$

$$= \sum_{k=1}^{s} \left\{ \sum_{j=1}^{p} \sum_{i=1}^{n} [1 - \exp(-CDI_{ij} \times SF_{ij})] \right\}$$

where:
TCR = probability of an individual developing cancer (dimensionless)
CDI_{ij} = chronic daily intake for the ith chemical and j^{th} route (mg/kg-day)
SF_{ij} = slope factor for the ith chemical and j^{th} route ([mg/kg-day]$^{-1}$)
n = total number of carcinogens
p = total number of pathways or exposure routes
s = total number for multiple sources of exposures to receptor
(e.g., dietary, drinking water, occupational, residential, recreational, etc.)

$$B_g = \sum_{i=1}^{N} B_{gi} = \sum_{i=1}^{N} (R_{gi} \times P_g) \qquad (8.7)$$

and the total population burden, B, is represented by:

$$B = \sum_{g=1}^{G} B_g = \sum_{g=1}^{G} \left\{ \sum_{i=1}^{N} B_{gi} \right\} = \sum_{g=1}^{G} \left\{ \sum_{i=1}^{N} (R_{gi} \times P_g) \right\} \qquad (8.8)$$

Insofar as possible, cancer risk estimates are expressed in terms of both individual and population risk. For the population risk, the individual upper-bound estimate of excess lifetime cancer risk for an average exposure scenario is simply multiplied by the size of the potentially exposed population.

8.2.2. CARCINOGENIC RISK COMPUTATIONS: ILLUSTRATION OF THE PROCESSES FOR CALCULATING CARCINOGENIC RISKS

In accordance with the relationships presented earlier on in this chapter, the potential carcinogenic risks associated with chemical exposures may be calculated for all relevant exposure routes. Illustrative example evaluations for potential receptor groups purported to be exposed through inhalation, soil ingestion (i.e., incidental or pica behavior), and dermal contact are discussed in the proceeding sections. The examples shown below are used to demonstrate the computational mechanics for estimating chemical risks; the same set of units is maintained throughout as given above in related prior discussions.

Carcinogenic Effects for Contaminants in Water
The carcinogenic risk associated with a potential receptor exposure to chemical constituents in water can be estimated using the following annotated relationship:

$$\begin{aligned}
\text{Risk}_{water} &= [CDI_o \times SF_o] + [CDI_i \times SF_i] \\
&= [(CDI_{ing} + CDI_{der}) \times SF_o] + [CDI_i \times SF_i] \\
&= \{[(INGf \times C_w) + (DEXf \times C_w)] \times SF_o]\} + \{[(INHf \times C_w) \times SF_i]\} \qquad (8.9)
\end{aligned}$$

More generally, the carcinogenic risk may be calculated from 'first principles' as follows:

Risk_{water}

$$
= \left\{ SF_o \times C_w \times \frac{(IR_{adult} \times FI \times ABS_{gi} \times EF \times ED_{adult})}{(BW_{adult} \times AT \times 365 \text{ day/yr})} \right\}
$$

$$
+ \left\{ SF_o \times C_w \times \frac{(IR_{child} \times FI \times ABS_{gi} \times EF \times ED_{child})}{(BW_{child} \times AT \times 365 \text{ day/yr})} \right\}
$$

$$
+ \left\{ SF_o \times C_w \times \frac{(SA_{adult} \times K_p \times CF \times FI \times ABS_{gi} \times EF \times ED_{adult} \times ET_{adult})}{(BW_{adult} \times AT \times 365 \text{ day/yr})} \right\}
$$

$$
+ \left\{ SF_o \times C_w \times \frac{(SA_{child} \times K_p \times CF \times FI \times ABS_{gi} \times EF \times ED_{child} \times ET_{child})}{(BW_{child} \times AT \times 365 \text{ day/yr})} \right\}
$$

$$
+ \left\{ SF_i \times C_w \times \frac{(IR_{adult} \times FI \times ABS_{gi} \times EF \times ED_{adult})}{(BW_{adult} \times AT \times 365 \text{ day/yr})} \right\} + \ldots \ldots
$$

$$+ \{SF_i \times C_w \times \frac{(IR_{child} \times FI \times ABS_{gi} \times EF \times ED_{child})}{(BW_{child} \times AT \times 365 \text{ day/yr})} \} \qquad (8.10)$$

As an example, substitution of the exposure assumptions presented in Box 8.2 into the above equation yields the following reduced form of equation (8.10):

$$Risk_{water} = (SF_o \times C_w \times 0.0149) + (SF_o \times C_w \times 0.0325 \times K_p)$$
$$+ (SF_i \times C_w \times 0.0149) \qquad (8.11)$$

Consequently, by substituting the chemical-specific parameters in the reduced risk equation, potential carcinogenic risks associated with the particular constituent can be determined.

Box 8.2. Definitions and exposure assumptions for example risk computations associated with exposure to environmental contaminants in water and soil

Parameter	Parameter Definition & Exposure Assumption
SF_o	Oral cancer potency slope (obtained from literature, or Appendix C) ($[mg/kg\text{-}day]^{-1}$)
SF_i	Inhalation cancer potency slope (from the literature, or Appendix C) ($[mg/kg\text{-}day]^{-1}$)
C_w	Chemical concentration in water (obtained from the sampling and/or modeling) (mg/L)
C_s	Chemical concentration in soil (obtained from the sampling and/or modeling) (mg/kg)
C_a	Chemical concentration in air (obtained from the sampling and/or modeling) (mg/m^3)
K_p	Chemical-specific dermal permeability coefficient from water (obtained from the literature, e.g., DTSC, 1994) (cm^2/hr)
AF	Soil to skin adherence factor ($1 mg/cm^2$)
SA	Skin surface area available for water contact (adult = 23,000 cm^2; child = 7,200 cm^2); Skin surface area available for soil contact (adult = 5,800 cm^2; child = 2,000 cm^2)
IR	Average water intake rate – where intake from inhalation of volatile constituents may be assumed as equivalent to the amount of ingested water (adult = 2L/day; child = 1L/day)
SIR	Average soil ingestion rate (adult = 100 mg/day; child = 200 mg/day)
IR_a	Inhalation rate (adult = 20 m^3/day; child = 10 m^3/day)
CF	Conversion factor for water (1L/1,000 cm^3); Conversion factor for soil (10^{-6} kg/mg)
FI	Fraction ingested from contaminated source (1)
ABS_{gi}	Bioavailability/gastrointestinal [GI] absorption factor (100%)
ABS_s	Chemical-specific skin absorption fraction of chemical from soil (%)
EF	Exposure frequency for water (350 days/year); Exposure frequency for soil (soil ingestion = 350 days/year; dermal contact – adult = 100 days/year, child = 350 days/year)
ED	Exposure duration (adult = 24 years; child = 6 years)
ET	Exposure time during showering/bathing (adult = 0.25 hr/day; child = 0.14 hr/day)
BW	Body weight (adult = 70 kg; child = 15 kg)
AT	Averaging time (period over which exposure is averaged = 70 years or [70 x 365] days)

Carcinogenic Effects for Contaminants in Soils

The carcinogenic risk associated with a potential receptor exposure to chemical constituents in soils can be estimated using the following annotated relationship:

$$\text{Risk}_{\text{soil}} = [CDI_o \times SF_o] + [CDI_i \times SF_i]$$
$$= [(CDI_{\text{ing}} + CDI_{\text{der}}) \times SF_o] + [CDI_i \times SF_i]$$
$$= \{[(\text{INGf} \times C_w) + (\text{DEXf} \times C_w)] \times SF_o]\} + \{[(\text{INHf} \times C_w) \times SF_i]\} \qquad (8.12)$$

More generally, the carcinogenic risk may be calculated from 'first principles' as follows:

$\text{Risk}_{\text{soil}}$

$$= \{SF_o \times C_s \times \frac{(SIR_{\text{adult}} \times CF \times FI \times ABS_{gi} \times EF \times ED_{\text{adult}})}{(BW_{\text{adult}} \times AT \times 365 \text{ day/yr})}\}$$

$$+ \{SF_o \times C_s \times \frac{(SIR_{\text{child}} \times CF \times FI \times ABS_{gi} \times EF \times ED_{\text{child}})}{(BW_{\text{child}} \times AT \times 365 \text{ day/yr})}\}$$

$$+ \{SF_o \times C_s \times \frac{(SA_{\text{adult}} \times AF \times CF \times FI \times ABS_{gi} \times ABS_s \times EF \times ED_{\text{adult}})}{(BW_{\text{adult}} \times AT \times 365 \text{ day/yr})}\}$$

$$+ \{SF_o \times C_s \times \frac{(SA_{\text{child}} \times AF \times CF \times FI \times ABS_{gi} \times ABS_s \times EF \times ED_{\text{child}})}{(BW_{\text{child}} \times AT \times 365 \text{ day/yr})}\}$$

$$+ \{SF_i \times C_a \times \frac{(IR_{\text{adult}} \times FI \times ABS_{gi} \times EF \times ED_{\text{adult}})}{(BW_{\text{adult}} \times AT \times 365 \text{ day/yr})}\}$$

$$+ \{SF_i \times C_a \times \frac{(IR_{\text{child}} \times FI \times ABS_{gi} \times EF \times ED_{\text{child}})}{(BW_{\text{child}} \times AT \times 365 \text{ day/yr})}\} \qquad (8.13)$$

As an example, substitution of the exposure assumptions previously shown in Box 8.2 into the above equation yields the following reduced form of equation (8.13):

$$\text{Risk}_{\text{soil}} = (SF_o \times C_s \times [1.57 \times 10^{-6}]) + (SF_o \times C_s \times [1.88 \times 10^{-5}] \times ABS_s)$$
$$+ (SF_i \times C_a \times 0.149) \qquad (8.14)$$

Consequently, by substituting the chemical-specific parameters in the reduced risk equation, potential carcinogenic risks associated with the particular constituent can be determined.

8.3. Non-cancer Risk Effects: Estimation of Non-carcinogenic Hazards to Human Health

The potential non-cancer health effects resulting from a chemical exposure problem is usually expressed by the hazard quotient (HQ) and/or the hazard index (HI). The HQ is defined by the ratio of the estimated chemical exposure level to the route-specific reference dose, represented as follows (USEPA 1989d):

$$\text{Hazard Quotient, } HQ = \frac{E}{\text{RfD}} \qquad (8.15)$$

where E is the chemical exposure level or intake (mg/kg-day); and RfD is the reference dose (mg/kg-day).

As a simple illustrative example calculation of human health non-carcinogenic risk, consider a situation where an aluminum container is used for the storage of water meant for household consumption. Laboratory testing of the water revealed that some aluminum consistently gets dissolved in this drinking water – averaging concentrations of approximately 10 mg/L. The question then is: 'what is the individual non-cancer risk for a person who uses this source for drinking water?' Assuming the only exposure route of concern is associated with water ingestion (a reasonable assumption for this situation), and using a non-cancer toxicity index (i.e., a RfD) of 1.0 (obtained from Table C.1 in Appendix C), then the non-cancer risk attributable to this exposure scenario is calculated to be:

$$\text{Hazard Index} = (1/\text{RfD}_o) \times \text{CDI}_o$$
$$= (1/\text{RfD}_o) \times C_w \times 0.0639$$
$$= 1.0 \times 10 \text{ mg/L} \times 0.0639 = 0.6$$

Similar evaluations can indeed be carried out for the various media and exposure routes.

As noted previously in Section 8.1.3, for multiple chemical exposures to non-carcinogens and the non-carcinogenic effects of carcinogens, constituents are normally grouped by the same mode of toxicological action. Cumulative non-cancer risk is evaluated through the use of a hazard index that is generated for each health or toxicological 'endpoint'. Chemicals with the same endpoint are generally included in a hazard index calculation. Thus, for multiple non-carcinogenic effects of several chemical compounds and multiple exposure routes, the aggregate non-cancer risk for all exposure routes and all constituents associated with a potential chemical exposure problem can be estimated using the algorithm shown in Box 8.3.

Box 8.3. General equation for calculating non-carcinogenic risks to human health

$$\text{Total Hazard Index} = \sum_{j=1}^{p} \sum_{i=1}^{n} \frac{E_{ij}}{RfD_{ij}} = \sum_{j=1}^{p} \sum_{i=1}^{n} [HQ]_{ij}$$

and

Aggregate/Cumulative Total Hazard Index

$$= \sum_{k=1}^{s} \left\{ \sum_{j=1}^{p} \sum_{i=1}^{n} \frac{E_{ij}}{RfD_{ij}} \right\} = \sum_{k=1}^{s} \left\{ \sum_{j=1}^{p} \sum_{i=1}^{n} [HQ]_{ij} \right\}$$

where:

E_{ij} = exposure level (or intake) for the ith chemical and j^{th} route (mg/kg-day)

RfD_{ij} = acceptable intake level (or reference dose) for the ith chemical and j^{th} exposure route (mg/kg-day)

$[HQ]_{ij}$ = hazard quotient for the ith chemical and j^{th} route

n = total number of chemicals showing non-carcinogenic effects

p = total number of pathways or exposure routes

s = total number for multiple sources of exposures to receptor
           (e.g., dietary, drinking water, occupational, residential, recreational, etc.)

The combination of hazard quotients across exposure routes is based on the assumption that the same receptors would consistently experience the reasonable maximum exposure via the multiple routes. Thus, if specific sources do not affect the same individual or receptor group, hazard quotients should not be combined under those circumstances. Furthermore, and in the strictest sense, constituents should not be grouped together unless the physiologic/toxicological endpoint is known to be the same, otherwise the process will likely over-estimate and over-state potential health effects.

In accordance with general guidelines on the interpretation of hazard indices, for any given chemical, there may be potential for adverse health effects if the hazard index exceeds unity (1). It is noteworthy that, since the RfD incorporates a large margin of safety, it is possible that no toxic effects may occur even if this benchmark level is exceeded. However, as a rule-of-thumb in the interpretation of the results from HI calculations, a reference value of less than or equal to unity (i.e., HI ≤1) should be taken as the acceptable benchmark. For HI values greater than unity (i.e., HI >1), the higher the value, the greater is the likelihood of adverse non-carcinogenic health impacts. In fact, since populations may be exposed to the same constituents from sources unknown or unrelated to a case-problem, it is preferred that the estimated non-carcinogenic hazard index be well below the benchmark level of unity – in order to allow for additional margin of protectiveness for populations potentially at risk. Indeed, if any calculated hazard index exceeds unity, then the health-based criterion for the chemical mixture or multiple routes has been exceeded and the need for corrective measures must be given serious consideration.

8.3.1. CHRONIC *VERSUS* SUBCHRONIC NON-CARCINOGENIC EFFECTS

Human receptor exposures to chemicals can occur over long-term periods (i.e., chronic exposures), or over short-term periods (i.e., subchronic exposures). Chronic exposures for humans usually range in duration from about seven years to a lifetime; sub-chronic human exposures typically range in duration from about two weeks to seven years (USEPA, 1989a) – albeit shorter-term exposures of less than two weeks could also be anticipated. Appropriate chronic and subchronic toxicity parameters and intakes are used in the estimation of non-carcinogenic effects associated with the different exposure duration.

The chronic non-cancer hazard index is represented by the following modification to the general equation presented above:

$$\text{Total Chronic Hazard Index} = \sum_{j=1}^{p} \sum_{i=1}^{n} \frac{CDI_{ij}}{RfD_{ij}} \qquad (8.16)$$

where: CDI_{ij} is chronic daily intake for the ith constituent and j^{th} exposure route, and RfD_{ij} is chronic reference dose for i^{th} constituent and j^{th} exposure route.

The subchronic non-cancer hazard index is represented by the following modification to the general equation presented above:

$$\text{Total Subchronic Hazard Index} = \sum_{j=1}^{p} \sum_{i=1}^{n} \frac{SDI_{ij}}{RfD_{sij}} \qquad (8.17)$$

where: SDI_{ij} is subchronic daily intake for the ith constituent and j^{th} exposure route, and RfD_{sij} is subchronic reference dose for i^{th} constituent and j^{th} exposure route.

8.3.2. NON-CARCINOGENIC HAZARD COMPUTATIONS: ILLUSTRATION OF THE PROCESSES FOR CALCULATING NON-CARCINOGENIC HAZARDS

In accordance with the relationships presented earlier on in this chapter, the potential non-cancer risks associated with chemical exposures may be calculated for all relevant exposure routes. Illustrative example evaluations for potential receptor groups purported to be exposed through inhalation, soil ingestion (i.e., incidental or pica behavior), and dermal contact are discussed in the proceeding sections. The examples shown below for childhood exposure from infancy through age six are used to demonstrate the computational mechanics for estimating chemical risks; the same set of units is maintained throughout as given above in related prior discussions.

Non-carcinogenic Effects for Contaminants in Water
The non-carcinogenic risk associated with a potential receptor exposure to chemical constituents in water can be estimated using the following annotated relationship:

$Hazard_{water}$

$$= [CDI_o \times \frac{1}{RfD_o}] + [CDI_i \times \frac{1}{RfD_i}]$$

$$= [(CDI_{ing} + CDI_{der}) \times \frac{1}{RfD_o}] + [CDI_i \times \frac{1}{RfD_i}]$$

$$= \{[(INGf \times C_w) + (DEXf \times C_w)] \times \frac{1}{RfD_o}\} + \{[(INHf_i \times C_w) \times \frac{1}{RfD_i}]\} \qquad (8.18)$$

More generally, the non-cancer risk may be calculated from 'first principles' as follows:

$Hazard_{water}$

$$= \{\frac{1}{RfD_o} \times C_w \times \frac{(IR_{child} \times FI \times ABS_{gi} \times EF \times ED_{child})}{(BW_{child} \times AT \times 365 \text{ day/yr})}\}$$

$$+ \{\frac{1}{RfD_o} \times C_w \times \frac{(SA_{child} \times K_p \times CF \times FI \times ABS_{gi} \times EF \times ED_{child} \times ET_{child})}{(BW_{child} \times AT \times 365 \text{ day/yr})}\}$$

$$+ \{\frac{1}{RfD_i} \times C_w \times \frac{(IR_{child} \times FI \times ABS_{gi} \times EF \times ED_{child})}{(BW_{child} \times AT \times 365 \text{ day/yr})}\} \qquad (8.19)$$

As an example, substitution from the exposure assumptions presented in Box 8.4 into the above equation yields the following reduced form of equation (8.19):

$$\text{Hazard}_{water} = (\frac{1}{RfD_o} \times C_w \times 0.0639) + (\frac{1}{RfD_o} \times C_w \times 0.0644 \times K_p)$$
$$+ (\frac{1}{RfD_i} \times C_w \times 0.0639) \qquad (8.20)$$

Consequently, by substituting the chemical-specific parameters in the reduced risk equation, potential non-carcinogenic risks associated with the particular constituent can be determined.

Box 8.4. Definitions and exposure assumptions for the example hazard computations associated with exposure to environmental contaminants in water and soil

Parameter	Parameter Definition & Exposure Assumption
RfD_o	Oral reference dose (obtained from the literature, or Appendix C) ([mg/kg-day])
RfD_i	Inhalation reference dose (from the literature, or Appendix C) ([mg/kg-day])
C_w	Chemical concentration in water (obtained from the sampling and/or modeling) (mg/L)
C_s	Chemical concentration in soil (obtained from the sampling and/or modeling) (mg/kg)
C_a	Chemical concentration in air (obtained from the sampling and/or modeling) (mg/m^3)
K_p	Chemical-specific dermal permeability coefficient from water (obtained from the literature, e.g., DTSC, 1994) (cm^2/hr)
AF	Soil to skin adherence factor (1 mg/cm^2)
SA	Skin surface area available for water contact (child = 7,200 cm^2)
SA	Skin surface area exposed/available for soil contact (child = 2,000 cm^2)
IR	Average water intake rate – where intake from inhalation of volatile constituents may be assumed as equivalent to the amount of ingested water (child = 1 L/day)
SIR	Average soil ingestion rate (child = 200 mg/day)
IR_a	Inhalation rate (child = 10 m^3/day)
CF	Conversion factor for water (1 L/1,000 cm^3)
CF	Conversion factor for soil (10^{-6} kg/mg)
FI	Fraction ingested from contaminated source (1)
ABS_{gi}	Bioavailability/gastrointestinal [GI] absorption factor (100%)
ABS_s	Chemical-specific skin absorption fraction of chemical from soil (%)
EF	Exposure frequency (350 days/years)
ED	Exposure duration (child = 6 years)
ET	Exposure time during showering/bathing (child = 0.14 hr/day)
BW	Body weight (child = 15 kg)
AT	Averaging time (period over which exposure is averaged = 6 years or [6 × 365] days)

Non-carcinogenic Effects for Contaminants in Soils

The non-carcinogenic risk associated with a potential receptor exposure to chemical constituents in soils can be estimated using the following annotated relationship:

Hazard_{soil}

$$= [CDI_o \times \frac{1}{RfD_o}] + [CDI_i \times \frac{1}{RfD_i}]$$

$$= [(CDI_{ing} + CDI_{der}) \times \frac{1}{RfD_o}] + [CDI_i \times \frac{1}{RfD_i}]$$

$$= \{[(INGf \times C_w) + (DEXf \times C_w)] \times \frac{1}{RfD_o}\} + \{[(INHf_i \times C_w) \times \frac{1}{RfD_i}]\} \quad (8.21)$$

More generally, the carcinogenic risk may be calculated from 'first principles' as follows:

$Hazard_{soil}$

$$= \{ \frac{1}{RfD_o} \times C_s \times \frac{(SIR_{child} \times CF \times FI \times ABS_{gi} \times EF \times ED_{child})}{(BW_{child} \times AT \times 365 \text{ day/yr})} \}$$

$$+ \{ \frac{1}{RfD_o} \times C_s \times \frac{(SA_{child} \times AF \times CF \times FI \times ABS_{gi} \times ABS_s \times EF \times ED_{child})}{(BW_{child} \times AT \times 365 \text{ day/yr})} \}$$

$$+ \{ \frac{1}{RfD_i} \times C_a \times \frac{(IR_{child} \times FI \times ABS_{gi} \times EF \times ED_{child})}{(BW_{child} \times AT \times 365 \text{ day/yr})} \} \qquad (8.22)$$

As an example, substitution from the exposure assumptions presented in Box 8.4 into the above equation yields the following reduced form of equation (8.22):

$$Hazard_{soil} = (\frac{1}{RfD_o} \times C_s \times [1.28 \times 10^{-5}])$$

$$+ (\frac{1}{RfD_o} \times C_s \times [1.28 \times 10^{-4}] \times ABS_s)$$

$$+ (\frac{1}{RfD_i} \times C_a \times 0.639) \qquad (8.23)$$

Consequently, by substituting the chemical-specific parameters in the reduced risk equation, potential non-carcinogenic risks associated with the particular constituent can be determined.

8.4. Risk Presentation

Consider the following illustrative practical example. Routine air monitoring at a housing development downwind of a chemical recycling facility has documented air contamination for the following chemicals (at the indicated average concentrations): Acetone = 12 $\mu g/m^3$; Benzene = 0.5 $\mu g/m^3$; and PCE = 2 $\mu g/m^3$. Now, it is required to determine the total health risk to a 70-kg adult in this housing estate, assuming an inhalation rate of 0.83 m^3/h. The computation process is provided below for this example problem.

Step 1 – Intake Computations

The intakes for the non-carcinogenic risk contributions from Acetone and PCE are estimated as follows:

$$NCInh_{(adultR)} = \left[\frac{(CA \times IR \times RR \times ABS_s \times ET \times EF \times ED)}{(BW \times AT)} \right]$$

Substituting CA = 12 μg/m^3 = (12 x 10^{-3}) mg/m^3 [Acetone] & 2 μg/m^3 = (2 x 10^{-3}) mg/m^3 [PCE]; IR = 0.83 m^3/h; RR = 1; ABS$_s$ = 1; ET = 12 h/d; EF = 365 d/year; ED = 58 years; BW = 70 kg; and AT = (ED x 365) = (58 x 365) days yields:

For Acetone:

$$NCInh_{(adultR)} = \left[\frac{(12\times10^{-3}\times0.83\times12\times365\times58)}{(70\times58\times365)}\right] \cong 1.71\times10^{-3}\,mg\,/\,kg-day$$

For PCE:

$$NCInh_{(adultR)} = \left[\frac{(2\times10^{-3}\times0.83\times12\times365\times58)}{(70\times58\times365)}\right] \cong 2.85\times10^{-4}\,mg\,/\,kg-day$$

Now, the intakes for the carcinogenic risk contributions from Benzene and PCE are estimated as follows:

$$CInh_{(adultR)} = \left[\frac{(CA\times IR\times RR\times ABS_s\times ET\times EF\times ED)}{(BW\times AT)}\right]$$

Substituting CA = 0.5 μg/m^3 = (0.5 x 10^{-3}) mg/m^3 [Benzene] & 2 μg/m^3 = (2 x 10^{-3}) mg/m^3 [PCE]; IR = 0.83 m^3/h; RR = 1; ABS$_s$ = 1; ET = 12 h/d; EF = 365 d/year; ED = 58 years; BW = 70 kg; and AT = (70 x 365) = (70 x 365) days yields:

For Benzene:

$$CInh_{(adultR)} = \left[\frac{(0.5\times10^{-3}\times0.83\times12\times365\times58)}{(70\times70\times365)}\right] \cong 5.89\times10^{-5}\,mg\,/\,kg-day$$

For PCE:

$$CInh_{(adultR)} = \left[\frac{(2\times10^{-3}\times0.83\times12\times365\times58)}{(70\times70\times365)}\right] \cong 2.36\times10^{-4}\,mg\,/\,kg-day$$

Step 2 – Risk Computations

For the non-carcinogenic risk, assuming RfD$_i$ = 1.00E-01 [Acetone] & 1.00E-02 [PCE], the hazard quotients are calculated to be:

$$HQ_{(acetone)} = \left[\frac{NCInh_{(adultR)}}{RfD_i} \right] = \frac{1.71 \times 10^{-3}}{1.00 \times 10^{-1}} \cong 1.71 \times 10^{-2}$$

$$HQ_{(PCE)} = \left[\frac{NCInh_{(adultR)}}{RfD_i} \right] = \frac{2.85 \times 10^{-4}}{1.00 \times 10^{-2}} \cong 2.85 \times 10^{-2}$$

Thence, the total hazard index is given by:

$$HI = (1.71 \times 10^{-2}) + (2.85 \times 10^{-2}) = 4.56 \times 10^{-2} = 0.05$$

For the carcinogenic risk, assuming SF_i = 2.90E-02 [Benzene] & 2.10E-02 [PCE], the cancer risks are calculated to be:

$$CR_{(benzene)} = \left[CInh_{(adult)} \times SF_i \right] = [(5.89 \times 10^{-5}) \times (2.90 \times 10^{-2})] \cong 1.71 \times 10^{-6}$$

$$CR_{(PCE)} = \left[CInh_{(adult)} \times SF_i \right] = [(2.36 \times 10^{-4}) \times (2.10 \times 10^{-2})] \cong 4.96 \times 10^{-6}$$

Thence, the total cancer risk is given by:

$$TCR = (1.71 \times 10^{-6}) + (4.96 \times 10^{-6}) = 6.67 \times 10^{-6}$$

After going through all the requisite computational exercises, the risk values are often stated simply as a number – such as is expressed in the following statements:

- Risk probability of occurrence of additional cases of cancer – e.g., a cancer risk of 1×10^{-6}, which reflects the estimated number of excess cancer cases in a population.
- Hazard index of non-cancer health effects such as neurotoxicity or birth defects – e.g., a hazard index of 1, reflecting the degree of harm from a given level of exposure.

One of the most important points to remember in all cases of risk presentation, however, is that the numbers by themselves may not tell the whole story. For instance, a human cancer risk of 10^{-6} for an 'average exposed person' (e.g., someone exposed via food products only) may not necessarily be interpreted to be the same as a cancer risk of 10^{-6} for a 'maximally exposed individual' (e.g., someone exposed from living in a highly contaminated area) – despite the fact that the numerical risk values may be identical. In fact, omission of the qualifier terms – e.g., 'average' or 'maximally/most exposed' – could mean an incomplete description of the true risk scenarios, and this could result in poor risk management strategies and/or a failure in risk communication tasks. Thus, it is very important to know, and to recognize such apparently subtle differences in the risk summarization – or indeed throughout the risk characterization process.

To ensure an effective risk presentation, it must be recognized that the qualitative aspect of a risk characterization (which may also include an explicit recognition of all assumptions, uncertainties, etc.) may be as important as its quantitative component (i.e., the estimated risk numbers). The qualitative considerations are indeed essential to making judgments about the reliability of the calculated risk numbers, and therefore the confidence associated with the characterization of the potential risks.

8.4.1. GRAPHICAL PRESENTATION OF THE RISK SUMMARY INFORMATION

Several graphical representations may be employed in presenting a summary of the requisite risk information that has been developed from the risk characterization efforts. Examples of such graphical forms include the following:

- *Pie charts,* such as shown in Figures 8.1(a) and 8.1(b) to illustrate the risk contributions from different chemical exposure sources.
- *Horizontal bar charts,* such as shown in Figure 8.2 to illustrate the hazard index contributions associated with different exposure routes and receptor groups.
- *Vertical bar charts,* such as shown in Figures 8.3(a), 8.3(b), and 8.3(c) to illustrate the hazard index and cancer risk contributions from different exposure sources and CoPCs.
- *Variety of relational plots,* such as shown in Figures 8.4 through 8.6 to illustrate various graphical relationships used to characterize risk associated with chemical exposure problems.

This listing is by no means complete; other novel representations that may consist of variations or convolutions of the above may indeed be found to be more appropriate and/or useful for some case-specific applications.

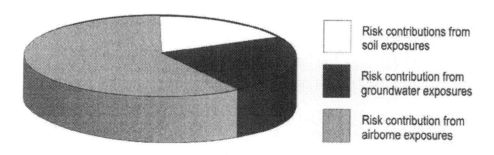

Risk contributions from soil exposures

Risk contribution from groundwater exposures

Risk contribution from airborne exposures

Figure 8.1(a). Pie chart illustration of risk summary results: a 3-D schematic

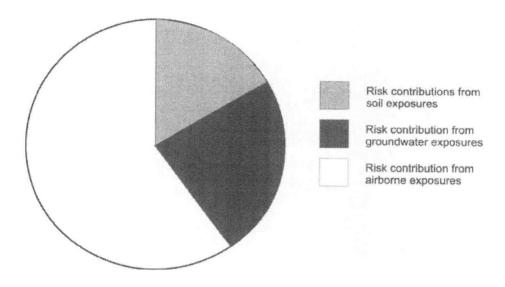

Figure 8.1(b). Pie chart illustration of risk summary results: a 2-D sketch

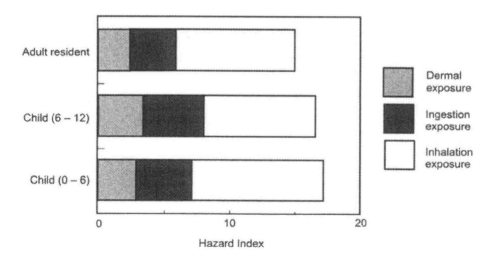

Figure 8.2. Horizontal bar chart illustration of risk summary results

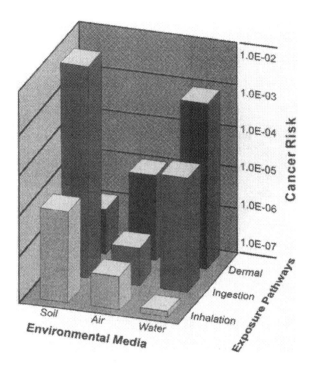

Figure 8.3. Vertical bar chart illustration of risk summary results

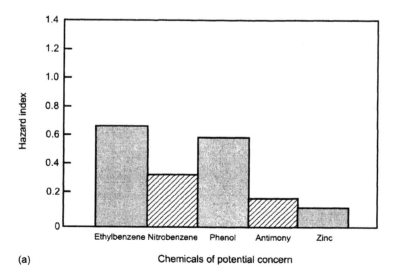

(a)

Figure 8.3(a). Vertical bar chart illustration of risk summary results:
Illustrative presentation of the relative contribution of individual chemicals to
overall hazard index estimates associated with a hypothetical public water supply system

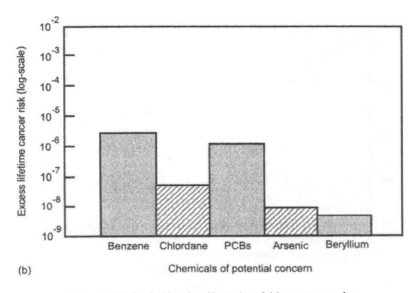

(b)

Figure 8.3(b). Vertical bar chart illustration of risk summary results:
Illustrative presentation of the relative contribution of individual chemicals to
overall cancer risk estimates associated with a hypothetical public water supply system (semi-log plot)

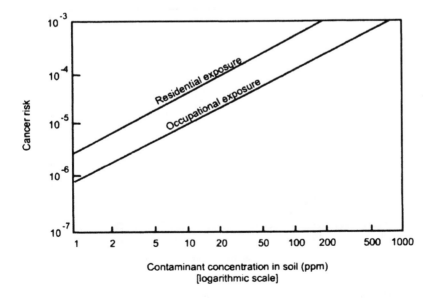

Figure 8.4. Illustrative sketch of the effects of choice of exposure scenarios on dose and risk estimates

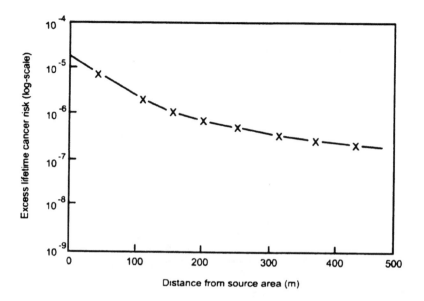

Figure 8.5. Illustrative sketch of the variation of estimated cancer risks with distance from contaminant source: A semi-log plot of cancer risk estimates from receptor exposures to benzene in groundwater at several different locations downgradient of a release source

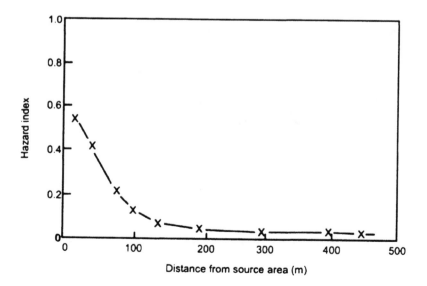

Figure 8.6. Illustrative sketch of the variation of estimated hazard index with distance from contaminant source: An arithmetic-scale plot of hazard index estimates from receptor exposures to ethylbenzene in groundwater at several different locations downgradient of a release source

8.5.   Uncertainty and Variability Issues in Risk Assessment

Risk assessments tend to be highly uncertain as well as highly variable. *Variability* (or *stochasticity*) arises from true heterogeneity in characteristics such as dose-response differences within a population, or differences in body weight, or differences in rates of food and water intakes/ingestion, or differences in chemical exposure levels in source materials. *Uncertainty* represents a lack of knowledge about factors such as adverse effects or chemical exposure levels – and this may be reduced with additional studies or investigations.

As an example of the intertwining relationships between uncertainty and variability, consider a situation involving the ingestion of contaminated drinking water; suppose that it is possible to measure an individual's daily water consumption (and indeed the contaminant concentration) in exact terms, thereby eliminating uncertainty in the measured daily dose. Even so, the daily dose still has an inherent day-to-day variability – due to changes in the individual's daily water intake or the contaminant concentration in the water. Ultimately, since the individual's true average daily dose is unknown, it is uncertain how close the estimate is to the true value. Thus, the variability across daily doses has been translated into uncertainty in the ADD. Consequently, although the individual's true ADD has no variability, the estimate of the ADD has some uncertainty associated with it (USEPA, 1997). In general, uncertainty can lead to inaccurate or biased estimates, whereas variability can affect the precision of the estimates and the degree to which they can be generalized.

Basically, 'variability' encompasses any aspect of the risk assessment process that can produce varying results. This includes the potential interpretations of the available data, the availability of different data sets collected under different experimental protocols, and the availability of different models and methods – albeit several of these may also be considered as sources of uncertainty (NRC, 1994; USEPA, 1992, 1997). Thus, the use of 'variability' to refer to differences attributable to diversity in biological sensitivity or exposure parameters means these differences can be better understood, but not reduced, by further research. On the other hand, 'uncertainty' – that refers to lack of knowledge about specific factors, parameters, or models – can generally be reduced through further study. Indeed, in principle uncertainty can be reduced through the acquisition of more information, whereas variability is irreducible.

In general, some parameters used in risk assessments may reflect both variability and uncertainty under different sets of circumstances or conditions. However, insofar as possible, stochastic variability and knowledge uncertainty should be segregated in the evaluation processes employed during the risk assessment.

8.5.1. TYPES AND NATURE OF VARIABILITY

Three fundamental types of variability may be defined for most risk assessment exercises, namely (USEPA, 1997):

1.   Spatial variability (i.e., variability across locations)
2.   Temporal variability (i.e., variability over time)
3.   Inter-receptor variability (i.e., variability amongst individual receptors)

Spatial variability can occur both at regional (macroscale) and local (microscale) levels. For example, fish intake rates can vary significantly depending on the region of a country – with higher consumption more likely to occur among populations located near large water bodies or coastal areas (USEPA, 1997). In general, higher exposures tend to be associated with receptors in closer proximity to the pollutant source.

Temporal variability refers to variations that occur over time – and this may relate to both long- and short-term situations. For example, seasonal fluctuations in weather, pesticide applications, use of wood-burning appliances, and fraction of time spent outdoors relate to longer-term variability; and shorter-term variability may include differences in individual or personal activities on weekdays *versus* weekends, or at different times of the day (USEPA, 1997).

Inter-receptor variability can be attributed/related to two major factors – namely, human characteristics (such as age or body weight) and human behaviors (such as location and activity patterns) – each of which in turn may be related to several underlying phenomena that vary as well (USEPA, 1997). For example, the natural variability in human weight is due to a combination of genetic, nutritional, and other lifestyle or environmental factors.

Variability may be confronted and evaluated in a variety of ways (see, e.g., NRC, 1994; USEPA, 1997) – albeit a strategy that involves using both the appropriate maximum and minimum parameter values seems to be favored in most chemical exposure and risk assessments. Such approach allows for the characterization of the variability by a range between the extreme values, as well as produces a measure of central tendency estimates.

8.5.2. TYPES AND NATURE OF UNCERTAINTY

The uncertainties that typically arise in risk assessments can be of three general types – namely (see, e.g., USEPA, 1992, 1997):

1. Uncertainties in parameter values (e.g., use of incomplete or biased values);
2. Uncertainties in parameter modeling (e.g., issue of model adequacy/inadequacy); and
3. Uncertainties in the degree of completeness (e.g., representativeness of evaluation scenarios).

Parameter uncertainties arise from the need to estimate parameter values from limited or inadequate data. Such uncertainties are inherent because the available data are usually incomplete, and the analyst must make inferences from a state of incomplete knowledge. Examples of uncertainties in parameter values relate to such issues as incomplete or biased data; applicability of available data to the particular case (i.e., generic vs. case-specific data); etc.

Modeling uncertainties stem from inadequacies in the various models used to evaluate hazards, exposures, and consequences, and also from the deficiencies of the models in representing reality. Examples of uncertainties in modeling relate to such issues as model adequacy; whether uncertainty is introduced by the mathematical or numerical approximations that are made for convenience; use of models outside its range of validity; etc.

Completeness/scenario uncertainties relate to the inability of the analyst to evaluate exhaustively all contributions to risk. They refer to the problem of assessing what may have been omitted in the analysis. Examples of uncertainties in the degree of completeness may relate to such questions as to whether the analyses have been taken to sufficient depth; whether all important hazard sources and exposure possibilities have been addressed; etc.

For all practical purposes, these uncertainties are propagated through the analysis. To the extent possible, a sensitivity analysis provides insight into the possible range of results. Sensitivity analysis entails the determination of how rapidly the output of an analysis changes with respect to variations in the input. Sensitivity studies do not usually incorporate the error range or uncertainty of the input – thus serving as a distinguishing element from uncertainty analyses.

Depending on the specific aspect or component of the risk assessment being performed, the type of uncertainty that dominates at each stage of the analysis can be different. Each type of uncertainty can be characterized either qualitatively or quantitatively. Various levels of uncertainty analysis can therefore be characterized by the degree to which each type of uncertainty is quantitatively analyzed. In any case, identification of the sources of uncertainty is an important first step in determining how to reduce that uncertainty. Furthermore, because the uncertainties tend to be fundamentally tied to a lack of knowledge concerning important evaluation factors/parameters, strategies for reducing uncertainty necessarily involve reduction or elimination of knowledge gaps (USEPA, 1997).

8.5.3. COMMON SOURCES OF UNCERTAINTY IN PUBLIC HEALTH ENDANGERMENT ASSESSMENTS

Inevitably, considerable uncertainty is inherent in the human risk assessment process; Box 8.5 identifies several major sources of uncertainty usually associated with human health risk assessments. In particular, uncertainties arise due to the use of several assumptions and inferences necessary to complete a risk assessment. For instance, human health risk assessments usually involve extrapolations and inferences to predict the occurrence of adverse health effects under certain conditions of exposure to chemicals present in the environment. The extrapolations and inferences are typically based on knowledge of the adverse effects that occur under a different set of exposure conditions (e.g., different dose levels and/or species). As a consequence of these types of extrapolations and projections, there is considerable uncertainty in the resulting conclusions – due in part to the several assumptions that is part of the evaluation process.

Box 8.5. Major sources of uncertainty in human health risk assessments

---

♦ Uncertainty in health effects/toxicity data
  ‣ Uncertainty in extrapolating from high dose to low dose
  ‣ Uncertainty in extrapolating data from experimental animals to humans
  ‣ Uncertainty due to differences between individuals
♦ Uncertainty in measuring or calculating exposure point concentrations
  ‣ Uncertainty in transposing chemical source concentrations into exposure point concentrations
  ‣ Uncertainty in assumptions used to model exposure point concentrations
♦ Uncertainty in calculating exposure dose
  ‣ Uncertainty in source terms (i.e., chemical source sampling and monitoring data)
  ‣ Uncertainty in estimating exposure dose using mathematical models

Indeed, for most chemical substances for which there are insufficient data in humans, a major uncertainty in the evaluation of potential health effects to humans is the reliance on animal studies. Such applications involve the use of high exposure in animals to predict human response at lower exposure. Furthermore, this is often carried out in the absence of an understanding of how an agent causes the observed toxicological effects in the animals, and in the face of the varying results frequently obtained with different animal species under different exposure conditions. Even when there are human data, there is uncertainty about average response at lower exposures and there is variability in individual response around this average.

In general, because of the various limitations and uncertainties, the results of a risk assessment cannot be considered as an absolutely accurate determination of risks. Commonly encountered limitations and uncertainties of considerable significance in relation to several components of the risk assessment process are enumerated below (see, e.g., Calabrese, 1984; Clewell and Andersen, 1985; Dourson and Stara, 1983; USEPA, 1989b).

- *Uncertainties in general extrapolations relevant to toxicity information.* Whereas some chemicals have been studied extensively under a variety of exposure conditions in several species (including humans), others may have only limited investigations done on them. This latter group will tend to have inherent limitations in toxicity data (arising for several reasons). Also, because data that specifically identify the hazards to humans as a result of their exposure to various chemicals of concern under the conditions of likely human exposure may not exist, it becomes necessary to infer such hazard effects by extrapolating from data obtained under different exposure conditions, usually in experimental animals. This introduces three major types of uncertainties – namely, that related to extrapolating from one species to another (i.e., uncertainties in interspecies extrapolation); those relating to extrapolation from a high-dose region curve to a low-dose region (i.e., uncertainties in intra-species extrapolation); and those related to extrapolating from one set of exposure conditions to another (i.e., uncertainties due to differences in exposure conditions).

- *Uncertainties from quantitative extrapolations and adjustments in dose-response evaluation.* Experimental studies to determine the carcinogenic effects due to low exposure levels often encountered in the environment generally are not feasible. This is because such effects are not readily apparent in the relatively short time frame over which it is usually possible to conduct such a study. Consequently, various mathematical models are used to extrapolate from the high doses used in animal studies to the doses encountered in exposure to ambient environmental concentrations. Extrapolating from a high dose (of animal studies) to a low dose (for human effects) introduces a level of uncertainty which could be significantly large, and which may have to be carefully addressed. For instance, in human health risk assessments, no-observed-adverse-effect-levels (NOAELs) and cancer potency slope factors (SFs) from animal studies are usually divided by a factor of 10 to account for extrapolation from animals to humans, and by an additional factor of 10 to account for variability in human responses (see Chapter 7). Given the recognized differences among species in their responses to toxic insult, and between strains of the same species, it is apparent that additional uncertainties will be introduced when this type of quantitative extrapolations and adjustments are made in the dose-response evaluation.

- *Uncertainty associated with the toxicity of chemical mixtures.* The effects of combining two chemicals may be synergistic (effect when outcome of combining two chemicals is greater than the sum of the inputs), antagonistic (effect when the outcome is less than the sum of the two inputs), or under potentiation (i.e., when one chemical has no toxic effect but combined with another chemical that is toxic, produces a much more toxic effect). Indeed, chemicals present in a mixture can interact to yield a new chemical; or one can interfere with the absorption, distribution, metabolism, or excretion of another. Notwithstanding all these possible scenarios, risk assessments often assume toxicity to be additive – resulting in an important source of uncertainty.

- *Limitations in model form.* Exposure scenarios and fate and behavior models usually can be a major contributor of uncertainty to risk assessments. Apart from general model imperfections, environmental and exposure models usually oversimplify reality, contributing one form of uncertainty or another. Also, the natural variability in environmental and exposure-related parameters causes variability in exposure factors, and therefore in exposure estimates developed on this basis. This therefore begs the question of how close to reality the model function and output are likely to be.

- *Consideration of ambient/'background' exposures.* For the most part, risk assessment methods used in practice tend to ignore background/ambient exposures; instead, the process considers only incremental risk estimates for the exposed populations. Consequently, such risk estimates do not address what constitutes the true health risks to the public – of which background or ambient exposures could be contributing in a very significant way.

- *Representativeness of sampling data.* Uncertainties may arise from random and systematic errors in the type of measurement and sampling techniques often used in environmental and exposure characterization activities. For instance, professional judgment (based on scientific assumptions) is frequently used for sampling design and also to make decisions on how to correct for data gaps – albeit this process has some inherent uncertainties associated with it.

In general, uncertainties are difficult to quantify, or at best, the quantification of uncertainty is itself uncertain. Thus, the risk levels generated in a risk assessment are useful only as a yardstick and as a decision-making tool for the prioritization of problem situations – rather than to be construed as actual expected rates of disease, or adversarial impacts in exposed populations. It is used only as an estimate of risks, based on current level of knowledge coupled with several assumptions. Quantitative descriptions of uncertainty, which could take into account random and systematic sources of uncertainty in potency, exposure, intakes, etc. would usually help present the spectrum of possible true values of risk estimates, together with the probability (or likelihood) associated with each point in the spectrum.

8.5.4. THE NEED FOR UNCERTAINTY AND VARIABILITY ANALYSES

A number of factors – directly or indirectly related to uncertainties and variabilities – may indeed cause an analysis to either under-estimate or over-estimate true risks that are associated with a chemical exposure problem. For instance, it is always possible that a chemical whose toxic properties have not been thoroughly tested may be more toxic than originally believed or anticipated; a chemical not tested for carcinogenicity or teratogenicity may in fact display those effects. Furthermore, a limitation of analysis for selected 'indicator chemicals' may have some limiting (even if insignificant) effects.

Indeed, all risk estimates involve some degree of uncertainty – especially because of the inability of the risk analyst to quantify all the requisite information necessary to complete a credible study. Uncertainty analysis should therefore become an integral part of all risk assessments, regardless of the scope or level of detail. In fact, it is prudent and essential to the credibility of the risk assessment, to describe the relevant uncertainties in as great a detail as possible. But, as one strives to be more scientifically credible, it is important not to attempt to infer levels of precision that clearly are not appropriate for quantitative risk assessments (Felter and Dourson, 1998). After all, the acknowledgment of 'inexactness' is very much in line with a cautionary note that Aristotle is quoted to have sounded, once upon a time – that: "It is the mark of an instructed mind to rest satisfied with the degree of precision which the nature of the subject permits and not to seek an exactness where only an approximation of the truth is possible."

Ultimately, the degree to which variability and uncertainty are addressed in a given study depends largely on the scope of the risk assessment and the resources available. For the study of variability, stochastic models are used as the more realistic representations of reality, rather than the use of deterministic models. In any case, as a guiding principle, the discussion of uncertainty and variability should, ideally, reflect the type and complexity of the risk assessment – with proportionate levels of effort dedicated to the risk assessment and the analysis or discussion of uncertainty and variability.

8.6. Characterization of Data Variability and Uncertainties

An uncertainty analysis consists of the process that translates uncertainties about models, variables and input data, and also the random variability in measured parameters into uncertainties in output variables (Calabrese and Kostecki, 1991; Finkel, 1990; Iman and Helton, 1988). The analysis of uncertainties will typically involve the following fundamental elements:

- Evaluation of uncertainties in the input of each of the relevant tasks;
- Propagation of input uncertainties through each task;
- Combination/convolution of uncertainties in the output from the various tasks; and
- Display and interpretation of the uncertainties in the final results.

The goal of an analysis of uncertainties is to provide decision-makers with the complete spectrum of information concerning the quality of an assessment – including the potential variability in the estimated parameters, the major data gaps, and the effect that such data gaps have on the accuracy and reasonableness of the estimates that are developed (Borgen, 1990; Covello *et al.*, 1987; Cox and Ricci, 1992; Finkel and Evans, 1987; Helton, 1993; Hoffmann and Hammonds, 1992; Morgan and Henrion, 1991; USEPA, 1989a). Analysis and presentation of the uncertainties allow analysts or decision-makers to better evaluate the risk assessment results in the context of other factors being considered. This, in turn, will generally result in a more sound and open decision-making process.

Overall, a premium should be placed on a critical evaluation and presentation of all environmental, biological, and statistical uncertainties in the final assessment. Furthermore, it may be useful to carefully reexamine the quality of the studies used to support all conclusions, and to compare data across similar studies that are relevant to specific assessments. When appropriate, policy makers may employ plausible ranges associated with default exposure, toxicological and other assumptions/policy positions. These may include, for instance, ranges of default values such as the range of pulmonary ventilation rates (e.g., of $8 - 20$ m^3/day), human body weight (e.g., of $10 - 60$ or 70 kg), or ranges based on the use of low-dose extrapolation models (such as logit, probit, multistage, etc. models).

An uncertainty analysis can be performed qualitatively or quantitatively. But whether qualitative or quantitative in nature, the analysis considers uncertainties in the database; uncertainties arising from assumptions in modeling; and the completeness of the analysis.

8.6.1. QUALITATIVE ANALYSIS OF UNCERTAINTIES

The qualitative analysis of uncertainties typically involves a determination of the general quality and reasonableness of the risk assessment data, parameters, and results. Qualitative analysis is usually most important to 'screening', 'preliminary', and 'intermediate' level assessments (USEPA, 1989a).

As part of the qualitative analysis, the cause(s) of uncertainty is initially determined. The basic cause of uncertainty is a lack of knowledge on the part of the analyst because of inadequate, or even nonexistent, experimental and operational data on key processes and parameters. The specific causes of uncertainty that are typically addressed here can be categorized as follows (USEPA, 1989a, 1992, 1997):

- Measurement errors (resulting from measurement techniques employed in the study that could yield imprecise or biased measurements)
- Sampling errors (arising from the degree of representativeness of sampled data to actual population – e.g., small or unrepresentative samples)
- Aggregation errors (such as results from spatial and temporal approximations)
- Incomplete analysis (such as results from overlooking an important exposure scenario)
- Natural variability – e.g., in time, space, or activities
- Model limitations (reflecting on how close to reality the models employed are)
- Application and quality of generic or indirect empirical data
- Professional/expert judgment (reflecting on the possible unreliability of scientific assumptions that may have been revoked/used – e.g., selection of an inappropriate model or surrogate data).

In general, once the causes of the uncertainties have been identified, the impact that these uncertainties have on the assessment results would then have to be determined. Insofar as possible, measures to minimize the impacts of such uncertainties on the results should be elaborated. Ultimately, the explicit presentation of the qualitative analysis results will transmit the level of confidence in the results to the decision-maker – facilitating the implementation of appropriate environmental and public health risk management actions.

8.6.2. QUANTITATIVE ANALYSIS OF UNCERTAINTIES

In addition to a qualitative analysis, most detailed risk assessments may also require quantitative uncertainty analysis techniques to be used in chemical exposure studies. The quantitative analysis of uncertainties, often employed in detailed assessments, usually will proceed via sensitivity analysis and/or probabilistic analysis (e.g., Monte Carlo simulation techniques). The technique of choice depends on the availability of input data statistics. But regardless of the technique of choice, the approach will generally allow for a deviation from the conservative and rather unrealistic approach of generating point estimates for risks, as has 'traditionally' been done in most risk assessment programs. Indeed, point estimates tend to confer a false sense of precision and population homogeneity – and subsequently disguises the basis for rational decision-making. On the other hand, techniques such as Monte Carlo simulation provides a more complete description of risks, allowing risk managers and other stakeholders to appreciate/understand the level of protection offered by various risk management alternatives in an explicit manner. Ultimately, the Monte Carlo simulation approach helps the risk manager avoid making decisions based on implausible and unrealistic risk estimates.

In general, quantitative analysis of uncertainty becomes very important and necessary when prior risk screening calculations indicate a potential problem, or when risk control actions may result in excessively high costs, or when it is necessary to establish the relative importance of chemicals and exposure routes. Conversely, if estimated chemical intakes or risks are most obviously small and/or if the consequence of a 'wrong' prediction/decision based on the calculated risk is negligible, then quantitative analysis of uncertainty may neither be necessary nor a worthwhile effort.

Probabilistic Analysis: The Application of Monte Carlo Simulation Techniques
Various probabilistic analysis techniques can be employed/used to quantify uncertainties in risk assessment (e.g., Burmaster, 1996; Finley and Paustenbach, 1994; Finley et al., 1994a,b; Lee and Kissel, 1995; Lee et al., 1995; Macintosh et al., 1994; Power and McCarty, 1996; Richardson, 1996; Smith et al., 1992). The driving force behind the development and use of probabilistic risk assessment techniques has been the desire to more completely reflect the complexity in exposure conditions and toxicological responses that are present in the real world (Boyce, 1998). The probabilistic risk analyses may serve several purposes, including being used to: propagate uncertainty in the estimate of exposure dose and risk; properly prioritize resources for risk reduction activities; and simulate stochastic variability among individuals in a population. Probabilistic analysis may indeed be applied to the evaluation of risks in order that uncertainties are accounted for systematically.

In general, probabilistic analyses require data on the range and probability function (or distribution) of each model parameter. In fact, a central part of probabilistic risk analyses is the selection of probability distributions for the uncertain input variables (Haas, 1997; Hamed and Bedient, 1997). Thus, it is usually recommended to undertake a formal selection among various distributional families, along with a formal statistical goodness-of-fit test, in order to obtain the most suitable family appropriate for characterizing the case data set.

The favored probabilistic approach for assessing uncertainty is via Monte Carlo Simulation (e.g., McKone, 1994; McKone and Borgen, 1991; Price *et al.*, 1996; Smith, 1994; Thompson *et al.*, 1992). Monte Carlo simulation is a statistical technique by which a quantity is calculated repeatedly, using randomly selected/generated scenarios for each calculation cycle – and typically presenting the results in simple graphs and tables. The results from the simulation process approximate the full range of possible outcomes, and the likelihood of each.

The Monte Carlo simulation process involves assigning a joint probability distribution to the input variables; the procedure yields a concomitant distribution that is strictly a consequence of the assumed distributions of the model inputs and the assumed functional form of the model (Figure 8.7). Several considerations may be important in the selection of appropriate probability distribution used to represent the relevant input parameters (Box 8.6) (Finley *et al.*, 1994a; USEPA, 1989a). Unless specific information on the relationships between the relevant parameters is available, values for the required input parameters will normally be assumed to be independent.

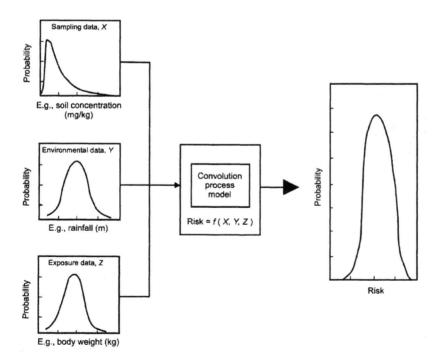

Figure 8.7. Conceptual illustration of the Monte Carlo simulation procedure

Box 8.6. Important considerations in the selection of appropriate probability distribution
in a Monte Carlo simulation

♦ A *uniform distribution* would be used to represent a factor/parameter when nothing is known about the factor except its finite range. The use of a uniform distribution assumes that all possible values within the range are equally likely.

♦ A *triangular distribution* would be used if the range of the parameter and its mode are known.

♦ A *Beta distribution* (scaled to the desired range) may be most appropriate if the parameter has a finite range of possible values and a smooth probability function is desired.

♦ A *Gamma, Log-Normal,* or *Weibull distribution* may be an appropriate choice if the parameter only assumes positive values. The Gamma distribution is probably the most flexible, especially because its probability function can assume a variety of shapes by varying its parameters, and it is mathematically tractable.

♦ A *Normal distribution* may be an appropriate choice if the parameter has an unrestricted range of possible values and is symmetrically distributed around its mode.

Monte Carlo simulations can indeed be used to develop numerical estimates of uncertainties that allows efficient ways to extend risk assessment methods to the estimation of both point values as well as distributions of risks posed by chemical exposure problems. In using Monte Carlo techniques, most or all input variables to the risk assessment models become random variables with known or estimated probability density functions (pdfs). Within this framework, a variable can take on a range of values with a known probability. In general, when Monte Carlo Simulation is applied to risk assessment, the risk presentation appears as frequency or probability distribution graphs – as illustrated by Figure 8.8 – from which the mean, median, variance, and/or percentile levels/values can be extracted.

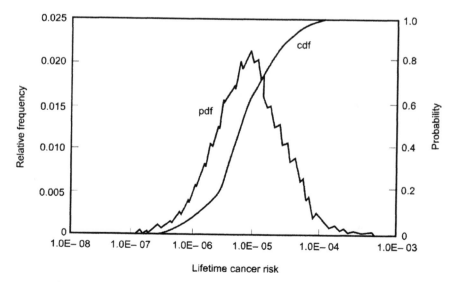

Figure 8.8. An illustrative sketch of a plot from a Monte Carlo simulation analysis (showing probability density function [pdf] and cumulative distribution function [cdf] for lifetime cancer risks from a contaminated site)

8.7. Presenting and Managing Uncertain Risks: The Role of Sensitivity Analyses

Inevitably, some degree of uncertainties remains in quantitative risk estimates in virtually all fields of applied risk analysis. A carefully executed analysis of uncertainties therefore plays a very important role in all risk assessments. On the other hand, either or both of a comprehensive qualitative analysis and a rigorous quantitative analysis of uncertainties will be of little value if the results of such analysis are not clearly presented for effective use in the decision-making process. A number of methods of approach have been suggested by investigators like Cox and Ricci (in Paustenbach, 1988) for presenting risk analysis results to decision-makers, including the following:

- Risk assessment results should be presented in a sufficiently disaggregated form (to show risks for different subgroups) so that key uncertainties and heterogeneities are not lost in the aggregation.
- Confidence bands around the predictions of statistical models can be useful, but uncertainties about the assumptions of the model itself should also be presented.
- Both individual (e.g., the typical and most threatened individuals in a population) and population/group risks should be presented, so that the equity of the distribution of individual risks in the population can be appreciated and taken into account.
- Any uncertainties, heterogeneities, or correlations across individual risks should be identified.
- Population risks can be described at the 'micro' level (namely, in terms of frequency distribution of individual risks), and/or at the 'macro' level (namely, by using decision-analytic models, in terms of attributes such as equivalent number of life-years).

It is noteworthy that, the uncertainty analysis can also be achieved via sensitivity analyses for key assumptions. This process serves to identify the sensitivity of the calculated result vis-à-vis the various input assumptions – and thus identify key uncertainties, as well as to bracket potential risks so that policy-makers can make more informed decisions or choices.

Sensitivity analysis, often a useful adjunct to the traditional uncertainty analysis, comprises of a process that examines the relative change or response of output variables caused by variation of the input variables and parameters (Calabrese and Kostecki, 1991; Iman and Helton, 1988; USEPA, 1992, 1997). It is a technique that tests the sensitivity of an output variable to the possible variation in the input variables of a given model. Typically, the performance of sensitivity testing requires data on the range of values for each relevant model parameter. The purpose of sensitivity analysis is then to identify the influential input variables, and to develop bounds on the model output. When computing the sensitivity with respect to a given input variable, all other input variables are held fixed at their nominal values. By identifying the influential or critical input variables, more resources can be directed to reduce their uncertainties, and thence reduce the output uncertainty.

Notwithstanding the added value of sensitivity analyses, several factors may still contribute to the over- or under-estimation of risks. For example, in human health risk assessments, some factors will invariably underestimate health impacts associated with the chemicals evaluated in the assessment. These may include: lack of potency data for some carcinogenic chemicals; risk contributions from compounds produced as transformation byproducts, but that are not quantified; and the fact that all risks are assumed to be additive, although certain combinations of exposure may have synergistic effects. Conversely, another set of factors would invariably cause the process to overestimate risks. These may include the fact that: many unit risk and potency factors are often considered plausible upper-bound estimates of carcinogenic potency, when indeed the true potency of the chemical could be considerably lower; exposure estimates are often very conservative; and possible antagonistic effects, for chemicals whose combined presence reduce toxic impacts, are not accounted for.

In conclusion, the results of deterministic risk assessments should be interpreted with caution, and never construed as absolute measures of risk. Notwithstanding, such point estimates of risk may still be useful in a qualitative sense for the ranking of different public health risk management programs or issues. In any event, probabilistic methods must be encouraged as the logical evolution of the risk assessment, and this should be accompanied by the development of risk management methods that can utilize the richness of information provided by Monte Carlo assessments and other similar techniques (Zemba *et al.*, 1996). In fact, it is believed that, the danger of mischaracterizing high-end, central tendency, and indeed other statistical exposure levels can only be properly alleviated via the development and utilization of full probabilistic analyses.

8.8. Suggested Further Reading

Goldman, M., 1996. Cancer risk at low-level exposure, *Science,* 271: 1821 – 1822

Goodrich, MT and JT McCord, 1995. Quantification of uncertainty in exposure assessment at hazardous waste sites, *Ground Water,* 33(5): 727 – 732

Law, AM and WD Kelton, 1991. Simulation Modeling and Analysis, McGraw-Hill, New York, NY

Maxwell, RM and WE Kastenberg, 1999. Stochastic environmental risk analysis: an integrated methodology for predicting cancer risk from contaminated groundwater, *Stochastic Environmental Research and Risk Assessment,* 13: 27 – 47

Nendza, M., 1997. *Structure-Activity Relationships in Environmental Sciences,* Kluwer Academic Publishers, Dordrecht, The Netherlands

Phelan, MJ, 1998. Environmental health policy decisions: the role of uncertainty in economic analysis, *Journal of Environmental Health,* 61(5): 8 – 13

Richards, D. and WD Rowe, 1999. Decision-Making with Heterogeneous Sources of Information, *Risk Analysis,* 19(1): 69-81

USEPA, 1997. *Guiding Principles for Monte Carlo Analysis,* EPA/630/R-97/001, Risk Assessment Forum, Office of Research and Development, US Environmental Protection Agency (USEPA), Washington, DC

USEPA, 1999. *Report of the Workshop on Selecting Input Distributions for Probabilistic Assessments,* EPA/630/R-98/004, Risk Assessment Forum, Office of Research and Development, US Environmental Protection Agency (USEPA), Washington, DC

West, GB, JH Brown, and BJ Enquist, 1997. A general model of the origin of allometric scaling laws in biology, *Science,* 276: 122 – 126

Chapter 9

DETERMINATION OF 'ACCEPTABLE' AND 'SAFE' LEVELS OF CHEMICAL EXPOSURE

An important and yet controversial issue that comes up in attempts to establish 'safe' or 'tolerable' levels for human exposure to chemical constituents relates to the notion of an 'acceptable chemical exposure level' (ACEL). The ACEL may be considered as the concentration of a chemical in a particular medium or product that, when exceeded, presents significant risk of adverse impact to potential receptors. In fact, in a number of situations, the ACEL concept tends to drive the public health risk management decision made about several consumer products. However, the ACELs may not always result in 'safe' or 'tolerable' risk levels – in part due to the nature of the critical exposure scenarios, receptor-specific factors, and other conditions that are specific to the particular hazard situation. Under such circumstances, it becomes necessary to develop more stringent and health-protective levels that will meet the 'safe' or 'tolerable' risk level criteria.

This chapter presents discussions of how risk assessment may facilitate a determination of what constitutes a reasonably 'safe' or 'acceptable' concentration of chemicals appearing in a variety of consumer products and in the human environments. This also includes an elaboration of a number of analytical relationships that can be adapted or used to estimate such 'safe' levels that are necessary for public health risk management decisions.

9.1. Requirements and Criteria for Establishing Risk-Based Chemical Exposure Levels

Risk-based chemical exposure levels (RBCELs) may generally be derived for various chemical sources by manipulating the exposure and risk models previously presented in Chapters 6 and 8. This involves a 'back-calculation' process that yields a media concentration predicated on health-protective exposure parameters; as an example, the RBCEL generally results in a cumulative non-cancer hazard index of ≤ 1 and/or a cumulative carcinogenic risk $\leq 10^{-6}$. In general, since risk is a function of both the exposure to a chemical and the toxicity of that chemical, a complete understanding of the exposure scenarios together with an accurate determination of the constituent toxicity is key to developing permissible exposure levels that will be protective of human health.

The target RBCELs are typically established for both the carcinogenic and non-carcinogenic effects of the constituents of concern – with the more stringent value usually being selected as a public health criterion (Figure 9.1); invariably, the carcinogenic limit tends to be more stringent in most situations where both values exist. Within the general procedural framework, the following criteria and general guidelines may additionally be used to facilitate the process of establishing media-specific RBCELs and/or public health goals:

- Assuming dose additivity, $\displaystyle\sum_{j=1}^{p} \sum_{i=1}^{n} \frac{CMAX_{ij}}{RBCEL_{ij}} < 1$

  where $CMAX_{ij}$ is the prevailing maximum concentration of constituent i in product or matrix j, and $RBCEL_{ij}$ is the risk-based chemical exposure level for constituent i in product or matrix j.
- In developing public health goals, it usually is necessary to establish a target level of risk for the constituents of concern; such standards are generally established within the risk range of 10^{-7} to 10^{-4} (with a lifetime excess cancer risk of 10^{-6} normally used as a point-of-departure) and a hazard index of 1.
- It is recommended that the cumulative risk posed by multiple chemical constituents not exceed a 10^{-4} cancer risk and/or a hazard index of unity.
- If sensitive populations (including vulnerable persons, such as children and the sick) are to be protected, then more stringent standards may be required.
- If nearby populations are exposed to hazardous constituents from other sources, lower target levels may generally be required than would ordinarily be necessary.
- If exposures to certain hazardous constituents occur through multiple routes, lower target levels should generally be prescribed.

Indeed, if/when these conditions are satisfied, then the RBCEL represents a maximum acceptable constituent level that will likely be sufficiently protective of public health. In general, exceeding the RBCEL will usually call for the development and implementation of a corrective action and/or public health risk management plan.

9.1.1. MISCELLANEOUS METHODS FOR ESTABLISHING ENVIRONMENTAL QUALITY GOALS

Several possibilities exist to use various analytical tools in the development of alternative and media-specific chemical exposure concentration limits and environmental quality goals. Some select general procedures commonly employed in establishing environmental quality goals are briefly annotated below. Broadly speaking, these approaches represent reasonably conservative ways of setting environmental quality goals. The use of such methods will generally ensure that risks are not underestimated – which tantamount to situations that result in the adequate protection of public health.

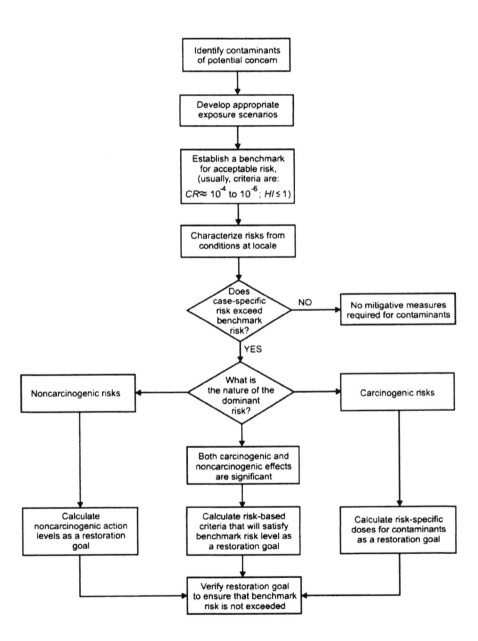

Figure 9.1. General protocol for developing risk-based chemical exposure levels and public health goals

Determination of Risk-specific Concentrations in Air
The estimation of health-protective concentrations of chemical constituents in air must take into account the toxicity of the chemicals of potential concern, as well as the potential exposure of individuals breathing the impacted air. By employing the risk assessment concepts and methodologies discussed in Chapters 6 through 8, risk-specific concentrations of chemicals in air may be estimated from the unit risk in air as follows:

$$\text{Air concentration } [\mu g/m^3] = \frac{[\text{specified risk level}] \times [\text{body weight}]}{SF_i \times [\text{inhalation rate}] \times 10^{-3}}$$

$$= \frac{[\text{specified risk level}]}{URF_i} = \frac{1 \times 10^{-6}}{URF_i} \qquad (9.1)$$

The assumptions generally used for such computations involve a specified risk level of 10^{-6}, a 70-kg body weight, and an average inhalation rate of 20 m^3/day.

Determination of Risk-specific Concentrations in Water
The estimation of health-protective concentrations of chemical constituents in drinking water must take into account the toxicity of the chemicals of potential concern, as well as the potential exposure of individuals using the water. By employing the risk assessment concepts and methodologies discussed in Chapters 6 through 8, risk-specific concentrations of chemicals in drinking water can be estimated from the oral slope factor. The water concentration corrected for an upper-bound increased lifetime risk of R $(=10^{-6})$ is given by:

$$\text{Water concentration } [mg/L] = \frac{[\text{specified risk level}] \times [\text{body weight}]}{SF_o \times [\text{ingestion rate}]}$$

or,

$$= \frac{\text{specified risk level}}{URF_o} \qquad (9.2)$$

The assumptions generally used for such computations involve a specified risk level of 10^{-6}, a 70-kg body weight, and an average water ingestion rate of 2 L/day – so that:

$$\text{Water concentration } [mg/L] = \frac{1 \times 10^{-6} \times 70 \text{ kg}}{SF_o \, (mg/kg/day)^{-1} \times 2 \text{ L/day}} = \frac{3.5 \times 10^{-5}}{SF_o}$$

It is noteworthy that, in general, the estimation of health-protective concentrations of chemical constituents in drinking water that is associated with negligible risks must also account for the fact that tap water is typically used directly as drinking water, as well as for preparing foods and beverages. The water is also used for bathing/showering, in washing clothes and dishes, flushing of toilets, and in a variety of other household uses – some of which could result in potential dermal and inhalation exposures as well. To allow for these additional exposures, therefore, the assumed daily volume of water consumed by an adult is typically increased from the default value of 2 L/day indicated above, to 3 L-equivalents/day (Leq/day).

9.2. Assessing the Safety of Chemicals in Consumer Products

Through the use of a variety of consumer products, numerous groups of peoples around the world are exposed to a barrage of chemical compounds on a daily basis. Typically, risk assessments (which allow the consumer exposures to be estimated by measurements and/or models) assist in the determination and management of potential health problems that could be expected or anticipated from the use of such consumer products. It is noteworthy, however, that the exposure assessment component of the processes involved tends to be particularly complicated, though not insurmountable – because of the huge diversity in usage and composition of consumer products. There is also the additional issue of intermittent exposures to variable amounts and types of products containing varying concentrations of chemical compounds (van Veen, 1996; Vermeire *et al.*, 1993). Notwithstanding, risk-based analyses can be carefully designed to help evaluate the safety of chemicals that appear in various consumer goods.

Consumer product safety is a function of exposure and toxicity, determined primarily based on the exposure patterns/rates and the toxicity of the chemical components of concern. This can be represented by the following conceptual expressions:

$$\text{Risk} = f(\text{Exposure}, \text{Toxicity}) \tag{9.3}$$

or,

$$\text{Safety} \ \alpha \ \frac{1}{\text{Risk}} \ = \ \frac{1}{f(\text{Exposure}, \text{Toxicity})} \tag{9.4}$$

For a particular consumer product to be classified as reasonably safe, the chemical-specific exposure dose should generally be less than the chemical's 'acceptable' daily intake – defined as the daily intake level for a chemical that represents no anticipated significant risk to the consumer or exposed individual.

9.2.1. DETERMINATION OF 'TOLERABLE' CHEMICAL CONCENTRATIONS

Chemicals in consumer products (including that occurring in dietary materials or foods) may be classified into two broad categories – carcinogenic and non-carcinogenic materials. The methods for deriving the 'acceptable' daily intakes and/or 'tolerable' concentrations for such chemicals are generally based on procedures/protocols presented earlier on in Chapters 6 through 8; the general concepts are briefly annotated below.

'Acceptable' Daily Intake and 'Tolerable' Concentration for Carcinogens
The 'acceptable' daily intake for carcinogenic materials appearing in a consumer product may be estimated by using the following approximate relationships:

$$\text{ADI}_{\text{carcinogen}} = \frac{[\text{TR x AT x 365 d/yr}]}{[\text{ED x EF x SF}]} \tag{9.5}$$

Thence, the 'tolerable' chemical concentration for carcinogens [$TC_{carcinogen}$] (mg/kg or mg/L) in the consumer product will be defined by,

$$TC_{carcinogen} = \frac{[ADI_{carcinogen} \times BW]}{[FR \times CR \times ABS]} \times CF \qquad (9.6)$$

where $ADI_{carcinogen}$ is the 'acceptable' daily intake for the carcinogenic materials (mg/kg-day); TR is the generally acceptable risk level (usually set at 10^{-6}); AT is the averaging time (years); ED is the exposure duration (yr); EF is the exposure frequency (d/yr); SF is the cancer potency or slope factor ($[mg/kg-d]^{-1}$); BW is the average body weight (kg); FR is the fraction of consumed material that is assumed to be contaminated; CR is the consumption rate (kg/d or L/d); ABS is the % absorption rate; and CF is a conversion factor to help maintain the dimensional tractability of the algorithm.

'Acceptable' Daily Intake and 'Tolerable' Concentration for Non-carcinogens
The 'acceptable' daily intake for non-carcinogenic materials appearing in a consumer product may be estimated by using the following approximate relationship:

$$ADI_{non-carcinogen} = \frac{[HQ \times AT \times 365 \ d/yr \times RfD]}{[ED \times EF]} \qquad (9.7)$$

Thence, the 'tolerable' chemical concentration for non-carcinogens [$TC_{non-carcinogen}$] (mg/kg or mg/L) in the consumer product will be defined by,

$$TC_{non-carcinogen} = \frac{[ADI_{noncarcinogen} \times BW]}{[FR \times CR \times ABS]} \times CF \qquad (9.8)$$

where $ADI_{non-carcinogen}$ is the 'acceptable' daily intake for the non-carcinogenic materials (mg/kg-day); HQ is the generally acceptable hazard level (usually set at 1); AT is the averaging time (years); ED is the exposure duration (yr); EF is the exposure frequency (d/yr); RfD is the non-cancer reference dose or acceptable daily intake (mg/kg-d); BW is the average body weight (kg); FR is the fraction of consumed material that is assumed to be contaminated; CR is the consumption rate (kg/d or L/d); ABS is the % absorption rate; and CF is a conversion factor to help maintain the dimensional tractability of the algorithm.

9.3. Determination of Risk-Based Chemical Exposure Levels

After defining the critical exposure routes and exposure scenarios appropriate for a given a chemical exposure problem, it generally becomes possible to estimate a corresponding RBCEL that would not pose significant risks to an exposed population. To determine the RBCEL for a chemical compound, algebraic manipulations of the hazard index and/or carcinogenic risk equations together with the exposure estimation equations discussed in Chapters 6 through 8 can be used to arrive at the appropriate analytical relationships. The step-wise computational efforts involved in this exercise consist of a 'back-calculation' process that yields a media concentration predicated on health-protective exposure parameters; as an example, the RBCEL generally results in a cumulative non-cancer hazard index of ≤ 1 and/or a cumulative carcinogenic risk $\leq 10^{-6}$. Indeed, for chemicals with carcinogenic effects, a target risk of (1×10^{-6}) is typically used in the 'back-calculation'; and a target hazard index of 1.0 is typically used for non-carcinogenic effects.

The processes involved in the determination of the RBCELs are summarized in the proceeding sections. For substances that are both carcinogenic and possess systemic toxicity properties, the lower of the carcinogenic or non-carcinogenic criterion should be used for the relevant public health risk management action or decision.

9.3.1. RBCELs FOR CARCINOGENIC CONSTITUENTS

As discussed in Chapters 6 through 8, the cancer risk (CR) for the significant human exposure routes (comprised of inhalation, ingestion, and dermal exposures) may be represented as follows:

$$CR = \{ \sum_{i=1}^{p} CDI_p \times SF_p \}$$

$$= [CDI_i \times SF_i]_{inhalation} + [CDI_o \times SF_o]_{ingestion} + [CDI_d \times SF_o]_{dermal\ contact}$$

$$\equiv C_m\{[INHf \times SF_i] + [INGf \times SF_o] + [DEXf \times SF_o]\} \qquad (9.9)$$

where the CDIs represent the chronic daily intakes, adjusted for absorption (mg/kg-day); INHf, INGf, and DEXf represent the inhalation, ingestion, and dermal contact 'intake factors', respectively (see Chapter 6); C_m is the chemical concentration in environmental/exposure matrix of concern; and the SFs are the route-specific cancer slope factors; and the subscripts i, o and d refer to the inhalation, oral ingestion, and dermal contact exposures, respectively.

The above model can be re-formulated to calculate the carcinogenic RBCEL (viz., $RBCEL_c$) for the environmental/exposure media of interest. This involves 'back-calculating' from the chemical intake equations presented in Chapter 6 for inhalation, ingestion, and dermal contact exposures. Hence,

$$RBCEL_c = C_m = \frac{CR}{\{[INHf \times SF_i] + [INGf \times SF_o] + [DEXf \times SF_o]\}} \qquad (9.10)$$

For illustrative purposes, let us assume that there is only one chemical constituent present in soils at a hypothetical contaminated land, and that only exposures via the dermal and ingestion routes contribute to, or at least dominate the total target carcinogenic risk (of, say CR = 10^{-6}). Then,

$$CDI = \frac{CR}{SF_O} = RSD$$

or,

$$(CDI_{ing} + CDI_{der}) = \frac{CR}{SF_O}$$

i.e.,

$$\frac{(RBC_c \times SIR \times CF \times FI \times ABS_{si} \times EF \times ED)}{(BW \times AT \times 365)}$$
$$+ \frac{(RBC_c \times CF \times SA \times AF \times ABS_{sd} \times SM \times EF \times ED)}{(BW \times AT \times 365)} = \frac{CR}{SF_O}$$

Consequently,

$$RBCEL_c = \frac{(BW \times AT \times 365) \times (RSD)}{(CF \times EF \times ED)\{(SIR \times FI \times ABS_{si}) + (SA \times AF \times ABS_{sd} \times SM)\}}$$

where RSD represents the risk-specific dose, defined by the ratio of the target risk to the slope factor.

Indeed, the estimated RBCEL may serve as surrogate for a health-based acceptable chemical exposure level (ACEL) – albeit some case-specific adjustments will usually be required, in order to arrive at a true ACEL used in public health risk management decisions.

Health-Based ACELs for Carcinogenic Chemicals
As health-based criteria, ACELs for carcinogens may be determined in a similar manner to the so-called 'virtually safe dose' (VSD) of a carcinogenic chemical constituent. A VSD is the daily dose of a carcinogenic chemical that, over a lifetime, will result in an incidence of cancer at a specified risk level; usually, this is calculated based on the appropriate *de minimis* risk level.

The governing equation for calculating ACELs for carcinogenic constituents is shown in Box 9.1. This model – developed from algorithms and concepts presented earlier on in Chapters 6 through 8 – assumes that there is only one chemical constituent involved in the problem situation. In other situations where several chemicals may be of concern, it is assumed (for simplification purposes) that each carcinogen has a different mode of biological action and target organs. Each of the carcinogens is, therefore, assigned 100% of the 'acceptable' excess carcinogenic risk (typically equal to $[1 \times 10^{-6}]$) in calculating the health-based ACELs; in other words, the excess carcinogenic risk is not allocated among the carcinogens.

Example Calculations. Consider a hypothetical situation whereby some human receptors may be consuming water contaminated with methylene chloride. Then, the allowable human exposure due to ingestion of 2 liters of the water containing methylene chloride (with oral SF = 7.5×10^{-3} $[mg/kg/day]^{-1}$) by a 70-kg weight adult over a 70-year lifetime is given by:

$$ACEL = \frac{(R \times BW \times LT \times CF)}{(SF \times I \times A \times ED)}$$

i.e.,

$$ACEL_{methchl} = \frac{[10^{-6} \times 70 \times 70 \times 1]}{[0.0075 \times 2 \times 1 \times 70]} \approx 0.005 \text{ mg/L} = 5 \text{ μg/L}$$

Thus, the health-based ACEL for methylene chloride, based on an acceptable excess lifetime cancer risk of 10^{-6}, is estimated to be 5 μg/L.

Next, consider another situation of a contaminated land impacting a multipurpose surface water-body due to overland flow. This surface water body is used both as a culinary water supply source and for recreational purposes. Assuming – in addition to the water intake – an average daily consumption of aquatic organisms, DIA, of 6.5g/day, and a BCF of 0.91 L/kg for methylene chloride, then the health-based exposure levels for the ingestion of both water and fish is determined from the following modified equation:

$$ACEL_{methchl} = \frac{[R \times BW \times LT \times CF]}{[SF \times (I + (DIA \times BCF)) \times A \times ED]}$$

$$= \frac{[10^{-6} \times 70 \times 70 \times 1]}{[0.0075 \times (2 + (0.0065 \times 0.91)) \times 1 \times 70]} \approx 0.005 \text{ mg/L} = 5 \text{ μg/L}$$

Thus, in this particular case, the allowable exposure levels for drinking water and eating aquatic organisms contaminated with methylene chloride is also approximately 5 μg/L.

Box 9.1. General equation for calculating acceptable chemical exposure levels for carcinogenic constituents

$$ACEL_c = \frac{(R \times BW \times LT \times CF)}{(SF \times I \times A \times ED)}$$

where:

$ACEL_c$ = acceptable chemical exposure level (equivalent to the VSD) in medium of concern (e.g., mg/kg in food; mg/L in water)

R = specified benchmark risk level, usually set at 10^{-6} (dimensionless)

BW = body weight (kg)

LT = assumed lifetime (years)

CF = conversion factor (equals 10^6 for ingestion exposure from solid materials; 1.00 for ingestion of fluids)

SF = cancer slope factor ($[\text{mg/kg-day}]^{-1}$)

I = intake assumption (mg/day for solid material ingestion rate; L/day for fluid ingestion)

A = absorption factor (dimensionless)

ED = exposure duration (years)

9.3.2. *RBCELs* FOR NON-CARCINOGENIC EFFECTS OF CHEMICAL CONSTITUENTS

As discussed in Chapters 6 through 8, the hazard index (HI) for the significant human exposure routes (comprised of inhalation, ingestion and dermal exposures) is given by:

$$HI = \{ \sum_{i=1}^{p} \frac{CDI_p}{RfD_p} \}$$

$$= [\frac{CDI_i}{RfD_i}] \text{inhalation} + [\frac{CDI_o}{RfD_o}] \text{ingestion} + [\frac{CDI_d}{RfD_o}] \text{dermal contact}$$

$$\equiv C_m \{ [\frac{INHf}{RfD_i}] + [\frac{INGf}{RfD_o}] + [\frac{DEXf}{RfD_o}] \} \tag{9.11}$$

where the CDIs represent the chronic daily intakes, adjusted for absorption (mg/kg-day); INHf, INGf, and DEXf represent the inhalation, ingestion, and dermal contact 'intake factors', respectively (see Chapter 6); C_m is the chemical concentration in environmental/exposure matrix of concern; and the RfDs are the route-specific reference doses; the subscripts *i, o* and *d* refer to the inhalation, oral ingestion and dermal contact exposures, respectively.

The above model can be re-formulated to calculate the non-carcinogenic RBCEL (viz., *RBCEL$_{nc}$*) for the environmental/exposure media of interest. This is derived by 'back-calculating' from the chemical intake equations presented in Chapter 6 for inhalation, ingestion, and dermal contact exposures. Hence,

$$RBCEL_{nc} = C_m = \frac{1}{\{ [\frac{INHf}{RfD_i}] + [\frac{INGf}{RfD_o}] + [\frac{DEXf}{RfD_o}] \}} \tag{9.12}$$

For illustrative purposes, assume that there is only one chemical constituent present in soils at a hypothetical contaminated land, and that only exposures via the dermal and ingestion routes contribute to, or at least dominate the total target hazard index (of HI = 1). Then,

$$CDI = RfD$$

or,

$$(CDI_{ing} + CDI_{der}) = RfD_o$$

i.e.,

$$\frac{(RBC_{nc} \times SIR \times CF \times FI \times ABS_{si} \times EF \times ED)}{(BW \times AT \times 365)}$$
$$+ \frac{(RBC_{nc} \times CF \times SA \times AF \times ABS_{sd} \times SM \times EF \times ED)}{(BW \times AT \times 365)} = RfD_o$$

Consequently,

$$RBCEL_{nc} = \frac{(BW \times AT \times 365) \times (RfD_O)}{(CF \times EF \times ED)\{(SIR \times FI \times ABS_{si}) + (SA \times AF \times ABS_{sd} \times SM)\}}$$

assuming a benchmark hazard index of unity.

Indeed, the estimated RBCEL may serve as surrogate for a health-based acceptable chemical exposure level (ACEL) – albeit some case-specific adjustments will usually be required, in order to arrive at a true ACEL used in public health risk management decisions.

Health-Based ACELs for Non-Carcinogenic Chemicals
As health-based criteria, ACELs for non-carcinogens may be determined in a similar manner to the so-called 'allowable daily intakes' (ADIs) of a non-carcinogenic chemical constituent. The ADI represents the threshold exposure limit below which no adverse effects are anticipated.

The governing equation for calculating ACELs for non-carcinogenic effects (i.e., the systemic toxicity) of chemical constituents is shown in Box 9.2. This model – derived from algorithms and concepts presented earlier on in Chapters 6 through 8 – assumes that there is only one chemical constituent involved. In situations where several chemicals may be of concern, it is assumed (for simplification purposes) that each chemical has a different organ-specific non-carcinogenic effect. Otherwise, the right hand side may be multiplied by a percentage factor to account for contribution to hazard index by each non-carcinogenic chemical subgroup.

Box 9.2. General equation for calculating acceptable chemical exposure levels for non-carcinogenic effects of systemic toxicants

$$ACEL_{nc} = \frac{(RfD \times BW \times CF)}{(I \times A)}$$

where:
$ACEL_{nc}$ = acceptable chemical exposure level in medium of concern (e.g., mg/kg in food; mg/L in water)
RfD = reference dose (mg/kg-day)
BW = body weight (kg)
CF = conversion factor (equals 10^6 for ingestion exposure from solid materials; 1.00 for fluid ingestion)
I = intake assumption (mg/day for solid material ingestion rate; L/day for fluid ingestion)
A = absorption factor (dimensionless)

Example Calculations. Consider a hypothetical situation whereby some human receptors may be consuming water contaminated with ethylbenzene. Then, the allowable human exposure concentration associated with the ingestion of 2 liters of water containing ethylbenzene (with RfD of 0.1 mg/kg/day) by a 70-kg weight adult is given by:

$$ACEL = \frac{[RfD \times BW]}{[DW \times A]}$$

i.e.,

$$ACEL_{ebz} = \frac{[0.1 \times 70]}{[2 \times 1]} = 3,500 \ \mu g/L$$

Next, consider another situation of a contaminated land impacting a multipurpose surface water-body due to overland flow. This surface water body is used both as a culinary water supply source and for recreational purposes. Assuming – in addition to the water intake – an average daily consumption of aquatic organisms, DIA, of 6.5g/day, and a BCF of 37.5 L/kg for ethylbenzene, then the health-based exposure levels for the ingestion of both water and fish is determined from the following modified equation:

$$ACEL_{ebz} \ [mg/L] = \frac{[RfD \times BW]}{[2 + (0.0065 \times BCF)] \times 1}$$

$$= \frac{[0.1 \times 70]}{[2 + (0.0065 \times 37.5)]} = 3,120 \ \mu g/L$$

Thus, the allowable exposure concentration (represented by the water ACEL) for drinking water and eating aquatic organisms contaminated with ethylbenzene is approximately 3,120 μg/L.

9.4.   Establishing Risk-Based Cleanup Limits for Contaminated Lands

Risk assessment has become particularly useful in determining the level of cleanup most appropriate for potentially contaminated lands. By utilizing methodologies that establish cleanup criteria based on risk assessment principles, corrective action programs can be conducted in a cost-effective and efficient manner. Once realistic risk reduction levels potentially achievable by various remedial alternatives are known, the decision-maker can then use other scientific criteria (such as implementability, reliability, operability, and cost) to select a final design alternative. Subsequently, an appropriate corrective action plan can be developed and implemented for the contaminated land. In fact, a major consideration in developing a remedial action plan for a contaminated land is the level of cleanup to be achieved – which could become the driving force behind remediation costs. The site cleanup limit concept generally facilitates decisions as to the effective use of limited funds to clean a site to a level appropriate/safe for its intended use. It is therefore be prudent to allocate adequate resources to develop the appropriate cleanup criteria.

In principle, the cleanup criteria selected for a potentially contaminated land may vary significantly from one site to another – due especially to the prevailing site-specific conditions. Similarly, mitigation measures may be case-specific for various hazardous situations and problems. In general, preliminary remediation goals (PRGs) are usually established as cleanup objectives early in a site characterization process. The development of PRGs requires site-specific data relating to the impacted media of interest, the chemicals of potential concern (CoPCs), and the probable future land uses.

The early determination of remediation goals facilitates the development of a range of feasible corrective action decisions, which in turn helps focus remedy selection on the most effective remedial alternative(s). It is noteworthy that an initial list of PRGs may have to be revised when new data becomes available during the site characterization process. In fact, PRGs are refined into final remediation goals throughout the process leading up to the final remedy selection. Consequently, it is important to iteratively review and re-evaluate the media and CoPCs, future land uses, and exposure assumptions originally identified during project formulation.

In addressing potentially contaminated land problems, soils can become the major focus of attention in the risk management decisions; this is because soils at such sites could serve as a major long-term reservoir for chemical contaminants – with the capacity to release contamination into several other environmental media. As such, the importance of soil cleanup for such contaminated lands cannot be over-emphasized. Indeed, the soil media typically requires a particularly close attention in most risk-based evaluations carried out for contaminated lands – albeit groundwater contaminant plumes underlying such sites are proving to be equally, if not more, problematic in some situations.

Consider, for illustrative purposes, a potentially contaminated land that is being considered for remediation so that it may be re-developed for either residential or industrial purposes. Contaminant levels in residential soils in which children might play (which allows for pica behavior in toddlers and other infants) must necessarily be lower than the same contaminant levels in soils present at a site designated for large industrial complexes (which effectively prevent direct exposures to contaminated soils). Also, the release potential of several chemical constituents will usually be different from sandy soils vs. clayey soils; this will affect the possible exposure scenarios, and therefore the acceptable soil contaminant levels that is designated for the different types of soils. Consequently, it is generally preferable to establish and use site-specific cleanup criteria for most contaminated land problems, especially where soil exposures is critical to the site restoration decisions.

To determine the risk-based cleanup level for a chemical compound present in soils at a contaminated land, algebraic manipulations of the hazard index and/or carcinogenic risk equations together with the exposure estimation equations discussed in Chapters 6 through 8 can be used to arrive at the appropriate analytical relationships. The step-wise computational efforts involved in this exercise consist of a 'back-calculation' process that yields an acceptable soil concentration (ASC) predicated on health-protective exposure parameters; as an example, the ASC generally results in a cumulative non-cancer hazard index of ≤ 1 and/or a cumulative carcinogenic risk $\leq 10^{-6}$. Indeed, for chemicals with carcinogenic effects, a target risk of $[1 \times 10^{-6}]$ is typically used in the 'back-calculation'; and a target hazard index of 1.0 is typically used for non-carcinogenic effects.

The processes involved in the determination of the ASCs are summarized in the proceeding sections. For substances that are both carcinogenic and possess systemic toxicity properties, the lower of the carcinogenic or non-carcinogenic criterion should be used for the relevant site restoration and/or risk management decisions.

9.4.1. SOIL CHEMICAL LIMITS FOR CARCINOGENIC CONTAMINANTS

Box 9.3 shows a general equation for calculating the risk-based site restoration criteria for a single carcinogenic chemical present in soils at a contaminated land. This has been derived by 'back-calculating' from the risk and chemical exposure equations associated with the inhalation of soil emissions, ingestion of soils, and dermal contact with soils. It is noteworthy that, where appropriate and necessary, this general equation may also be re-formulated to incorporate the receptor age-adjustment exposure factors developed and presented earlier on in Chapter 6.

Box 9.3. General equation for calculating risk-based soil cleanup level
for a carcinogenic chemical constituent

$$ASC_c =$$

$$\frac{TCR}{(\frac{EF \times ED \times CF}{BW \times AT \times 365}) \times \{[SF_i \times IR \times RR \times ABS_a \times AEF \times CF_a] + [(SF_o \times SIR \times FI \times ABS_{si}) + (SF_o \times SA \times AF \times ABS_{sd} \times SM)]\}}$$

$$= \frac{(TCR) \times (BW \times AT \times 365)}{(EF \times ED \times CF) \times \{[SF_i \times IR \times RR \times ABS_a \times AEF \times CF_a] + SF_o[(SIR \times FI \times ABS_{si}) + (SA \times AF \times ABS_{sd} \times SM)]\}}$$

where:

ASC_c	= acceptable soil concentration (i.e., acceptable risk-based cleanup level) of carcinogenic contaminant in soil (mg/kg)
TCR	= target cancer risk, usually set at 10^{-6} (dimensionless)
SF_i	= inhalation slope factor ($[mg/kg\text{-}day]^{-1}$)
SF_o	= oral slope factor ($[mg/kg\text{-}day]^{-1}$)
IR	= inhalation rate (m^3/day)
RR	= retention rate of inhaled air (%)
ABS_a	= percent chemical absorbed into bloodstream (%)
AEF	= air emissions factor, i.e., PM_{10} particulate emissions or volatilization (kg/m^3)
CF_a	= conversion factor for air emission term (10^6)
SIR	= soil ingestion rate (mg/day)
CF	= conversion factor (10^{-6} kg/mg)
FI	= fraction ingested from contaminated source (dimensionless)
ABS_{si}	= bioavailability absorption factor for ingestion exposure (%)
ABS_{sd}	= bioavailability absorption factor for dermal exposures (%)
SA	= skin surface area available for contact, i.e., surface area of exposed skin (cm^2/event)
AF	= soil to skin adherence factor, i.e., soil loading on skin (mg/cm^2)
SM	= factor for soil matrix effects (%)
EF	= exposure frequency (days/year)
ED	= exposure duration (years)
BW	= body weight (kg)
AT	= averaging time (i.e., period over which exposure is averaged) (years)

An Illustrative Example

In a simplified example of the application of the above equation (for calculating media-specific ASC for a carcinogenic chemical), consider a hypothetical site located within a residential setting where children might become exposed to site contamination during recreational activities. It has been found that soil at this playground for young children in the neighborhood is contaminated with methylene chloride. It is expected that children aged 1 to 6 years could be ingesting approximately 200 mg of the contaminated soils per day during outdoor activities at the impacted playground. The ASC associated with the *ingestion only exposure* of 200 mg of soil (contaminated with methylene chloride, with an oral SF of 7.5×10^{-3} [mg/kg-day]$^{-1}$) on a daily basis, by a 16-kg child, over a 5-year exposure period is conservatively estimated to be:

$$ASC_{mc} = \frac{[10^{-6} \times 16 \times 70 \times 365]}{[0.0075 \times 200 \times 1 \times 1 \times 365 \times 5 \times 10^{-6}]} \approx 149 \text{ mg/kg}$$

That is, the allowable exposure concentration (represented by the ASC) for methylene chloride in soils within this residential setting, assuming a benchmark excess lifetime cancer risk level of 10^{-6}, is estimated to be approximately 149 mg/kg. Thus, if environmental sampling and analysis indicates contamination levels in excess of 149 mg/kg at this residential playground, then immediate risk control action (such as restricting access to the playground as an interim measure) should be implemented. It is noteworthy that, other potentially significant exposure routes (e.g., dermal contact and inhalation) as well as other sources of exposure (e.g., via drinking water and food) have not been accounted for in this illustrative example. However, all such other exposure routes and sources may require the need to further lower the calculated ASC for any site restoration decisions. Indeed, regulatory guidance would probably require reducing the contaminant concentration, ASC_{mc}, to only a fraction (e.g., 20%) of the calculated value in view of the fact that there could be other sources of exposure (e.g., air, food, etc.). This thinking should generally be factored into the overall risk management decisions about contaminated land management problems.

9.4.2. SOIL CHEMICAL LIMITS FOR THE NON-CARCINOGENIC EFFECTS OF SITE CONTAMINANTS

Box 9.4 shows a general equation for calculating the risk-based site restoration criteria for the non-carcinogenic effects of a single chemical constituent found in soils at a contaminated land. This has been derived by 'back-calculating' from the hazard and chemical exposure equations associated with the inhalation of soil emissions, ingestion of soils, and dermal contact with soils.

An Illustrative Example
In a simplified example of the application of the ASC equation (for calculating media-specific ASC for the non-carcinogenic effects of a chemical constituent), consider a hypothetical site located within a residential setting where children may be exposed to site contamination during recreational activities. It has been found that soil at this playground for young children in the neighborhood is contaminated with ethylbenzene. It is expected that children aged 1 to 6 years could be ingesting approximately 200 mg of contaminated soils per day during outdoor activities at the impacted playground. The ASC associated with the *ingestion only exposure* of 200 mg of soil (contaminated with ethylbenzene, with an oral RfD of 0.1 mg/kg-day) on a daily basis, by a 16-kg child, over a 5-year exposure period is conservatively estimated to be:

$$ASC_{ebz} = \frac{0.1 \times [1 \times 16 \times 5 \times 365]}{[200 \times 1 \times 1 \times 365 \times 5 \times 10^{-6}]} \approx 8{,}000 \text{ mg/kg.}$$

Box 9.4. General equation for calculating risk-based soil cleanup level
for the non-carcinogenic effects of a chemical constituent

$$ASC_{nc} = \frac{\text{Target Hazard Quotient}}{(\frac{EF \times ED \times 10^{-6}}{BW \times AT \times 365}) \times \{[\frac{IR \times RR \times ABS_a}{RfD_i} \times AEF \times CF_a] + [(\frac{SIR}{RfD_o} \times FI \times ABS_{si})] + [\frac{SA \times AF \times ABS_{sd} \times SM}{RfD_o}]\}}$$

$$= \frac{(THQ) \times (BW \times AT \times 365)}{(EF \times ED \times CF) \times \{[\frac{IR \times RR \times ABS_a}{RfD_i} \times AEF \times CF_a] + \frac{1}{RfD_o}[(SIR \times FI \times ABS_{si}) + (SA \times AF \times ABS_{sd} \times SM)]\}}$$

where:

ASC_{nc}	= acceptable soil concentration (i.e., acceptable risk-based cleanup level) of non-carcinogenic contaminant in soil (mg/kg)
THQ	= target hazard quotient (usually equal to 1) (unitless)
RfD_i	= inhalation reference dose (mg/kg-day)
RfD_o	= oral reference dose (mg/kg-day)
IR	= inhalation rate (m^3/day)
RR	= retention rate of inhaled air (%)
ABS_a	= percent chemical absorbed into bloodstream (%)
AEF	= air emission factor, i.e., PM_{10} particulate emissions or volatilization (kg/m^3)
CFa	= conversion factor for air emission term (10^6)
SIR	= soil ingestion rate (mg/day)
CF	= conversion factor (10^{-6} kg/mg)
FI	= fraction ingested from contaminated source (dimensionless)
ABS_{si}	= bioavailability absorption factor for ingestion exposure (%)
ABS_{sd}	= bioavailability absorption factor for dermal exposures (%)
SA	= skin surface area available for contact, i.e., surface area of exposed skin (cm^2/event)
AF	= soil to skin adherence factor, i.e., soil loading on skin (mg/cm^2)
SM	= factor for soil matrix effects (%)
EF	= exposure frequency (days/year)
ED	= exposure duration (years)
BW	= body weight (kg)
AT	= averaging time (i.e., period over which exposure is averaged, equals ED for non-carcinogens) (years)

That is, the allowable exposure concentration (represented by the ASC) for ethylbenzene in soils within this residential setting is estimated to be approximately 8,000 mg/kg. Thus, if environmental sampling and analysis indicates contamination levels in excess of 8,000 mg/kg at this residential playground, then immediate risk control action (such as restricting access to the playground as an interim measure) should be implemented. It is noteworthy that, other potentially significant exposure routes (e.g., dermal contact and inhalation) as well as other sources of exposure (e.g., via drinking water and food) have not been accounted for in this illustrative example.

However, all such other exposure routes and sources may require the need to further lower the calculated ASC for any site restoration decisions. Indeed, regulatory guidance would probably require reducing the contaminant concentration, ASC_{ebz}, to only a fraction (e.g., 20%) of the calculated value in view of the fact that there could be other sources of exposure (e.g., air, food, etc.). This thinking should generally be factored into the overall risk management decisions about contaminated land management problems.

9.5.   Establishing Risk-Based Cleanup Limits for Contaminated Waters

To determine the risk-based cleanup level for a chemical compound present in water, algebraic manipulations of the hazard index and/or carcinogenic risk equations together with the exposure estimation equations discussed in Chapters 6 through 8 can be used to arrive at the appropriate analytical relationships. The step-wise computational efforts involved in this exercise consist of a 'back-calculation' process that yields an acceptable water concentration (AWC) predicated on health-protective exposure parameters; as an example, the AWC generally results in a cumulative non-cancer hazard index of ≤ 1 and/or a cumulative carcinogenic risk $\leq 10^{-6}$. Indeed, for chemicals with carcinogenic effects, a target risk of $[1 \times 10^{-6}]$ is typically used in the 'back-calculation'; and a target hazard index of 1.0 is typically used for non-carcinogenic effects.

The processes involved in the determination of the AWCs are summarized in the proceeding sections. For substances that are both carcinogenic and possess systemic toxicity properties, the lower of the carcinogenic or non-carcinogenic criterion should be used for the relevant corrective action and/or risk management decisions.

9.5.1. WATER CHEMICAL LIMITS FOR CARCINOGENIC CONTAMINANTS

Box 9.5 shows a general equation for calculating the risk-based restoration criteria for a single carcinogenic constituent present in potable water. This has been derived by 'back-calculating' from the risk and chemical exposure equations associated with the inhalation of contaminants in water (for volatile constituents only), ingestion of water, and dermal contact with water. It is noteworthy that, where appropriate and necessary, this general equation may also be re-formulated to incorporate the receptor age-adjustment exposure factors developed and presented earlier on in Chapter 6.

An Illustrative Example
In a simplified example of the application of the above equation (for calculating media-specific AWC for a carcinogenic chemical), consider the case of a contaminated site that is impacting an underlying water supply aquifer as a result of contaminant migration into groundwater. This groundwater resource is used for culinary water supply purposes. The AWC associated with the *ingestion only exposure* to 2 liters of water (contaminated with methylene chloride, with an oral SF of 7.5×10^{-3} [mg/kg-day]$^{-1}$) on a daily basis, by a 70-kg adult, over a 70-year lifetime is given by the following approximation:

$$AWC_{mc} = \frac{[10^{-6} \times 70 \times 70 \times 365]}{[0.0075 \times 2 \times 1 \times 365 \times 70]} \approx 0.005 \text{ mg/L} = 5\mu\text{g/L}$$

That is, assuming a benchmark excess lifetime cancer risk level of 10^{-6}, the allowable exposure concentration for methylene chloride (represented by the AWC) is estimated at 5 µg/L. Obviously, the inclusion of other pertinent exposure routes (such as inhalation of vapors, and dermal contacts during showering/bathing activities, etc.) would likely call for a lower AWC in any aquifer restoration decision. Indeed, regulatory guidance would probably require reducing the contaminant concentration, AWC_{mc}, to only a fraction (e.g., 20%) of the calculated value in view of the fact that there could be other sources of exposure (e.g., air, food, etc.). This thinking should generally be factored into the overall risk management decisions about contaminated water management problems.

Box 9.5. General equation for calculating risk-based water cleanup level
for a carcinogenic chemical constituent

$$AWC_C =$$

$$\frac{TCR}{(\frac{EF \times ED}{BW \times AT \times 365}) \times \{[SF_i \times IR_w \times RR \times ABS_a \times CF_a] + [(SF_o \times WIR \times FI \times ABS_{si}) + (SF_o \times SA \times K_p \times ET \times ABS_{sd} \times CF)]\}}$$

$$= \frac{TCR \times (BW \times AT \times 365)}{(EF \times ED) \times \{[SF_i \times IR_w \times RR \times ABS_a \times CF_a] + SF_o[(WIR \times FI \times ABS_{si}) + (SA \times K_p \times ET \times ABS_{sd} \times CF)]\}}$$

where:

AWC_C	= acceptable water concentration (i.e., acceptable risk-based cleanup level) of carcinogenic contaminant in water (mg/L)
TCR	= target cancer risk, usually set at 10^{-6} (dimensionless)
SF_i	= inhalation slope factor ($[\text{mg/kg-day}]^{-1}$)
SF_o	= oral slope factor ($[\text{mg/kg-day}]^{-1}$)
IR_w	= intake from the inhalation of volatile compounds (sometimes equivalent to the amount of ingested water) (m^3/day)
RR	= retention rate of inhaled air (%)
ABS_a	= percent chemical absorbed into bloodstream (%)
CF_a	= conversion factor for volatiles inhalation term (1,000 L/1 m^3 = 10^3 L/m^3)
WIR	= water ingestion rate (L/day)
CF	= conversion factor (1 L/1,000 cm^3 = 10^{-3} L/cm^3)
FI	= Fraction ingested from contaminated source (unitless)
ABS_{si}	= bioavailability absorption factor for ingestion exposure (%)
ABS_{sd}	= bioavailability absorption factor for dermal exposures (%)
SA	= skin surface area available for contact, i.e., surface area of exposed skin (cm^2/event)
K_p	= chemical-specific dermal permeability coefficient from water (cm^2/hr)
ET	= exposure time during water contacts (e.g., during showering/bathing activity) (hr/day)
EF	= exposure frequency (days/years)
ED	= exposure duration (years)
BW	= body weight (kg)
AT	= averaging time (i.e., period over which exposure is averaged) (years).

9.5.2. WATER CHEMICAL LIMITS FOR THE NON-CARCINOGENIC EFFECTS OF SITE CONTAMINANTS

Box 9.6 shows a general equation for calculating the risk-based restoration criteria for a single non-carcinogenic constituent present in potable water. This has been derived by 'back-calculating' from the risk and chemical exposure equations associated with the inhalation of contaminants in water (for volatile constituents only), ingestion of water, and dermal contact with water.

Box 9.6. General equation for calculating risk-based water cleanup level
for non-carcinogenic effects of a chemical constituent

$$AWC_{nc} = \frac{THQ}{\left(\dfrac{EF \times ED}{BW \times AT \times 365}\right) \times \left\{\left[\dfrac{IR_w \times RR \times ABS_a \times CF_a}{RfD_i}\right] + \left[\left(\dfrac{WIR}{RfD_o} \times FI \times ABS_{si}\right)\right] + \left[\dfrac{SA \times K_p \times ET \times ABS_{sd} \times CF}{RfD_o}\right]\right\}}$$

$$= \frac{THQ \times (BW \times AT \times 365)}{(EF \times ED) \times \left\{\left[\dfrac{IR_w \times RR \times ABS_a \times CF_a}{RfD_i}\right] + \dfrac{1}{RfD_o}\left[(WIR \times FI \times ABS_{si}) + (SA \times K_p \times ET \times ABS_{sd} \times CF)\right]\right\}}$$

where:
AWC_{nc} = acceptable water concentration (i.e., acceptable risk-based cleanup level)
        of non-carcinogenic contaminant in water (mg/L)
THQ   = target hazard quotient (usually equal to 1)
RfD_i = inhalation reference dose (mg/kg-day)
RfD_o = oral reference dose (mg/kg-day)
IR_w = inhalation intake rate (m^3/day)
RR   = retention rate of inhaled air (%)
ABS_a = percent chemical absorbed into bloodstream (%)
CF_a = conversion factor for volatiles inhalation term ($1,000\ L/1\ m^3 = 10^3\ L/m^3$)
WIR   = water intake rate (L/day)
CF   = conversion factor ($1\ L/1,000\ cm^3 = 10^{-3}\ L/cm^3$)
FI   = fraction ingested from contaminated source (dimensionless)
ABS_{si} = bioavailability absorption factor for ingestion exposure (%)
ABS_{sd} = bioavailability absorption factor for dermal exposures (%)
SA   = skin surface area available for contact, i.e., surface area of exposed skin (cm^2/event)
K_p = chemical-specific dermal permeability coefficient from water (cm^2/hr)
ET   = exposure time during water contacts (e.g., during showering/bathing activity) (hr/day)
EF   = exposure frequency (days/years)
ED   = exposure duration (years)
BW   = body weight (kg)
AT   = averaging time (i.e., period over which exposure is averaged) (years).

An Illustrative Example

In a simplified example of the application of the above equation (for calculating media-specific AWC for a non-carcinogenic chemical), consider the case of a contaminated site that is impacting an underlying water supply aquifer as a result of contaminant migration into groundwater. This groundwater resource is used for culinary water supply purposes. The AWC associated with the *ingestion only exposure* to 2 liters of water (contaminated with ethylbenzene, with an oral RfD of 0.1 mg/kg-day) on a daily basis, by a 70-kg adult is approximated by:

$$AWC_{ebz} = \frac{0.1 \times [1 \times 70 \times 70 \times 365]}{[2 \times 1 \times 1 \times 365 \times 70]} \approx 3,500 \ \mu g/L$$

Thus, the allowable exposure concentration (represented by the AWC) for ethylbenzene is estimated to be 3,500 µg/L. Of course, additional exposures via inhalation and dermal contacts during showering/bathing and washing activities may also have to be incorporated to yield an even lower AWC, in order to arrive at a more responsible water restoration decision. Indeed, regulatory guidance would probably require reducing the contaminant concentration, AWC_{ebz}, to only a fraction (e.g., 20%) of the calculated value in view of the fact that there could be other sources of exposure (e.g., air, food, etc.). This thinking should generally be factored into the overall risk management decisions about contaminated water management problems.

9.6.   A 'Preferable' Health-Protective Chemical Level

Oftentimes, the RBCEL that has been established based on an acceptable risk level or hazard index are for a single contaminant in one environmental matrix or exposure media. Therefore the risk and hazard associated with multiple contaminants in a multi-media setting are not fully accounted for during the 'back-modeling' process used to establish the RBCELs. In contrast, the evaluation of risks associated with a given chemical exposure problem usually involves a set of equations designed to estimate hazard and risk for several chemicals, and for a multiplicity of exposure routes. Under this latter type of scenario, the computed 'acceptable' risks could indeed exceed the health-protective limits; consequently, it becomes necessary to establish a modified RBCEL for the requisite environmental or public health risk management decision. To obtain the modified RBCEL, the 'acceptable' chemical exposure level is estimated in the same manner as elaborated in the preceding sections – but with the cumulative effects of multiple chemicals being taken into account through a process of apportioning target risks and hazards among all the CoPCs.

9.6.1. THE MODIFIED *RBCEL* FOR CARCINOGENIC CHEMICALS

A modified RBCEL for carcinogenic constituents may be derived by the application of a 'risk disaggregation factor' – that allows for the apportionment of risk amongst all CoPCs. That is, the new RBCEL may be estimated by proportionately aggregating – or rather disaggregating – the target cancer risk amongst the CoPCs, and then using the corresponding target risk level in the equation presented earlier on in Boxes 9.1. The assumption used for apportioning the excess carcinogenic risk may be that all carcinogens have the same mode of biological actions and target organs; otherwise, excess carcinogenic risk is not apportioned among carcinogens, but rather each assumes the same value in the computational efforts. A more comprehensive approach to 'partitioning' or combining risks would involve more complicated mathematical manipulations, such as by the use of linear programming algorithms.

In general, the acceptable risk level may be apportioned between the chemical constituents contributing to the overall target risk by assuming that each constituent contributes equally or proportionately to the total acceptable risk. The 'risk fraction' obtained for each constituent can then be used to derive the modified RBCEL – by working from the relationships established previously for the computation of RBCELs (Section 9.3); by using the approach to estimating media RBCELs, the modified RBCEL is derived in accordance the following approximate relationship:

$$RBCEL_{c\text{-}mod} = \frac{[\%] \times CR}{\{[INHf \times SF_i] + [INGf \times SF_o] + [DEXf \times SF_o]\}} \qquad (9.13)$$

All the terms are the same as defined previously in Section 9.3 and [%] represents the proportionate contribution from a specific chemical constituent to the overall target risk level. One may also choose to use weighting factors in apportioning the chemical contributions to the target risk levels; for instance, this could be based on carcinogenic classes – such that class A carcinogens are given twice as much weight as class B, etc.; or chemicals posing carcinogenic risk via all exposure routes are given more weight than those presenting similar risks via specific routes only. Overall, the use of the modified RBCEL approach will ensure that the sum of risks from all the chemicals involved over all exposure pathways is less than or equal to the set target *de minimis* risk (e.g., $\leq 10^{-6}$).

9.6.2. THE MODIFIED *RBCEL* FOR NON-CARCINOGENIC CONSTITUENTS

A modified RBCEL for non-carcinogenic constituents may be derived by application of a 'hazard disaggregation factor' – that allows for the apportionment of target hazard index amongst all CoPCs. That is, the new RBCEL may be estimated by proportionately aggregating – or rather disaggregating – the non-cancer hazard index amongst the CoPCs, and then using the corresponding target hazard level in the equation presented earlier on in Box 9.2.

In general, the acceptable hazard level may be apportioned between the chemical constituents contributing to the overall hazard index by assuming that each constituent contributes equally or proportionately to the total acceptable hazard index. The 'hazard fraction' obtained for each constituent can then be used to derive the modified RBCEL – by working from the relationships established previously for the computation of RBCELs (Section 9.3). By using the approach to estimating media RBCELs, the modified RBCEL is derived in accordance the following approximate relationship for non-carcinogenic effects of chemicals having the same toxicological endpoints:

$$RBCEL_{nc\text{-}mod} = \frac{[\%] \times 1}{\{[\frac{INHf}{RfD_i}] + [\frac{INGf}{RfD_o}] + [\frac{DEXf}{RfD_o}]\}} \qquad (9.14)$$

All the terms are the same as defined previously in Section 9.3 and [%] represents the proportionate contribution from a specific chemical constituent to the overall target hazard index for the non-carcinogenic effects of chemicals with same physiologic endpoint. Overall, the use of the modified RBCEL approach will ensure that the sum of hazard quotients over all exposure pathways for all chemicals (with the same physiologic endpoints) is less than or equal to the hazard index criterion of 1.0.

9.6.3. INCORPORATING DEGRADATION RATES INTO THE ESTIMATION OF ENVIRONMENTAL QUALITY CRITERIA

The effect of chemical degradation is generally not incorporated into estimated RBCELs. However, since exposure scenarios used in calculating the RBCELs or similar criteria usually make the assumption that exposures could be occurring over long time periods (up to a lifetime of 70 years), it is prudent, at least in a detailed analysis, to consider the fact that degradation or other transformation of the CoPC could occur. Under such circumstances, the degradation properties of the CoPCs should be carefully evaluated. Subsequently, an adjusted RBCEL (or its equivalent) can be estimated – that is based on the original RBCEL (or equivalent), a degradation rate coefficient, and the specified exposure duration. The new adjusted RBCEL is then given by:

$$RBCEL_a = \frac{RBC}{\text{degradation factor (DGF)}} \qquad (9.15)$$

where $RBCEL_a$ is the adjusted RBCEL or its equivalent, and that incorporates a degradation rate coefficient. Assuming first-order kinetics, as an example, an approximation of the degradation effects can be obtained as follows:

$$DGF = \frac{(1 - e^{-kt})}{kt} \qquad (9.16)$$

where k is a chemical-specific degradation rate constant (days^{-1}), and t is time period over which exposure occurs (days). For a first-order decaying substance, k is estimated from the following relationship:

$$T_{1/2} \text{ [days]} = \frac{0.693}{k} \quad \text{or} \quad k \text{ [days}^{-1}] = \frac{0.693}{T_{1/2}} \qquad (9.17)$$

where $T_{1/2}$ is the half-life, which is the time after which the mass of a given substance will be one-half its initial value. Consequently,

$$RBCEL_a = RBCEL \times \frac{kt}{(1 - e^{-kt})} \qquad (9.18)$$

This relationship assumes that a first-order degradation/decay is occurring during the complete exposure period; decay/degradation is initiated at time, $t = 0$ years; and the RBCEL is the average allowable concentration over the exposure period. In fact, if significant degradation is likely to occur, the $RBCEL_a$ calculations become much more complicated; in that case, predicated source chemical levels must be calculated at frequent intervals and summed over the exposure period.

9.7. Public Health Goals *vs.* Risk-Based Chemical Exposure Levels

Pre-established public health goals (PHGs) are often used to define acceptable chemical exposure limits for human exposure – i.e., if they are determined to represent 'safe' or 'tolerable' benchmark levels for the case-specific situation. However, such generic PHGs may not always be available, or may not even offer adequate public health protection under certain circumstances. For instance, the presence of multiple constituents, multiple exposure routes, or other extraneous factors could result in 'unacceptable' aggregate risk being associated with a PHG for the particular situation. Under such circumstances, a new 'acceptable' or 'safe' level may be better represented by the RBCEL – that are derived for the various exposure routes, and from elaborately defined exposure scenarios. As the preferred risk-based benchmark, the RBCEL can then be used as a surrogate or replacement for the PHG of the CoPC.

Typically, risk-based benchmarks are developed via 'back-modeling' from a target risk level that produces an acceptable RBCEL – which can serve as a surrogate PHG. Invariably, the type of exposure scenarios envisioned as well as the exposure assumptions used may predicate the new benchmark level. It is noteworthy that, when the calculated RBCEL based on non-cancer toxicity is less protective of public health than the cancer-based value, the surrogate PHG for the CoPC is set at the lower of the two – usually the one based on the cancer effects. In any case, even for a criteria predicated on the cancer toxicity, the adopted PHG is considered to contain an adequate margin of safety for the potential non-carcinogenic adverse effects, such as adverse effects on the renal, neurological and reproductive systems.

In general, the risk-based benchmarks predicated on RBCELs may be used to: determine the degree of chemical exposures; evaluate the need for intervention and receptor monitoring; provide guidance on the need for risk control and/or corrective actions; establish safer PHGs; and verify the adequacy of possible remedial/corrective actions. Overall, the use of risk assessment principles to establish case-specific benchmarks for chemical exposure problems represent an even better and more sophisticated approach to designing cost-effective public health risk management programs – in comparison with the use of generic benchmarks. Ultimately, the use of such an approach aids in the development and/or selection of appropriate public health risk management strategies capable of achieving a set of performance goals – such that public health is not jeopardized.

9.8. Suggested Further Reading

Beck, LW, AW Maki, NR Artman, and ER Wilson, 1981. Outline and Criteria for Evaluating the Safety of New Chemicals, *Regul. Toxicol. Pharmacol.*, 1:19 – 58

Bowers, TS, NS Shifrin, and BL Murphy, 1996. Statistical approach to meeting soil cleanup goals, *Environmental Science and Technology*, 30(5): 1437 – 1444

FDA, 1994. *Action Levels for Poisonous or Deleterious Substances in Human Food and Animal Feed*, US Food and Drug Administration (FDA), Washington, DC

Ginevan, ME and DE Splitstone, 1997. Improving remediation decisions at hazardous waste sites with risk-based geostatistical analysis, *Environmental Science and Technology*, 31(2): 92A-96A

Krewski, D., C. Brown, and D. Murdoch, 1984. Determining safe levels of exposure: safety factors for mathematical models, *Fundam. Appl. Toxicol.*, 4: S383 – S394

Lu, FC, 1988. Acceptable daily intake: inception, evolution, and application, *Regul. Toxicol. Pharmacol.*, 8: 45 – 60

NRC, 1996. *Linking Science and Technology to Society's Environmental Goals,* National Forum on Science and Technology Goals: Environment, National Research Council (NRC), National Academy Press: Washington, DC

Watson, DH (ed.), 1998. *Natural Toxicants in Food*, CRC Press LLC, Boca Raton, FL

Wolt, JD, 1999. Exposure endpoint selection in acute dietary risk assessment, *Regulatory Toxicology and Pharmacology,* 29(3): 279 – 286

Chapter 10

DESIGN OF PUBLIC HEALTH
RISK MANAGEMENT PROGRAMS

Risk management is a decision-making process that entails weighing policy alternatives and then selecting the most appropriate regulatory action. This is accomplished by integrating the results of risk assessment with scientific data as well as with social, economic and political concerns, in order to arrive at an appropriate decision on a potential hazard situation (Cohrssen and Covello, 1989; NRC, 1994; Seip and Heiberg, 1989; van Leeuwen and Hermens, 1995). Risk management may also include the design and implementation of policies and strategies that result from this decision process.

Risk management programs are typically directed at risk reduction (i.e., taking measures to protect humans and/or the environment against previously identified risks); risk mitigation (i.e., implementing measures to remove risks); and/or risk prevention (i.e., instituting measures to completely prevent the occurrence of risks). Indeed, risk management, mitigation, and preventative programs generally can help facilitate an increase in the level of protection to public health and safety, as well as aid in the reduction of liability. This chapter elaborates the key steps in the effectual design of typical public health risk assessment and risk management programs.

10.1. Risk Assessment as a Cost-Effective Tool in the Formulation of Public Health and Environmental Management Decisions

Risk assessment is a systematic technique that can be used to make estimates of significant and likely risk factors associated with chemical exposure problems. Oftentimes, risk assessment is used as a management tool to facilitate effective decision-making on the control of chemical exposure problems. In fact, the chief purpose of risk assessment is to aid decision-making and this focus should be maintained throughout an environmental or public health risk management program. In particular, the application of risk assessment to chemical exposure problems can indeed remove some of the ambiguities in the decision-making process. It can also aid in the selection of prudent, technically feasible, and scientifically justifiable risk control or corrective actions that will help protect public health and the environment in a cost-effective manner.

It is noteworthy that risk assessments performed for chemical exposure problems usually will depend on an understanding of the fate and behavior of the chemical constituents of concern. Consequently, the fate and behavior issues in the various exposure settings should be carefully analyzed with the best available scientific tools. In general, the procedures utilized must reflect current state-of-the-art methods for conducting risk assessments. For all intents and purposes, the following are noteworthy recommendations in the exercises typically involved:

- Performing risk assessment to incorporate all likely scenarios envisaged rather than for the 'worst-case' alone allows better comparison to be made between risk assessments performed by different scientists and analysts whose views on what represents a 'worst-case' may be very subjective and therefore may vary significantly.

- Exposure scenarios and chemical fate and behavior models may contribute significant uncertainty to the risk assessment. The uncertainties, heterogeneities, and similarities should be identified and well documented throughout the risk assessment.

- Whenever possible, the synergistic, antagonistic, and potentiation (i.e., the case of a non-hazardous situation becoming hazardous due to its combination with others) effects of chemicals and other hazardous situations should be carefully evaluated for inclusion in the risk decisions.

- It is prudent to assess what the 'baseline' (no-action) risks are for a potentially hazardous situation or chemical exposure problem. This will provide a reflection on what the existing situation is, which can then be compared against future improved situations.

- An evaluation of the 'post-remedy' risks (i.e., residual risks remaining after the implementation of corrective actions) for a potentially hazardous situation or chemical exposure problem should generally be carried out for alternative mitigation measures. This will give a reflection of what the anticipated improved situation is, compared with the prior conditions associated with the problem situation.

Risk assessments do indeed provide decision-makers with scientifically defensible information for determining whether a chemical exposure problem poses a significant threat to human health or the environment. It may be conducted to assist in the development of cost-effective strategies for the management of chemical exposure problems. Ultimately, the risk assessment efforts can help minimize or eliminate potential long-term problems or liabilities that could result from hazards associated with chemical exposure problems.

The risk assessment process can be used to define the level of risk, which will in turn aid in determining the level of analysis and the type of risk management actions to adopt for a given chemical exposure or environmental management problem. The level of risk considered in such applications can be depicted in a risk-decision matrix (Figure 10.1) – that will help distinguish between imminent health hazards and risks. In general, this can be used as an aid for policy decisions, in order to develop variations in the scope of work necessary for case-specific public health risk management programs.

In general, the benefits of risk assessment designed to facilitate public health risk management decisions outweigh any possible disadvantages; still, it must be recognized that this process will not be without tribulations. Indeed, risk assessment is by no means a panacea. Its use, however, is an attempt to widen and extend the decision-maker's knowledge-base and thus improve the decision-making capability. Overall, the method deserves the effort required for its continual refinement as a public health risk management tool.

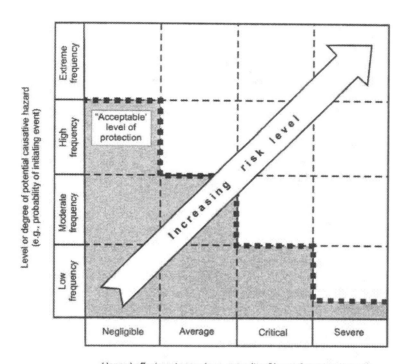

Figure 10.1. A conceptual representation defining risk profiles in a risk-decision matrix

10.2. Comparative Risk Analysis: Application of Environmental Decision Analysis Methods to Public Health Risk Management Programs

Decision analysis is a management tool comprised of a conceptual and systematic procedure for rationally analyzing complex sets of alternative solutions to a problem, in order to improve the overall performance of the decision-making process. Decision theory provides a logical and systematic framework to structure the problem objectives and to evaluate and rank alternative potential solutions to the problem. Environmental decision analyses typically involve the use of a series of techniques to comprehensively develop risk control or corrective action plans, and to evaluate appropriate mitigative alternatives in a technically defensible manner.

As part of a corrective action assessment program, it is almost inevitable that the policy analyst will often have to make choices between alternative remedial options. These are based on an evaluation of risk tradeoffs and relative risks among available decision alternatives; evaluation of the cost-effectiveness of corrective action plans; or a risk-cost-benefit comparison of several management options. In fact, comparing risks, benefits, and costs amongst various risk management strategies can become very important in most environmental and public health risk management programs. A number of analytical tools may generally be used to assist the processes involved (see, e.g., Bentkover *et al.*, 1986; Haimes, 1981; Haimes *et al.*, 1990; Keeney, 1990; Lind *et al.*, 1991; Nathwani *et al.*, 1990; Seip and Heiberg, 1989; USEPA, 1984). Examples of the relevant tools are annotated below.

10.2.1. COST-EFFECTIVENESS ANALYSIS

Cost-effectiveness analysis involves a comparison of the costs of alternative methods to achieve some set goal(s) of risk reduction, such as an established benchmark risk or environmental cleanup criteria. The process compares the costs associated with different methods of achieving a specific risk management goal. The analysis can be used to allocate limited resources among several risk abatement programs, aimed at achieving the maximum positive results per unit cost. The procedure may also be used to project and compare total costs of several risk management plans.

In the application of cost-effective analyses to risk management actions, a fixed goal is established, and then policy options are evaluated on the ability to achieve that goal in a most cost-effective manner. The goal generally consists of a specified level of 'acceptable' risk, and the risk management options are compared on the basis of the monetary costs necessary to reach the benchmark risk. Cost constraints can also be imposed so that the options are assessed on their ability to control the risk most effectively for a fixed cost. The efficacy of the risk management action alternatives in the hazard reduction process can subsequently be assessed, and the most cost-effective course of action (i.e., one with minimum cost meeting the constraint of a benchmark risk/hazard level) can then be implemented. This would then guarantee the objective of meeting the constraints at the lowest feasible cost.

10.2.2. RISK-COST-BENEFIT OPTIMIZATION

Subjective and controversial as it might appear to express certain hazards in terms of cost, especially where public health and/or safety is concerned, it nevertheless has been used to provide an objective way of evaluating risk management action problems. This is particularly true where risk factors are considered in the overall study.

Risk-cost-benefit analysis is a generic term for techniques encompassing risk assessment and the inclusive evaluation of risks, costs, and benefits of alternative projects or policies. In performing risk-cost-benefit analysis, one attempts to measure risks, costs and benefits, to identify uncertainties and potential tradeoffs, and then to present this information coherently to decision-makers. A general form of objective function for use in a risk-cost-benefit analysis that treats the stream of benefits, costs, and risks in a net present value calculation is given by (Crouch and Wilson 1982; Massmann and Freeze 1987):

$$\Phi = \sum_{t=0}^{T} \frac{1}{(1+r)^t} \left[B(t) - C(t) - R(t) \right] \qquad (10.1)$$

where: Φ = objective function (\$); t = time, spanning 0 to T (yrs); T = time horizon (yrs); r = discount rate; B(t) = benefits in year t (\$); C(t) = costs in year t (\$); R(t) = risks in year t (\$). The risk term is defined as the expected cost associated with the probability of significant impacts or failure, and is a function of the costs due to the consequences of failure in year t. In general, tradeoff decisions made in the process will be directed at improving both short- and long-term benefits of the program.

10.2.3. MULTI-ATTRIBUTE DECISION ANALYSIS AND UTILITY THEORY APPLICATIONS

Multi-attribute decision analysis and utility theory have been suggested (e.g., Keeney and Raiffa 1976; Lifson 1972) for the evaluation of problems involving multiple conflicting objectives, such as is the case for decisions on environmental and public health risk management programs. In such situations, the decision-maker is faced with the problem of having to trade-off the performance of one objective for another. In addressing these types of problem, a mathematical structure may be developed around utility theory that presents a deductive philosophy for risk-based decisions (Keeney 1984; Keeney and Raiffa 1976; Lifson 1972; Starr and Whipple 1980).

The use of structured decision support systems has proven to be efficient and cost-effective in making sound environmental and public health risk management decisions. Such tools can indeed play vital roles in improving the decision-making process. It should be acknowledged, however, that despite the fact that decision analysis presents a systematic and flexible technique that incorporates the decision-maker's judgment, it does not necessarily provide a complete analysis of the public's perception of risk.

Risk tradeoffs between increased expenditure of a risk management action and the hazard reduction achieved upon implementation may be assessed by the use of multi-attribute decision analysis and utility theory methods. Multi-attribute decision analysis and utility theory can indeed be applied in the investigation and management of environmental contamination and chemical exposure problems, in order to determine whether one set of risk management action alternatives is more or less desirable than another set. With such a formulation, an explicitly logical and justifiable solution can be assessed for the complex decisions involved in environmental and public health risk management programs. In using expected utility maximization, the preferred alternative will be the one that maximizes the expected utility – or equivalently, the one that minimizes the loss of expected utility. In a way, this is a nonlinear generalization of cost-benefit or risk-benefit analysis.

Even though utility theory offers a rational procedure for evaluating environmental and public health risk management measures, it may transfer the burden of decision to the assessment of utility functions. Also, several subjective assumptions are used in the application of utility functions that are a subject of debate. The details of the paradoxes surrounding conclusions from expected utility applications are beyond the scope of this elaboration and are not discussed here.

Utility-Attribute Analysis

Attributes measure how well a set of objectives is being achieved. Through the use of multiple attributes scaled in the form of utilities, and weighted according to their relative importance, a decision analyst can describe an expanded set of consequences associated with an environmental or public health risk management program. Adopting utility as the criterion of choice among alternatives allows a multifaceted representation of each possible consequence. Hence, in its application to environmental contamination and chemical exposure management problems, both hazards and costs can be converted to utility values, as measured by the relative importance that the decision-maker attaches to either attribute.

The utility function need not be linear since the utility is not necessarily proportional to the attribute. Thus, curves of the forms shown in Figure 10.2 can be generated for the utility function. An arbitrary value [e.g., *0* or *1*] of *1* can be assigned to the 'ideal' situation (i.e., a 'no hazard/no cost scenario'), and the 'dooms-day' scenario (i.e., 'high hazard/high cost') is then assigned a corresponding relative value [e.g., *-1* or *0*] of *0*. The shape of the curves is determined by the relative value given each attribute. The range in utilities is the same for each attribute, and attributes should, strictly speaking, be expressed as specific functions of system characteristics.

In assigning utility value to hazard, it is a commonplace to rely on various social and environmental or public health goals that can help determine the threats posed by the hazard, rather than use the direct concept of hazard. These utility values can then be used as the basis for selection among the environmental and public health risk management action alternatives.

Preferences and Evaluation of Utility Functions
Evaluation of utility functions requires skill, and when the utility function represents the preferences of a particular interest group, additional difficulties arise. Nonetheless, risk tradeoffs may be determined by applying weighting factors of preferences in a utility-attribute analysis.

Preferences are directly incorporated in the utility functions by assigning an appropriate weighting factor to each utility term. The weighting factors are changed to reflect varying tradeoff values associated with alternative decisions. For instance, if minimizing hazards is *k* times as important as minimizing costs, then weighting factors of [*k/(k+1)*] and [*1/(k+1)*] would be assigned to the hazard utility and the cost utility, respectively. These weighting factors would reflect, or give a measure of, the preferences for a given utility function. Past decisions can help provide empirical data that can be used for quantifying the tradeoffs and therefore the *k* values. The given utilities are weighted by their preferences, and are summed over all the objectives. For *n* alternatives, the value of the *i*-th alternative would be determined as follows:

$$V_i = \frac{k}{(k+1)} \, U(H_i) + \frac{1}{(k+1)} \, U(\$_i) \qquad (10.2)$$

where:

V_i = the total relative value for the *i*-th alternative
$U(H_i)$ = the hazard utility, *H*, for the *i*-th alternative
$U(\$_i)$ = the cost utility, *$*, associated with alternative *i*.

In general, the largest total relative value would ultimately be selected as the best alternative.

(a) Utility function for hazards

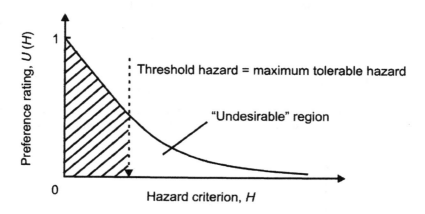

(b) Utility function for costs

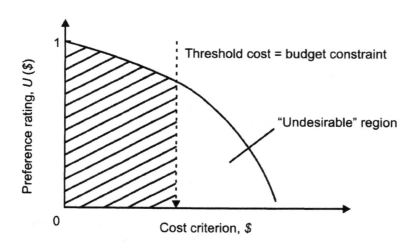

Figure 10.2. Utility functions giving the relative values of hazards and costs in similar (dimensionless) terms

Utility Optimization

To facilitate the development of an optimal risk management program, the total relative value can be plotted against the cost (Figure 10.3). From this plot, the optimum cost is that cost value which corresponds to the maximum total relative value. The optimum cost is equivalently obtained, mathematically, as follows:

$$\frac{dV}{(d\$)} = \frac{d}{(d\$)} \left[\frac{k}{(k+1)} U(H) + \frac{1}{(k+1)} U(\$) \right] = 0$$

or,   (10.3)

$$k \frac{dU(H)}{(d\$)} = -\frac{dU(\$)}{(d\$)}$$

where $\frac{dU(H)}{(d\$)}$ is the derivative of hazard utility relative to cost, and $\frac{dU(\$)}{(d\$)}$ is the derivative of cost utility relative to cost. The optimum cost is obtained by solving this equation for $\$$; this would represent the most cost-effective option for project execution.

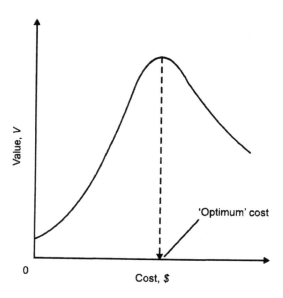

Figure 10.3. Value function for costs

In an evaluation similar to the one presented above, a plot of total relative value against hazard provides a representation of the 'optimum hazard' (Figure 10.4). Again, this result can be evaluated in an analytical manner similar to that presented above for cost; the 'optimum hazard' is given, mathematically, by:

$$\frac{dV}{(dH)} = \frac{d}{(dH)} \left[\frac{k}{(k+1)} U(H) + \frac{1}{(k+1)} U(\$) \right] = 0$$

or   (10.4)

$$k \frac{dU(H)}{(dH)} = -\frac{dU(\$)}{(dH)}$$

where $\frac{dU(H)}{(dH)}$ is the derivative of hazard utility relative to hazard, and $\frac{dU(\$)}{(dH)}$ is the derivative of cost utility relative to hazard. Solving for H yields the 'optimum' value for the hazard.

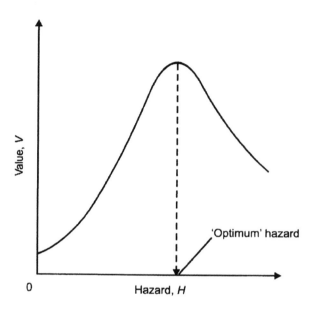

Figure 10.4. Value function for hazards

10.3. The General Nature of Risk Management Programs

The management of chemical exposure problems usually involves competing and contradictory objectives – with the prime objective to minimize both hazards and risk management action costs under multiple constraints. Typically, once a minimum acceptable and achievable level of protection has been established via hazard assessment, alternative courses of action can be developed that weigh the magnitude of adverse consequences against the cost of risk management actions. In general, reducing hazards would require increasing costs, and cost minimization during hazard abatement will likely leave higher degrees of unmitigated hazards (Figure 10.5). Typically, a decision is made based on the alternative that accomplishes the desired objectives at the least total cost – total cost here being the sum of hazard cost and risk management cost.

Risk management uses information from hazard analyses and/or risk assessment – along with information about technical resources; social, economic, and political values; and regulatory control or response options – to determine what actions to take in order to reduce or eliminate a risk. It is comprised of actions evaluated and implemented to help in risk reduction policies, and may include concepts for prioritizing the risks, as well as an evaluation of the costs and benefits of proposed risk reduction programs.

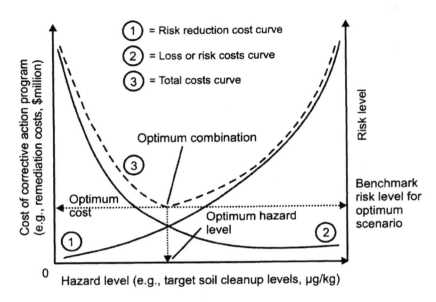

Figure 10.5. Risk reduction vs. costs: a schematic of corrective action costs (e.g., cleanup or remediation costs) for varying hazard levels (e.g., chemical concentrations in environmental media or residual risk)

Examples of risk management actions that are commonly encountered in environmental and public health risk management programs include:

- Deciding on how much of a chemical a manufacturing company may discharge into a river;
- Deciding on which substances may be handled at a hazardous waste treatment, storage, and disposal facilities (TSDF);
- Deciding on the extent of cleanup warranted at a hazardous waste site;
- Setting general permit levels for discharge, storage, or transport of hazardous materials;
- Establishing levels for air contaminant emissions for air pollution control purposes;
- Determining allowable levels of contamination in drinking water or food;
- Deciding on the use of specific chemicals in manufacturing processes and related industrial activities;
- Determining hazardous waste facility design and operation requirements;
- Consideration of the harmful effect of chemical pollutants that need to be controlled; and
- Determining the relinquished benefits of using a pesticide or other toxic chemical.

Risk management decisions associated with these types of issues are made based on inputs from a prior risk assessment conducted for the applicable case-specific problem.

In fact, risk assessment results, serving as input to risk management, generally help in the setting of priorities for a variety of chemical exposure problems – further to producing more efficient and consistent risk reduction policies.

Risk management does indeed provide a context for balanced analysis and decision-making. Public health risk management programs are generally designed with the goal to minimize potential negative impacts associated with chemical exposure problems.

10.4. A Framework for Risk Management Programs

Risk management decisions generally are complex processes that involve a variety of technical, political, and socioeconomic considerations. Notwithstanding the complexity and the fuzziness of the issues involved, the ultimate goal of public health risk management programs is to protect public health. The application of risk assessment can remove some of the ambiguity in the decision-making process; it can also aid in the selection of prudent, technically feasible, and scientifically justifiable risk management actions that will help protect public health in a cost-effective manner. To successfully apply the risk assessment process to a potential chemical exposure problem, however, the process must be tailored to the case-specific conditions and relevant regulatory constraints. Based on the results of a risk assessment, decisions can then be made relating to the types of risk management actions needed for a given chemical exposure problem. If unacceptable risk levels are identified, the risk assessment process can further be employed in the evaluation of remedial or risk control action alternatives. This will ensure that net risks to human health are truly reduced to acceptable levels via the remedial or risk management action of choice.

Figure 10.6 provides a framework that may be used to facilitate the environmental and public health risk management decision-making process involved in chemical exposure programs. The process will generally incorporate a consideration of the complex interactions existing between the exposure setting, regulatory policies, and technical feasibility of risk management options. Ultimately, the tasks involved should help public health risk analysts identify, rank/categorize, and monitor the status of potential chemical exposure problems; identify field data needs and decide on the best investigation or sampling strategy; establish appropriate public health goals; and choose the risk management action that is most cost-effective in controlling or abating the risks associated with the chemical exposure problem.

In the arena of chemical exposure problems, it is noteworthy that, public health risk management decisions should be based on a wide range of issues relevant to risk analysis – including medical opinion, epidemiology, and professional judgment, along with socioeconomic factors and technical feasibility. It is also imperative to systematically identify hazards throughout an entire public health risk management system, assess the potential consequences due to any associated hazards, and examine corrective measures for dealing with the case-specific type of problem. Risk management – used in tandem with risk assessment – offers the necessary mechanism for achieving such goals.

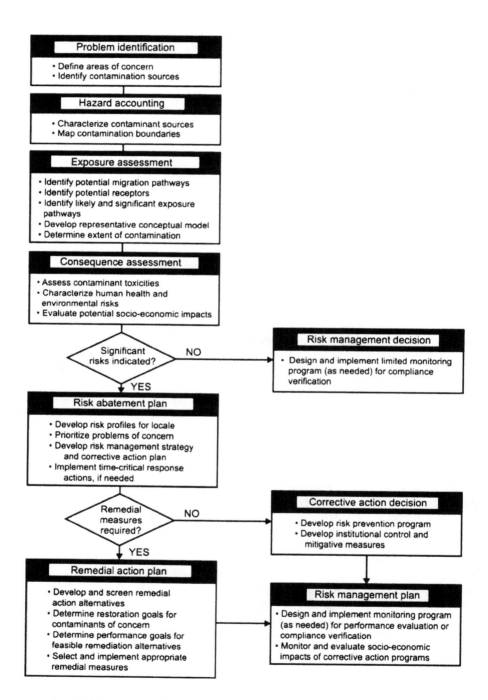

Figure 10.6. A risk management decision framework for the management of chemical exposure problems

10.5. Risk Communication as a Facilitator of Risk Management

Risk management combines socioeconomic, political, legal, and scientific approaches to manage risks. Risk assessment information is used in the risk management process to help in deciding how to best protect public health. Thus, essentially, risk assessment provides *information* on the risks, and risk management develops and implements an *action* based on that information. This means that, risk assessment can in principle be carried out objectively, whereas risk management usually involves preferences and attitudes, and should therefore be considered a subjective activity (Seip and Heiberg, 1989; NRC, 1983; USEPA, 1984). The subjectivity of the risk management task calls for the use of very effective facilitator tools/techniques – with good risk communication being the logical choice. Risk communication is an interactive process or exchange of information and opinions among interested parties or stakeholders concerning risk, potential risk, or perceived risk.

Risk communication has formally been defined as the process of conveying or transmitting information among interested parties about the following types of issues: levels of health and environmental risks; the significance or meaning of health or environmental risks; and decisions, actions, or policies aimed at managing or controlling health or environmental risks (Cohrssen and Covello, 1989). It offers a forum at which various stakeholders discuss the nature, magnitude, significance, or control of risks and related consequences with one another. Effective risk communication is indeed important for the implementation of an effectual risk management program. It is therefore quite important to give adequate consideration to risk communication issues when developing a risk management agenda. In fact, in many a situation, even credible risk assessment and risk management decisions may never be implemented unless they are effectively communicated to all interested stakeholders.

One goal of risk communication is to improve the agreement between the magnitude of a risk and the public's political and behavioral response to this risk – necessitating researchers to investigate a number of message characteristics and risk communication strategies (Weinstein *et al.*, 1996; Weinstein and Sandman, 1993). The process involved provides information to a concerned public about potential health risks from exposure to toxic chemicals or similar environmental hazards. In fact, because the perception of risks often differs widely, risk communication typically requires a sensitive approach and should involve genuine dialogue (van Leeuwen and Hermens, 1995). Among several other factors, trust and credibility are believed to be key determinants in the realization of any risk communication goals. Apparently, defying a negative stereotype is key to improving perceptions of trust and credibility (Peters *et al.*, 1997).

Indeed, risk communication is vital to the risk assessment and risk management processes – and, ultimately, to the success of most risk management actions. To be able to design an effectual risk management program, a variety of qualitative issues become equally important in addition to any prior risk quantification. Risk communication may indeed dictate public perception, and therefore public acceptance of risk management strategies and overall environmental and public health risk management decisions.

Only limited presentation on the risk communication topic is given in this volume – with more detailed elaboration/discussions to be found in the literature elsewhere (e.g., Cohrssen and Covello, 1989; Covello, 1992, 1993; Covello and Allen, 1988;

Fisher and Johnson, 1989; Freudenburg and Pastor, 1992; Hance *et al.*, 1990; Kasperson and Stallen, 1991; Laird, 1989; Leiss, 1989; Leiss and Chociolko, 1994; Lundgren, 1994; Morgan and Lave, 1990; NRC, 1989; Pedersen, 1989; Peters *et al.*, 1997; Renn, 1992; Silk and Kent, 1995; Slovic, 1993; van Leeuwen and Hermens, 1995; Vaughan, 1995; Weinstein *et al.*, 1996; Weinstein and Sandman, 1993). The literature available on the subject addresses several important elements/issues – including checklists for improving both the process and content of risk communication efforts.

10.5.1. DESIGNING AN EFFECTUAL RISK COMMUNICATION PROGRAM

Several rules and guidelines have been suggested/proposed to facilitate effective risk communication (e.g., Cohrssen and Covello, 1989; Covello and Allen, 1988) – albeit there are no easy prescriptions. In fact, it is very important that risk communication should consider and embrace several important elements (Box 10.1), in order to minimize or even prevent suspicion/outrage from a usually cynical public. Serious consideration of the relevant elements should help move a potentially charged atmosphere to a responsible one, and one of cooperation and dialog. Ultimately, a proactive, planned program of risk communication will – at the very least – usually place the intended message in the public eye in advance of negative publicity and sensational media headlines. A reliable tool and channel of communication should be identified to ensure effective transmittal of all relevant information.

Overall, a systematic evaluation using structured decision methods, such as the use of the event tree approach (see, e.g., Asante-Duah, 1998), can greatly help in this direction. The event tree illustrates the cause and effect ordering of event scenarios, with each event being shown by a branch of the event tree in the context of the decision problem. The event tree model structure can indeed aid risk communicators in improving the quality and effectiveness of their performance and presentations. Ultimately, scientific information about health and environmental risks is generally communicated to the public through a variety of channels – ranging from warning labels on consumer products to public meetings/forums involving representatives from government, industry, the media, the populations potentially at risk, and other sectors of the general public (Cohrssen and Covello, 1989). Important traditional techniques of risk communication usually consist of community public health education programs, 'fact sheets', newsletters, public notices, workshops, focus groups, public meetings, and similar forum types – all of which seem to work very well if rightly implemented or utilized.

Irrespective of the approach or technique adopted, however, it must be acknowledged that hazard perception and risk thresholds tend to be quite different in different parts of the world. In fact, there could also be variations within different sectors of a community or society within the same locale or region. Even so, such variances should not affect the general design principles when one is developing a risk communication program.

Box 10.1. Important strategic considerations in developing an effective risk communication program

- Accept and involve the public as a legitimate partner – especially all parties that have an interest or direct stake in the particular risk situation.
- Involve all stakeholders as early as possible – via taking a proactive stance, based on a coherent strategy, sound tactics, and careful planning of community relations and actions.
- Listen to your audience – recognizing communication is a two-way activity, and take note of the public's specific concerns.
- Ensure an effective two-way discourse/dialogue, to ensure adequate flow of information in both directions between the risk communication team and the interested public/parties. That is, flow of information should be from, and to all stakeholders.
- Be honest, frank, and open – since lost trust and credibility are almost impossible to regain.
- Focus should be on what the risks are, and what is already being done to keep these risks as low as reasonably possible.
- Have an even greater focus on long-term implication of risk management decisions/strategies, without necessarily discounting potential short-term consequences.
- Have an elaborate evaluation of alternative choice of proposed risk management strategies.
- Anticipate or investigate the affected party's likely perception to the prevailing or expected risks, since this could be central to any response to proposed actions or risk management strategies.
- Speak clearly and with compassion – especially minimizing excessive use of technical language and jargon.
- Avoid use of unnecessary jargon and excessive technical details, in order to allow the community to focus on the real/practical issues of interest to the PARs.
- Anticipate controversy, request for changes to proposed risk management plans, and then offer positive response that may form a basis for consensus between all stakeholders.
- Keep the needs and perceptions of the 'outsider' stakeholders in perspective, to ensure a balanced and equitable program needs.
- Focus more on psychological needs of community, rather than economic realities/interests of project.
- Plan carefully and evaluate performance – to help re-focus, if necessary.
- Coordinate and collaborate with other credible sources – such as by issuing communication jointly with other trustworthy sources like credible university scientists, area physicians, trusted local officials, and opinion leaders.
- Meet the needs of the media – recognizing that they tend to play a critical role in setting agendas and determining outcomes.
- Clarify who the risk assessment protects – i.e., the community and/or stakeholders
- Identify potentially overlooked stakeholder project knowledge – and incorporate such useful information in the overall strategic plan
- Give credit for stakeholder roles and contributions in a decision – also explaining how and why stakeholder input has or has not been used in the process

10.6. The Use of Contemporary Risk Mapping Tools: GIS in Public Health Risk Management Applications

A Geographic Information System (GIS) is a computer-based tool used to capture, manipulate, process, and display spatial or geo-referenced data (Bernhardsen, 1992; Gattrell and Loytonen, 1998; Goodchild *et al.*, 1993, 1996). It is a tool that can serve a wide range of research and surveillance purposes; it allows the layering of health, demographic, environmental and other traditional data sources to be analyzed by their location on the earth's surface. Indeed, the GIS technology has become an important tool for public health professionals – as this generally allows for the mapping and analysis of public health and environmental management data.

It is apparent that, recent advances in the application of GIS technology have improved, and will continue to revolutionize the spatial analysis of diseases, environmental contamination, and social/demographic information. In a way, the unlimited future of GIS in public health risk management is derived from comparable situations from centuries ago, whereby public health surveillance activities by health professionals have relied on maps to locate and identify changes in patterns of human disease. The GIS of today provides a relatively easy tool for overlaying and analyzing disparate data sets that relate to each other by location on the earth's surface. The growing availability of health, demographic, and environmental databases containing local, regional, national, and international information are propelling major advances in the use of GIS and computer mapping with spatial statistical analyses.

In general, understanding and communicating the association between environmental hazards and disease incidence are essential requirements of an effective environmental health policy. For instance, in the United States, routine public health surveillance programs generate massive amounts of information. Environmental monitoring and simulation modeling projects also provide an equally large volume of data. But, due to the lack of a coordinated framework to organize, manage, analyze, and display the data, majority of this information has been poorly utilized in the past. This is one reason why the advances in the geospatial information technologies, such as GIS, could become very useful in a whole wide range of public health risk management functions.

10.6.1. UTILIZATION OF GIS IN RISK ASSESSMENT AND ENVIRONMENTAL MANAGEMENT PROGRAMS

The GIS can process geo-referenced data and provide answers to such questions as the particulars of a given location, the distribution of selected phenomena and their temporal changes, the impact of a specific event, or the relationships and systematic patterns of a region (Bernhardsen, 1992). In fact, it has been suggested that, as a planning and policy tool, the GIS technology could be used to 'regionalize' the risk analysis process – moving it from its traditional focus on a micro-scale (i.e., site-specific problems) to a true macro-scale (e.g., urban or regional risk analysis, comparative risk analysis, risk equity analysis) (D. Rejeski, in Goodchild *et al.*, 1993).

GIS is indeed a rapidly developing technology for handling, analyzing, and modeling geographic information. It is not a source of information *per se*, but only a way to manipulate information. When the manipulation and presentation of data relates to geographic locations of interest, then our understanding of the real world is enhanced. Example application types for the utilization of GIS in environmental and public health risk management decisions are briefly discussed below.

- *Exposure Assessment of Population Groups.* The exposure assessment of population groups is usually made through the linkage of environmental and health data. However, environmental and exposure data are generally referred to scattered and instantaneous samples, while epidemiological data integrate periods of time within administrative territories. GIS can be used as an organizing tool of health and environmental data sets. For example, in a case for a water supply system, potential health risk in a locality or supply area can be examined via the overlay of information layers containing data on the presence and quality of water supply service *vs.* the primary georeferenced data such as: the census tract which contains information on the manner of how the household is supplied; the water distribution system; and the water quality data derived from a monitoring program. The population groups potentially at risk can then be identified, and then appropriate corrective actions can subsequently be undertaken.

- *Development of Thematic Map Layers for Risk Management.* By combining population density information with ambient concentrations for a specific chemical ($\mu g/m^3$), and the unit risk factor for that chemical (risk/$\mu g/m^3$), a map showing the risk per unit of population may be produced – that could be used to facilitate risk management decisions.

- *Data Improvement for Source/Pathway Characterization.* It is recognized that, in general, an adequate characterization of the exposure pathways that affect the fate and behavior of risk-inducing agents can help improve risk estimates significantly. Thus, more spatial data could be added to a GIS model to empirically describe the environmental medium and its effects on the distribution or dispersion of risk agents.

- *Community Health and Environmental Assessments.* GIS is an effective tool that can be used by local environmental community health departments to perform health and environmental assessments, improve public access to environmental health information, and increase organizational effectiveness and efficiency. In fact, in the daily work activities required to protect public health and welfare, environmental health departments usually collect large amounts of useful data. Much of this data has a geographic component. These data, while integral to daily environmental health tasks, can have many additional applications – especially as the field of environmental health grows more assessment-oriented. Indeed, a trend for regulatory and public health agencies to track their activities with GIS offers environmental health departments a unique opportunity to join in sharing information while serving the public's interest in health and environmental needs.

- *Potential Risk Cataloging System.* A potential risk cataloging system may be designed to serve as a screening methodology that ranks key areas of concern (e.g., hazardous facilities, industrial sectors, etc.) vis-à-vis multimedia chemical releases, chemical toxicities, and selected demographics of surrounding populations. The system can utilize GIS technologies to display vast quantities of data to assist users in cumulative risk analysis and other decision-making processes. A 'vulnerability index' is used to characterize potentially exposed populations – highlighting those that may be more vulnerable. In this manner, the screening can include population characteristics without using broad assumptions about exposure conditions. The 'chemical release index' and the 'vulnerability index' can then be overlaid to identify potential incidence of highly toxic and large release combinations within areas with relatively high percentage of vulnerable populations, or the like.

- *Characterization and mapping of public health concerns and problems associated with various industrial sectors in a region.* An important aspect of public health risk management with growing interest relates to the coupling of environmental health data and environmental models with information systems – such as GIS – in order to allow for effectual risk mapping of a study area. GIS can indeed be used to map the location and proximity of risk to identified or selected populations. The GIS can process geo-referenced data and provide answers to such questions as the distribution of selected phenomena and their temporal changes, the impact of a specific event, or the relationships and systematic patterns of a region. In fact, it has been suggested that, as a planning and policy tool, the GIS technology could be used to 'regionalize' a risk analysis process. For the characterization and mapping of public health concerns and problems associated with various industrial sectors, a typical study could consist of a survey of workers and communities within major industrial bases, and an investigation of the pattern of likely environmental health problems borne by such communities. GIS can then be used to examine the spatial distribution of risks around toxic sources (such as hazardous waste sites or incinerators and smelters). This would incorporate the development of thematic map layers for public health risk management.

- *Design of Risk Reduction Strategies.* Once risks have been mapped using GIS, it may be possible to match estimated risks to risk reduction strategies, and also to delineate spatially, the regions where resources should be invested, as well as the appropriate strategies to adopt for various geographical dichotomies.

- *Utilization in Remediation Planning and Design.* GIS may be used to examine the spatial distribution of risks around toxic sources, such as hazardous waste sites or incinerators and smelters. The ability of GIS to aid in the calculation of volumes, for instance, would allow soil removal and transport costs to be estimated, and for an optimal remedial action decision to be made.

- *Investigation of Risk Equity Assurance Issues.* Considering the fact that the notion of environmental justice, based on an equitable distribution of risks, is emerging as a critical theme in environmental and public health risk decision-making, GIS could become a powerful tool of choice for exploring such risk equity issues. This type of application moves beyond simply calculating risks based on somehow abstract and subjective probabilities and actually present comparative analyses for all stakeholders and 'contesting' regions or neighborhoods.

In general, the evaluation of possible exposures to environmental chemicals requires the integration of information from several and various sources. Typically, environmental sampling and chemical exposure data is analyzed to determine the magnitude and extent of contamination and/or human exposures. GIS provides a means of viewing spatial characteristics of such contamination or exposures. The sampling data may be overlayed with geographical features such as roads, streams, schools, and census data. During the assessment of possible exposures, a user may define locations of concern (such as areas within a groundwater plume, or areas with contaminated soil) by drawing a polygon. Environmental sampling data contained within the polygon is then summarized, and population characteristics for residents within the polygon are estimated from census data. The user may also define an exposure pathway by tying together the sampling data and population characteristics, along with other information such as the exposure route (e.g., oral ingestion), the period of time over which the exposure occurred, and characteristics of different groups within the exposed population (e.g., the body weight of children). All this information may then be used to estimate exposure doses, and subsequently, the potential for adverse health effects to occur. GIS has indeed proven to be a valuable tool during this type of evaluation. It provides the means to visually assess large quantities of environmental and public health information, and to relate that information to locations where exposures might occur. Overall, the GIS technology provides a way to manage and analyze information, and allows users to visualize information using combinations of different map layers.

10.6.2. THE ROLE OF GIS APPLICATIONS IN ENVIRONMENTAL MANAGEMENT

Every field of environmental and exposure modeling is increasingly using spatially distributed approaches, and the use of GIS methods will likely become widespread. Even so, it is noteworthy that models lacking a spatial component clearly have no use for GIS. The specific role of environmental and exposure models integrated with GIS would largely be in their ability to communicate effectively – i.e., via the use of maps as a well-understood and accepted form of information display; and by generating a widely accepted and familiar format for the sharing of information (Goodchild *et al.*, 1993). In general, the GIS describes the spatial environment, and the environmental and risk modeling simulates the functioning of exposure processes (Goodchild *et al.*, 1996). Thus, GIS can serve as a common data and analysis framework for environmental and exposure models.

It must be acknowledged that, although linkage of environmental and exposure models with GIS has frequently been encountered in the past, in the majority of the cases, the GIS and environmental/risk models had not been truly integrated – but simply used together (Goodchild *et al.*, 1993). The GIS has often used as pre-processors to prepare spatially distributed input data, and as post-processors to display and possibly analyze model results further. Compared to maps, however, GIS has the inherent advantage that the data storage and data presentation aspects are separate. Consequently, data may be presented and viewed in a variety of ways.

In general, the integration of GIS into risk assessment and environmental management programs can result in the following particularly important uses for GIS: as a tool in environmental and exposure modeling; as a tool for hazard, exposure and risk mapping; and as a tool for risk communication. Indeed, exposure analysis as an overlay of sources and receptor represents an almost classical GIS application. GIS have the ability to integrate spatial variables into risk assessment models, yielding maps that are powerful visual tools to communicate risk information (Goodchild *et al.*, 1993). In principle, the conceptual mapping of risk makes it much easier to communicate hazard and risk levels to potentially affected society and other stakeholders.

In typical application scenarios, the GIS will generally form a central framework and integrating component that provides a variety of map types for use in an overall environmental and public health risk management system. Maps or overlays include simple line features (such as residential boundaries) or complex topical maps serving as background for the spatially distributed environmental models. A simple example of such a representation is concentration fields of pollutants from air, groundwater or surface water models stored as grid cell files.

10.7. Suggested Further Reading

Bates, DV, 1994. *Environmental Health Risks and Public Policy*, Univ. of Washington Press, Seattle, Washington

Hilts, SR, 1996. A co-operative approach to risk management in an active lead/zinc smelter community, *Environmental Geochemistry and Health*, 18: 17 – 24

McKone, TE, WE Kastenberg, and D. Okrent, 1983. The use of landscape chemical cycles for indexing the health risks of toxic elements and radionuclides, *Risk Analysis*, 3(3): 189 – 205

Moeller, DW, 1992. *Environmental Health*, Harvard Univ. Press, Cambridge, Mass

Mohamed, AMO and K. Côté, 1999. Decision analysis of polluted sites – a fuzzy set approach, *Waste Management,* 19: 519 – 533

Mitchell, P. and D. Barr, 1995. The nature and significance of public exposure to arsenic: a review of its relevance to South West England, *Environmental Geochemistry and Health*, 17: 57 – 82

Russell, M. and M. Gruber, 1987. Risk assessment in environmental policy-making, *Science,* 236: 286 – 290

Suter II, GW, BW Cornaby, *et al.*, 1995. An approach for balancing health and ecological risks at hazardous waste sites, *Risk Analysis*, 15: 221-231

Williams, FLR and SA Ogston, 2002. Identifying populations at risk from environmental contamination from point sources, *Occupational and Environmental Medicine*, 59(1): 2 – 8

Chapter 11

THE ROLE OF RISK ASSESSMENT
IN PUBLIC HEALTH POLICY DECISIONS

Risk assessment seems to be gaining wider grounds in making public health policy decisions on the control of risks associated with human exposures to chemicals. This situation may be attributed to the fact that, the very process of performing a risk assessment can lead to a better understanding and appreciation of the nature of the risks inherent in a study, and further helps develop steps that can be taken to reduce these risks. Overall, the application of risk assessment to chemical exposure problems helps identify critical receptor exposure routes, and other extraneous factors contributing most to total risks. It also facilitates the determination of cost-effective risk reduction policies. Ultimately, based on the results of a risk assessment, a more effectual decision can be made in relation to the types of risk management actions necessary to address a given chemical exposure problem or a hazardous situation.

Indeed, the risk assessment process is intended to give the risk management team the best possible evaluation of all available scientific data – in order to arrive at justifiable and defensible decisions on a wide range of issues. For example, to ensure public safety in chemical exposure situations, receptors must not exceed some stipulated risk-based exposure levels or acceptable public health goals – typically established through a risk assessment process. In general, it is apparent that, some form of risk assessment is inevitable if public health and environmental management programs are to be conducted in a sensible and deliberate manner.

Inevitably, risk-based decision-making will generally result in the design of better environmental and public health risk management programs. This is because risk assessment can produce more efficient and consistent risk reduction policies. It can also be used as a screening device for the setting of policy priorities. This chapter enumerates the general scope for the application of risk assessment, as pertains to the management of potential chemical exposure problems; it also discusses specific practical example situations for the utilization of the risk assessment paradigm.

11.1. General Scope of Public Health Risk Assessment Practice

Risk assessment has several specific applications that could affect the type of decisions to be made in relation to environmental and public health risk management programs.

A number of practical examples of the potential application of risk assessment principles, concepts, and techniques – including the identification of key decision issues associated with specific problems – abound in the literature of risk analysis. Some of the broad applications often encountered in chemical exposure situations include uses:

- Analysis of human health impacts from chemical residues found in food products (such as contaminated fish and pesticide-treated produce), as well as a variety of consumer products – including cosmetics and pharmaceuticals.
- Addressing the health and safety issues associated with environmental chemicals – i.e., to determine 'safe' exposure limits for toxic chemicals used or found in the workplace and residences.
- Facilitation of decisions about the use of specific chemicals in manufacturing processes and industrial activities.
- Implementation of general risk management and risk prevention programs for public health and environmental management planning.
- Evaluation and management of potential risks due to toxic air emissions from industrial facilities and incinerators.
- Evaluation of potential risks associated with the migration of contaminant vapors into building structures.
- Facilitation of property transactions by assisting developers, lenders, and buyers in the 'safe' acquisition of both residential and commercial properties.
- Determination of potential risks associated with industrial, commercial, and residential properties – to facilitate land-use decisions and/or restrictions.

The application of the risk assessment process to chemical exposure problems will generally serve to document the fact that risks to human health and the environment have been evaluated and incorporated into a set of appropriate response actions. In fact, almost invariably, every process for developing effectual environmental and public health risk management programs should incorporate some concepts or principles of risk assessment. In particular, all decisions on corrective action plans for potential chemical exposure problems will include, implicitly or explicitly, some elements of risk assessment.

Appropriately applied, risk assessment techniques can indeed be used to estimate the risks posed by chemical hazards under various exposure scenarios, and to further estimate the degree of risk reduction achievable by implementing various scientific remedies. Overall, a risk assessment will generally provide the decision-maker with scientifically defensible procedures for determining whether or not a potential chemical exposure problem could represent a significant adverse health and environmental risk, and if it should therefore be considered a candidate for mitigative actions. In fact, several issues that – directly or indirectly – affect public health and environmental management programs may be addressed by using some form of risk assessment.

11.1.1. ILLUSTRATIVE EXAMPLES OF
PUBLIC HEALTH RISK ASSESSMENT IN PRACTICE

In the applications of risk assessment, it is important to adequately characterize the exposure and physical settings for the problem situation, in order to allow for a proper application of appropriate risk assessment methods of approach. Unfortunately, there tends to be several unique complexities associated with real-life chemical exposure scenarios, and this can seriously overburden the overall process. Also, the populations potentially at risk from chemical exposure problems are usually heterogeneous – and this can greatly influence the anticipated impacts/consequences. Critical receptors should therefore be carefully identified with respect to numbers, location (areal and temporal), sensitivities, etc., so that risks are neither underestimated nor conservatively overestimated.

The determination of potential risks associated with chemical exposure problems invariably plays an important role in public health risk mitigation and/or risk management strategies – as demonstrated by the hypothetical example problems that follow below. It is noteworthy that, risk assessments may be formulated quite differently for differing situations or circumstances – such as one that is purely qualitative in nature, through a completely quantitative evaluation.

Evaluation of Human Health Risks Associated with
Airborne Exposures to Asbestos
This section presents a discussion of the investigation and assessment of the human health risks associated with worker exposures to asbestos in the ventilation systems of a commercial/office building.

There are two sub-divisions of asbestos: the serpentine group containing only chrysotile (which consists of bundles of curly fibrils); and the amphibole group containing several minerals (which tend to be more straight and rigid). Asbestos is neither water-soluble nor volatile, so that the form of concern is microscopic fibers (usually reported as, or measured in the environment in units of fibers per m^3 or fibers per cc).

Processed asbestos has typically been fabricated into a wide variety of materials that have been used in consumer products (such as cigarette filters, wine filters, hair dryers, brake linings, vinyl floor tiles, and cement pipe), and also in a variety of construction materials (e.g., asbestos-cement pipe, flooring, friction products, roofing, sheeting, coating and papers, packaging and gaskets, thermal insulation, electric insulation, etc.). Notwithstanding the apparent useful commercial attributes, asbestos has emerged as one of the most complex, alarming, costly, and tragic environmental health problems (Brooks *et al.*, 1995). A case in point, asbestos materials are frequently removed and discarded during building renovations and demolitions. To ensure safe ambient conditions under such circumstances, it often becomes necessary to conduct an asbestos sampling and analysis – which results can be used to support a risk assessment.

Study Objective. The primary concern of the risk assessment for the ventilation systems in the case building is to determine the level of asbestos exposures that potential receptors (especially workers cleaning the ventilation systems) could experience, and whether such exposure constitutes potential significant risks.

Summary Results of Environmental Sampling and Analysis. Standard air samples are usually collected on a filter paper and fibers >5μm long are counted with a phase contrast microscope; alternative approaches include both scanning and transmission electron microscopy and X-ray diffraction. It is generally believed that fibers that are 5μm or longer are of potential concern (USEPA, 1990a,b).

Following an asbestos identification survey of the case structure, air samples collected from suspect areas in the building's ventilation systems were analyzed using phase contrast microscopy (PCM), and highly suspect ones further analyzed by using transmission electron microscopy (TEM). The TEM analytical results are important because they serve as a means/methods for distinguishing asbestos particles from other fibers or dust particles.

The PCM analysis produced concentration of asbestos fibers in the range of <0.002 to a maximum of 0.008 fibers/cm^3. From the TEM, chrysotile asbestos was determined to be at <0.004 structures per cm^3 (str/cm) in all the environmental air samples.

The Risk Estimation. For asbestos fibers to cause any disease in a potentially exposed population, they must gain access to the potential receptor's body. Since they do not pass through the intact skin, their main entry routes are by inhalation or ingestion of contaminated air or water (Brooks *et al.*, 1995) – with the inhalation pathway apparently being the most critical in typical exposure scenarios. That is, for asbestos exposures, inhalation is expected to be the only significant exposure pathway. Consequently, intake is based on estimates of the asbestos concentration in air, the rate of contact with the contaminated air, and the duration of exposure. Subsequently, the intake is integrated with the toxicity index to determine the potential risks associated with any exposures.

Individual excess cancer risk is a function of the airborne contaminant concentration, the probability of an exposure causing risk, and the exposure duration. By using the cancer risk equations presented earlier in Chapter 8, the cancer risk from asbestos exposures may be estimated in accordance with the following relationship:

Cancer Risk = [airborne fiber concentration (fibers/m^3)]
            x [exposure constant (unitless)]
            x [inhalation unit risk ((100 PCM fibers/m^3)$^{-1}$)]   (11.1)

or,

Risk Probability = Intake x UR = [C$_a$ x INHf] x UR   (11.2)

The following exposure assumptions are used to facilitate the intake computation for this particular problem identified above:

- It is assumed that workers cleaning the ventilation system will complete this task within two weeks for a 5-day work-week. Hence, the maximum exposure duration is taken as, ED = 10 days – in comparison to a 70-year lifetime daily exposure.
- Assumed exposure time is 40 minutes per working hour, for an 8-hour work-day.
- Inhalation rate is 20 m^3/day (or 0.83 m^3/hr).

The exposure evaluation utilizes the information obtained from the airborne fiber samples collected and analyzed for during the prior air sampling activities; to be conservative, the maximum concentrations measured from the analytical protocols are used in the risk estimation. Thence, the fraction of an individual's lifetime for which exposure occurs – represented by the inhalation factor – is estimated to be:

$$INHf = (40/60) \times (8/24) \times (10/365) \times (1/70) = 8.7 \times 10^{-5}$$

Next, asbestos is considered carcinogenic with a unit risk of approximately 1.9×10^{-4} $(100 \text{ PCM fibers/m}^3)^{-1}$ (see, e.g., DTSC/Cal EPA, 1994). Consequently, potential risk associated with the 'possible' but unlikely (represented by an evaluation based on the PCM analysis results) and the reasonable/likely (represented by an evaluation based on TEM analysis results) asbestos concentrations are determined, respectively, as follows:

- Risk represented by results of the PCM analyses is estimated by integrating the following information,

  ◆ PCM-based airborne fiber concentration (maximum)
    $= 0.008 \text{ fibers/cc} = 8 \times 10^3 \text{ fibers/m}^3$
  ◆ $INHf = 8.7 \times 10^{-5}$
  ◆ $UR = 1.9 \times 10^{-4} (100 \text{ PCM fibers/m}^3)^{-1} \equiv 1.9 \times 10^{-6} \text{ per fibers/m}^3$

Hence,

$$\text{Cancer Risk (based on PCM concentration)} = 1.32 \times 10^{-6}$$

- Risk represented by results of the TEM analyses is estimated by integrating the following information,

  ◆ TEM-based airborne asbestos concentration (maximum)
    $= 0.004 \text{ structures/cc} = 4 \times 10^3 \text{ str/m}^3$
  ◆ $INHf = 8.7 \times 10^{-5}$
  ◆ $UR = 1.9 \times 10^{-4} (100 \text{ PCM fibers/m}^3)^{-1} \equiv 1.9 \times 10^{-6} \text{ per fibers/m}^3$

Hence,

$$\text{Cancer Risk (based on TEM concentration)} = 6.6 \times 10^{-7}$$

A Risk Management Decision. All risk estimates indicated here are near the lower end of the generally acceptable risk range/spectrum (i.e., 10^{-4} to 10^{-6}). Thence, it may be concluded that asbestos in the case building should represent minimal potential risks of concern for workers entering the ventilation system to clean it up. Nonetheless, it is generally advisable to incorporate adequate worker protection through the use of appropriate respirators. In general, any asbestos abatement or removal program should indeed conform to strict health and safety requirements – with on-site enforcement of the specifications being carried out by a qualified health and safety officer or industrial hygienist.

A Human Health Risk Assessment Associated with
PCB Release into the Environment
PCBs (polychlorinated biphenyls) are mixtures of synthetic organic chemicals. Different mixtures can take on forms ranging from oily liquids to waxy solids. Although their chemical properties vary widely, different mixtures can have many common components. Because of their non-inflammability, chemical stability, and

insulating properties, commercial PCB mixtures had been used in many industrial applications, especially in capacitors, transformers, and other electrical equipment. These chemical properties, however, also contribute to the persistence of PCBs after they are released into the environment. In fact, because of evidence that PCBs persist in the environment and cause harmful effects, the manufacture of commercial mixtures was stopped in the late1970s – albeit existing PCBs continued in use.

Problem Scenario. Consider a release of PCBs onto the ground near a lake. Potential pathways of human exposure have been determined to include vapor inhalation, drinking water, fish ingestion, and skin contact with ambient water and contaminated soil.

The population of interest includes anglers who consume an average of two 105g portions of local fish each week; this translates into 30g of fish ingestion per day (i.e., [2 x 105g per week]/7days per week = [210/7] = 30g per day). They also spend most of their time in the area, on average, breathing 20m^3 of air and drinking 2L of water each day. Skin contact with ambient water and soil is negligible for this population. A 30-year human exposure duration is assumed, with a representative lifespan of 70years, and an average body weight of 70kg.

Environmental samples indicate long-term average concentrations of 0.01 μg/m^3 in ambient air, 5 μg/L in drinking water, and 110 μg/kg in the edible portion of local fish. Issues pertaining to dust in ambient air and sediment in drinking water are considered negligible.

PCBs persist in the body, providing a continuing source of internal exposure after external exposure stops. There may be greater-than-proportional effects from less-than-lifetime exposure, especially for persistent mixtures and for early-life exposures. PCBs are absorbed through ingestion, inhalation, and dermal exposure, after which they are transported similarly through the circulation. This provides a reasonable basis to expect similar internal effects from different routes of human exposure. Indeed, joint consideration of cancer studies and environmental processes leads to a conclusion that environmental PCB mixtures are highly likely to pose a risk of cancer to humans. Apart from the cancer effects, PCBs also have significant human health effects other than cancer – including neurotoxicity, reproductive and developmental toxicity, immune system suppression, liver damage, skin irritation, and endocrine disruption. Toxic effects have indeed been observed from acute and chronic exposures to PCB mixtures with varying chlorine content.

The Exposure Scenarios. Three different exposure pathways are assumed for this case problem – namely, vapor inhalation, water ingestion, and fish consumption. Because of partitioning, transformation, and bioaccumulation, different fractions of the original mixture are encountered through these pathways – and hence different potency values are appropriate. Vapor inhalation is associated with 'low risk' (because evaporating congeners tend to have low chlorine content and be inclined to metabolism and elimination), so the low end of the range (upper-bound slope of 0.4 per mg/kg-d) is used for vapor inhalation (USEPA, 1996). Similarly, ingestion of water-soluble congeners is associated with 'low risk' (because dissolved congeners tend to have low chlorine content and be inclined to metabolism and elimination) – so the low end (of 0.07 per mg/kg-d) is also used for drinking water (USEPA, 1996). (It is noteworthy that, if ambient air or drinking water had contained significant amounts of contaminated dust or sediment, the high-end potency values would be more appropriate, as adsorbed congeners tend to be of high chlorine content and persistence.) Finally, food chain exposure is more realistically associated with 'high risk' (because aquatic organisms and fish selectively accumulate congeners of high chlorine content and persistence that are resistant to metabolism and elimination) – and thus, the high end of the range (upper-bound slope of 2 per mg/kg-d) is used for fish ingestion (USEPA, 1996).

Risk Calculation. The lifetime average daily dose (LADD) is calculated as the product of concentration C, intake rate IR, and exposure duration ED divided by body weight BW and lifetime LT, as follows:

$$\text{Pathway Exposure, LADD} = [C \times IR \times ED]/[BW \times LT] \tag{11.3}$$

Thence,

Vapor Inhalation LADD
$$= [0.01\mu g/m^3 \times 20m^3/d \times 30yr]/[70kg \times 70yr] = 1.2 \times 10^{-6} \text{ mg/kg-d}$$

Drinking Water LADD
$$= [5.0\mu g/L \times 2L/d \times 30 \text{ yr}]/[70kg \times 70yr] = 6.1 \times 10^{-5} \text{ mg/kg-d}$$

Fish Ingestion LADD
$$= [110\mu g/kg \times 30g/d \times 30yr]/[70kg \times 70yr] = 2.0 \times 10^{-5} \text{ mg/kg-d}$$

Subsequently, for each pathway, the lifetime average daily dose is multiplied by the appropriate slope factor to arrive at the estimated risk, as follows:

$$\text{Pathway Risk} = [\text{LADD}] \times [\text{Cancer Slope Factor}] \tag{11.4}$$

Thence,

Vapor Inhalation Risk $= 1.2 \times 10^{-6}$ mg/kg-d $\times$ 0.4 per mg/kg-d $= 4.8 \times 10^{-7}$

Drinking Water Risk $= 6.1 \times 10^{-5}$ mg/kg-d $\times$ 0.07 per mg/kg-d $= 4.3 \times 10^{-6}$

Fish Ingestion Risk $= 2.0 \times 10^{-5}$ mg/kg-d $\times$ 2 per mg/kg-d $= 4.0 \times 10^{-5}$

Thus,
$$\text{Total LADD} = 8.2 \times 10^{-5} \text{ mg/kg-d}$$
and
$$\text{Total Risk} = 4.5 \times 10^{-5}$$

A Risk Management Decision. The above evaluation leads to a conclusion that fish ingestion is the principal pathway contributing to risk, and that drinking water and vapor inhalation are of lesser consequence. Indeed, it would be advisable to examine variability in fish consumption rates and fish tissue concentrations to determine whether some individuals are at much higher risk. In any case, it also is important to recognize that, this specific site exposure adds to a background level of exposure from other sources.

11.2. The Public Health Risk Assessment Paradigm

In our attempts to shape public health risk management policy decisions, one must appreciate what Rachel Carson notes in her book, *Silent Spring*, that: "As the tide of chemicals born of the Industrial Age has arisen to engulf our environment, a drastic change has come about in the nature of the most serious public health problems" (Carson, 1962; 1994). Indeed, chemicals have become an integral part of modern ways of life – with the capacity to improve as well as endanger public health. The general population is typically exposed to chemicals in air, water, foods, cosmetics, household products, and a variety of therapeutic drugs. In everyday life, a person may experience a multitude of exposure to potentially toxic substances, singly and in combination, and both synthetic and natural. Levels of exposure tend to vary and may or may not pose a hazard – depending on the dose, route, and duration of exposure. The consequences of human exposure to chemicals have therefore become (or should become) a very important driving force in public health policy decisions. To effectively address this situation, the traditional approach to dealing with public health risk management issues may not suffice in this day and age. Contemporary risk assessment methods of approach may therefore be used to facilitate the design of more reliable public health risk management strategies/schemes. But it must also be recognized that, a given risk assessment provides only a snapshot in time (and indeed in space as well) of the estimated risk of a given toxic agent at a particular phase of our understanding of the issues and problems. As Moeller (1997) notes, unless care is exercised and all interacting factors are considered, risk assessments directed at single issues, followed by ill-conceived risk management strategies, can create problems worse than those the management strategies were designed to correct. The single-issue approach can also create public myopia by excluding the totality of alternatives and consequences needed for an informed public choice (Moeller, 1997). Indeed, to be truly instructive and constructive, therefore, risk assessments will usually be conducted on an iterative basis – being updated as new information and knowledge become available. Ultimately, it is quite important to examine the total system to which a given risk assessment is being applied.

It is expected that, there will be growing applications of the risk assessment paradigm to several specific chemical exposure problems, and this could affect the type of decisions made in relation to public health risk management programs. Such applications may cover a wide range of diverse problem situations – as exemplified by the illustrative application scenarios annotated below. This listing of public health study designs is by no means complete and exhaustive, since variations or even completely different and unique problems may be resolved by use of one form of risk assessment principle and methodology or another.

- *Investigation of blood lead (Pb) distribution amongst population groups.* This type of public health risk assessment study may be used to help determine the likely impacts of Pb exposures on various population groups.

    - *Study Rationale.* For young children, Pb can cause lower levels of intelligence, behavioral problems, and school failures. In fact, recent studies conducted by the Harvard School of Public Health in the United States have determined that, a woman's lead exposure during pregnancy can threaten the fetus' nervous system and other developing organs. Also, it has been noted that women of child-bearing age who were exposed to lead as children usually will have this lead accumulated in their bones, threatening the health of their babies many years later. The pre-natal exposure scenario is: if a little girl is exposed to lead, the lead is stored in her bones as she grows, and when she becomes a pregnant adult, the lead moves from her bones – exposing her fetus to lead. Furthermore, lead acquired pre-natally can contribute to the lead burden in young children, implying a potential concern for pregnant women. Further yet, recent studies (e.g., Tellez-Rojo *et al.*, 2002) designed to evaluate the impact of breast-feeding on the mobilization of lead from bone seem to confirm the hypothesis that lactation stimulates lead release from bone to blood. Lactation has indeed been recognized as a powerful stimulus for bone resorption; thus, Pb accumulated in bone from past exposures may be released into the bloodstream and excreted in breast milk, constituting an important source of lead exposure for the breast-fed infant (see, e.g., Silbergeld, 1991; Tellez-Rojo *et al.*, 2002). Also of significant interest, recent studies document the impact of low level lead exposure on blood pressure in adults (Schwartz, 1991). The significance of investigating lead exposure to a community can therefore not be underestimated, since it will ultimately threaten infant development as well as adult welfare.

♦ *Scientific Design.* A study may be designed to determine the presence of, and the degree of population group exposures and impacts in different regions. The study may document results with respect to gender differences, age categories and even the different socioeconomic classes of a community.

♦ *Significance of Study.* Considering the frequent occurrence of lead in several environmental settings, and in view of the dangers associated with lead exposures – especially to children – it is important to adequately document this kind of information, and then help develop strategies to deal with likely problems associated with the lead contamination and exposure situations. In fact, Pb is naturally occurring, but often is released into the environment from human-made sources; it has been mined, smelted, refined, and used for hundreds of years. For example, Pb has been used as an additive in paint and gasoline, and in leaded pipes, solder, crystal, and ceramics. Mining, smelting, and refining activities have resulted in substantial increases in Pb levels in the environment, especially near mining and smelting sites, near some types of industrial and municipal facilities, and adjacent to highways.

Pb particles in the environment can attach to dust and be carried long distances in the air. Such Pb-containing dust can be removed from the air by rain and deposited on surface soil, where it may remain for many years. In addition, heavy rains may cause Pb in surface soil to migrate into ground water and eventually into water systems. Given its widespread distribution, everyone is exposed to 'background' levels of Pb. In fact, there are many possible ways to be exposed to Pb, including ingestion of Pb-contaminated water, soil, paint chips, and dust; inhalation of Pb-containing particles of soil or dust in air; and ingestion of foods that contain Pb from soil or water. Pb poisoning is a particularly insidious public health threat because there may be no unique signs or symptoms. Early symptoms of Pb exposure may include persistent fatigue, irritability, loss of appetite, stomach discomfort, reduced attention span, insomnia, and constipation. Failure to treat Pb poisoning in the early stages can cause long-term or permanent health damage, but because of the general nature of symptoms at this stage, Pb poisoning is often not suspected.

In adults, Pb poisoning can cause irritability, poor muscle coordination, and nerve damage to the sense organs and nerves controlling the body. It may cause increased blood pressure, hearing and vision impairment, and reproductive problems (such as a decreased sperm count). It also can retard fetal development even at relatively low levels of Pb. In children, Pb poisoning can cause brain damage, mental retardation, behavioral problems, anemia, liver and kidney damage, hearing loss, hyperactivity, developmental delays, other physical and mental problems, and in extreme cases, death. Although the effects of Pb exposure are a potential concern for all humans, young children (0 to 7 years old) are the most at risk. This increased vulnerability results from a combination of the following factors:

▶ Children typically have higher intake rates per unit body weight for environmental media such as soil, dust, food, water, air, and paint, than adults since they are more likely to play in dirt and to place their hands and other objects in their mouths;

▶ Children tend to absorb a higher fraction of ingested Pb from the gastrointestinal tract than adults;

▶ Children tend to be more susceptible to the adverse neurological and developmental effects of Pb than adults; and

▶ Nutritional deficiencies of iron or calcium, which are prevalent in children, may facilitate Pb absorption and exacerbate the toxic effects of Pb.

The current/typical blood Pb level of concern in children is 10 micrograms (μg) of Pb per deciliter (dL) of blood (i.e., 10μg/dL). However, since adverse effects may occur at lower levels than previously thought, various agencies are considering whether this level should be lowered further.

• *Assessment of risks from chemical contaminants in nursing mothers' breast milk.* This type of public health risk assessment study may be designed to help determine the potential risks associated with the breast-feeding of infants.

◆ *Study Rationale.* It apparent that human breast milk is a primary source of infant nutrition – especially in most developing countries. But the investigation of the levels of chemical contaminants in human milk in some locales elucidate the potential risks of such contaminants to the health of breast-fed infants – and who are indeed more susceptible to chemical exposure effects. In view of the fact that welfare establishments have been encouraging mothers to breast-feed more often, the question needs to be raised as to whether this is always 'safe' for all the children – considering the fact that some nursing mothers may have high accumulation of dangerous environmental chemicals in their breast milk? For example, pre-natal exposures to PCBs in foods, etc. can be passed on to newborn infants.

- ♦ *Scientific Design.* A study may be so-designed to identify and quantify the presence and levels of select chemicals (e.g., Pb, PCBs, dioxins, and organochlorine pesticides such as DDT, dieldrin, lindane, aldrin, hexachlorobenzene, chlordane, etc.) in mothers' milk for nursing mothers in different locales, and from different socioeconomic classes of a community. Subsequently, the toxicological implications to the health status of both the nursing mother and the breast-fed infants can be determined.

- ♦ *Significance of Study.* The toxicological implications derived from this study can become a very important guide for health care providers – especially at post-natal health care facilities.

- *Analysis of human health impacts from chemical residues in food and consumer products.* This type of public health risk assessment study may be designed to help determine the potential risks associated with population exposures to chemical residues in food (including contaminated fish and pesticide-treated produce) and a variety of consumer products.

  - ♦ *Study Rationale.* It is almost indisputable that all peoples around the world consume a whole variety of plant and animal products, some of which have – directly or indirectly – been exposed to chemical substances at one time or another. Also, a variety of consumer products abound on the world markets today – some having questionable origins, and several containing potentially toxic chemicals.

  - ♦ *Scientific Design.* A study may be so-designed that consists of an investigation into the occurrence, and the measurement of the levels of selected chemical contaminants (including organochlorine pesticides residues and various inorganic chemicals) that may be present in typical staple food products of the average person in the case-study area – especially for fish, meat, and chicken eggs. In implementing this program, food samples can be collected and analyzed for selected chemical contaminants warranting investigation for the particular setting/location. Also, information can be collected on food consumption patterns of the various communities to facilitate realistic risk assessments. The relevant studies may also consist of an investigation into the types of toxic chemicals found in selected consumer products (such as hair products, skin care products, processed and canned food items, etc.) commonly found on the market for a case-study area, and that have widespread usage among various sectors of society. Based on an examination of the consumer use patterns and exposures, risks to human health from the widely used consumer products may be determined. As part of this study, a comparative look can be made for the urban populace *vs.* the rural dweller.

  - ♦ *Significance of Study.* Results of the investigation can help develop an effectual public health education and awareness program about the potential harms from toxic chemicals – such as PCBs and DDT – potentially present in common food and consumer products finding widespread use in a region. It can also help national or regional governments in establishing long-term national food contamination monitoring program – as part of the comparable United Nations/WHO programs.

- *Investigating the health impacts of mining activities on a community.* This type of public health risk assessment study may be used to help determine the potential impacts associated with mining activities.

  ♦ *Study Rationale.* As part of the recent economic rejuvenation programs in a number of nations, mining activities seem to have become one of the most popular ventures attracting a wide spectrum of investors. But even long before the recent additions to the mining sector, in some regions (especially in the newly emerging economies), environmental degradation from mining activities has always been a critical but neglected issue.

  ♦ *Scientific Design.* A comprehensive tiered study may be so-designed that consists of both general and specific investigations of the health implications of the various mining sectors in the case region. This study may also include a look at the distribution of likely health problems associated with the different mining sectors and communities. For the mining communities selected for more detailed investigations, attention could be focused on issues such as water quality problems, levels of various chemicals in the blood of the target populations, etc. that are associated with particular mining activities. For example, mercury will be of particular interest in the gold-mining areas (since it is often used to extract gold), especially when fishing streams can be found within the watershed. Also, the study may be so-designed to cover all seasons, in order to ensure accurate measurements and that will ensure effective policy decisions. The complete seasonal investigation is deemed necessary, in part because recent studies by some Canadian scientists in the Brazilian Amazon found a link between the seasons and methyl mercury (a highly toxic form of mercury created when the metal is released into streams and modified by bacteria) levels found in a village population in that region; in that case, it was proven that the contamination levels in the Amazon were highest during the rainy season. Whereas this seasonal variation in contaminant levels may not be universally true, it will seem prudent to extend such types of investigation to cover all the seasons – at least to give representative database for statistical analysis purposes.

  For practicality, this type of somewhat extensive project can be carried out in phases – albeit each phase can be designed to yield project outputs/results that can individually be used to make important public health policies and to guide risk management programs. For example, a 'Phase 1' may consist of a survey to identify the prominent types of toxic chemicals found to originate from mining activities; the collection of chemical and exposure data, based on the nature of mining activities; and a statistical compilation of common health problems in the target communities. The information from this initial phase can be analyzed (both qualitatively and quantitatively) in order to determine the potential risks to the exposed populations. A 'Phase 2' may aim at collecting specific material samples that can undergo appropriate laboratory analyses (to facilitate a more accurate quantification of the likely risks to populations exposed to mining-related environmental chemicals under a variety of conditions). The 'Phase 2' sampling program may involve taking blood and other biomarker samples from representative residents of the target community, to be analyzed for the suspect chemicals of concern, and also the sampling and analysis of potable water supplies and selected dietary/farm produce of the target communities.

♦   *Significance of Study*. Results from each of the project phases can be used to design appropriate mitigative and public health risk management programs in relation to the impacts of mining activities on various sectors of the populations in the study locale. The results of the study can also be used to help develop effectual corrective measures for current and future mining practices in a locale or region. Ultimately, such a project can be expected to help improve risk mitigation and public health risk management programs associated with mining activities in a region.

•   *Health implications of pesticide use in agricultural communities.* This type of public health risk assessment study may be used to facilitate public policy decisions on pesticide applications.

♦   *Study Rationale.* The use of a wide range of chemical pesticides seems to be an important aspect of agricultural practice worldwide. But then, such increased use of pesticides is of grave concern due to their potential effects on human health. In fact, the situation is particularly worrisome in the newly emerging economies, where lesser protective measures are generally taken and also where reliable data on the population exposures to pesticides (and indeed several other environmental and 'social' chemicals) is lacking. As has been affirmed by the World Health Organization (WHO) and the United Nations Environment Programme (UNEP), pesticides currently in use involve a wide variety of chemicals, with great differences in their mode of action, uptake by the body, metabolism, elimination from the body, and toxicity to humans. Also, it is an undisputed fact that, the health effects will depend on the health status of the individual exposed. Thus, malnutrition and dehydration – situations fairly prevalent in developing economies in particular – are likely to increase sensitivity to pesticides. In addition, several environmental factors – such as temperature and humidity – will tend to affect the absorption of pesticides by exposed individuals.

♦   *Scientific Design.* A typical research study may be designed that consists of both general and specific investigations of the health implications of pesticide use in the selected agricultural (or other pesticide applicator) communities. In particular, the levels of pesticides in farmers can be investigated. The study may also cover such parameters as miscarriages and the prevalence of birth defects, in order to determine any possible association with pesticide usage in particular communities.

As part of the overall program, information can be collected on the pesticide use/application patterns for major farming communities. Blood samples can be collected from farmers and analyzed for organochlorine pesticide residues, and indeed other related chemical contaminants warranting investigation for the particular setting/location. The study can also include an investigation of possible pesticide contamination of rural drinking water supplies located near farmlands.

Indeed, such type of project may be carried out in phases – albeit each phase can be designed to yield project outputs/results that can individually be used to make important public health policies and to guide risk management programs. For example, in a two-phase design, 'Phase 1' may consist of a survey to identify the prominent types of pesticides in general use; the collection of chemical and exposure data, based on the pesticide use patterns; and a statistical compilation of common health problems in the target communities. The information from this initial phase can be analyzed (both qualitatively and quantitatively) in order to determine the potential risks to the exposed populations. 'Phase 2' may then aim at collecting specific material samples that can undergo appropriate laboratory analyses (to facilitate a more accurate quantification of the likely risks to populations exposed to pesticides under a variety of conditions). The 'Phase 2' sampling program may, for example, involve taking blood and other biomarker samples from farmers, to be analyzed for organochlorine pesticide residues, and also the sampling and analysis of potable water supplies and selected dietary/farm produce of the target communities. Results from each of the project phases can then be used to design appropriate mitigative and public health risk management programs in relation to pesticide usage in the region.

◆ *Significance of Study.* Results from such a study can be used to help develop an effectual public education program about the potential harms from pesticide usage in region, by providing a basis for public education on pesticides use. Ultimately, such a project can be expected to help improve risk mitigation actions and public health risk management programs involving pesticide usage in a region, particularly for the 'pesticide applicator' areas.

• *Evaluation of selected urban occupational worker exposure risks.* So many innocent and unsuspecting workers in various countries are exposed to a variety of toxic substances on a daily basis – but this represents a preventable situation, achievable by using appropriate protective equipment and clothing. In fact, chronic worker exposure for select categories of workers who often work with dangerous chemicals without personal protective tools/equipment/clothes is becoming an even more serious problem in a number developing and newly industrializing countries – especially with the mushrooming of several small businesses and mini-industries.

◆ *Study Rationale.* A study may be undertaken that consists of an investigation into the types of toxic chemicals widely used in consumer products (such as hair products, select food items, etc.); examination of the consumer use patterns and exposures (to include chronic worker exposure for select categories of workers who often work with dangerous chemicals without personal protective tools/equipment/clothes); and the assessment of human health risks from the use of various consumer products commonly found on the market in the region of concern, and that have widespread use among various sectors of society.

- ◆ *Scientific Design.* Typically, the study will focus on select 'high-risk' urban occupational groups – to include, e.g., hairdressers and beauticians (for various chemicals found in cosmetics, etc.), auto mechanics/automotive shop workers (for various solvents and metals), and fuel station attendants (especially for benzene and possible fuel additives) – all of whom are exposed to specific chemicals on an almost daily basis. The study may consist of an investigation into occupational hazards associated with the types of toxic chemicals widely used by these worker groups.

- ◆ *Significance of Study.* Results from this study can help develop an effectual public education program and worker protection campaign about the potential harms from toxic chemicals often encountered by various worker groups.

- • *Morbidity effect of particulate matter (PM) exposures: an epidemiologic study of the human health effects from ambient particulate matter.* Epidemiologic studies that link population ambient PM exposures to adverse health effects can provide an indication of the measurable excesses in pulmonary function decrements, respiratory symptoms, hospital and emergency department admissions, and indeed mortality associated with ambient levels of $PM_{2.5}$, PM_{10}, and other indicators of PM exposures.

- ◆ *Study Rationale.* Epidemiology studies can be used to establish causal inferences about PM health effects. Subsequently, the causal inference methodology should play a key role in evaluating the effectiveness of proposed interventions (such as changes in regulatory standards for ambient PM). Ultimately, this would help clarify the predicted effect of reductions in ambient PM on public health.

- ◆ *Scientific Design.* Typically, epidemiologic studies are divided into *morbidity studies* and *mortality studies*. The morbidity studies would include a wide range of health endpoints – such as changes in pulmonary function test, reports of respiratory symptoms, self-medication in asthmatics, medical visits, low birth-weight infants, and hospitalization. (By the way, mortality studies from many causes tend to provide the most unambiguous evidence of a clearly adverse endpoint.)

    A typical study design may consist of the so-called 'cross-sectional' studies – that evaluates subjects at a 'point' in time, where measurements of health status, pollution exposure, and individual covariates are observed simultaneously. In general, studies with individual-level outcome data, covariates, and PM exposure indices should be preferred – albeit individual-level exposure data are the most commonly missing component.

    In this type of study, the hypothesis being tested will consist of the null hypothesis, H_0: exposure to ambient PM at current levels cannot cause adverse health effects in susceptible sub-populations or individuals *vs.* the alternate hypothesis, H_A: exposure to ambient PM or some component at current levels is associated with adverse health effects in some susceptible sub-populations or individuals.

♦ *Significance of Study.* Past epidemiologic studies strongly implicate respirable particles in increased morbidity and mortality in the general population. Specific epidemiologic studies can provide information on issues such as the following:

▸ short-term PM exposure effects on lung function and respiratory symptoms in asthmatics and non-asthmatics;

▸ long-term PM exposure effects on lung function and respiratory symptoms;

▸ relationships of short-term PM exposure to the incidence of respiratory and other medical visits, as well as hospitalization (i.e., hospital admissions over limited to extended duration); and

▸ effects of ambient PM exposure on acute cardiovascular morbidity

For most of the types of applications identified above, studies using various epidemiological designs would be employed; the various epidemiological studies may include both observational and experimental study designs – and that may consist of ordinary descriptive study designs, case control studies, prospective studies, and indeed various types of experimental studies. Overall, these types of public health risk assessment studies will seek to increase understanding and preventative strategies to be adopted by public health policy makers and also community healthcare providers in chemical exposure problem situations. Also, the specific projects or investigations could plan to involve students from the appropriate institutions dedicated to teaching health-related sciences, and thus help bring early awareness to the would-be healthcare providers at an earlier stage of their training. Ultimately, such projects are expected to help improve risk mitigation and public health risk management programs associated with various chemical exposure problems.

Finally, it must be recognized that a clear understanding and effective communication about the association between environmental hazards and disease incidence are indeed essential requirements of an effective environmental and public health risk management policy. Results from the types of projects exemplified here will help develop preventative strategies that can be adopted by public health policy makers as well as community healthcare providers.

11.3. Suggested Further Reading

American Academy of Pediatrics, 1997. Breastfeeding and the use of human milk, *Pediatrics*, 100:1035 – 1039

ATSDR, 1990 – Present. *ATSDR Case Studies in Environmental Medicine,* Agency for Toxic Substances and Disease Registry (ATSDR), US Dept. of Health and Human Service, Atlanta, GA

ATSDR, 1994. *Priority Health Conditions: An Integrated Strategy to Evaluate the Relationship Between Illness and Exposure to Hazardous Substances,* Agency for Toxic Substances and Disease Registry (ATSDR), US Dept. of Health and Human Service, Atlanta, GA

Awasthi, S., HA Glick, RH Fletcher, and N. Ahmed, 1996. Ambient air pollution and respiratory symptoms complex in preschool children, *Indian J. Med. Res.*, 104: 257 – 262

Bates, DV, 1992. Health indices of the adverse effects of air pollution: the question of coherence, *Environmental Research*, 59: 336 – 349

Bates, DV, 1994. *Environmental Health Risks and Public Policy*, Univ. of Washington Press, Seattle, Washington

Benedetti, M., I. Lavarone, and P. Comba, 2001. Cancer risk associated with residential proximity to industrial sites: a review, *Archives of Environmental Health*, 56(4): 342 – 349

Brunekreef, B., 1997. Air pollution and life expectancy: is there a relation?, *Occup. Environ. Med.*, 54: 781 – 784

Churchill, JE and WE Kaye, 2001. Recent chemical exposures and blood volatile organic compound levels in a large population-based sample, *Archives of Environmental Health*, 56(2): 157 – 166

Dykeman, R., *et al.*, 2002. Lead exposure in Mexican radiator repair workers, *American Journal of Industrial Medicine*, 41(3): 179 – 187

Goyer, RA, 1996. Results of lead research: prenatal exposure and neurological consequences, *Environmental Health Perspectives*, 104: 1050 – 1054

Hauptmann, M., H. Pohlabeln, *et al.*, 2002. The exposure-time-response relationship between occupational asbestos exposure and lung cancer in two German case-control studies, *American Journal of Industrial Medicine*, 41(2): 89 – 97

Heikkila, P., R. Riala, *et al.*, 2002. Occupational exposure to bitumen during road paving, *AIHA Journal*, 63(2): 156 – 165

Hill, AB, 1965. The environment and disease: association or causation?, *Proc. R. Soc. Med.*, 58: 295 – 300

Holtta, P., H. Kiviranta, *et al.*, 2001. Developmental dental defects in children who reside by a river polluted by dioxins and furans, *Archives of Environmental Health*, 56(6): 522 – 528

Johnson, DL, K. McDade, and D. Griffith, 1996. Seasonal variation in paediatric blood lead levels in Syracuse, NY, USA, *Environmental Geochemistry and Health*, 18: 81 – 88

Karmaus, W., J. Kuehr, and H. Kruse, 2001. Infections and atopic disorders in childhood and organochlorine exposure, *Archives of Environmental Health*, 56(6): 485 – 492

Onalaja, AO and L. Claudio, 2000. Genetic susceptibility to lead poisoning, *Environmental Health Perspectives*, 108 (Suppl.1): 23 – 28

Park, R., F. Rice, L. Stayner, *et al.*, 2002. Exposure to crystalline silica, silicosis, and lung disease other than cancer in diatomaceous earth industry workers: a quantitative risk assessment, *Occupational and Environmental Medicine*, 59(1): 36 – 43

Pocock, SJ, M. Smith and P. Baghurst, 1994. Environmental lead and children's intelligence: a systematic review of the epidemiological evidence, *British Medical Journal*, 309: 1189 – 97

Russell, M. and M. Gruber, 1987. Risk assessment in environmental policy-making, *Science*, 236: 286 – 90

Schwartz, J., 1996. Air pollution and hospital admissions for respiratory disease, *Epidemiology*, 7: 20 – 28

Seaton, A., 1996. Particles in the air: the enigma of urban air pollution, *Jour. Royal Society of Medicine*, 89: 604 – 607

List of References and Bibliography

Abbaspour, KC, R. Schulin, E. Schläppi, and H. Flühler, 1996. A Bayesian approach for incorporating uncertainty and data worth in environmental projects, *Environmental Modelling and Assessment*, 1(3/4): 151-158

ACS and RFF, 1998. *Understanding Risk Analysis (A Short Guide for Health, Safety and Environmental Policy Making)*, A Publication of the American Chemical Society (ACS) and Resources for the Future (RFF) – written by M. Boroush, ACS, Washington, DC

AIHC, 1994. *Exposure Factors Sourcebook*, American Industrial Health Council (AIHC), Washington, DC

Allan, M. and GM Richardson, 1998. Probability density functions describing 24-hour inhalation rates for use in human health risk assessments, *Human and Ecological Risk Assessment*, 4(2): 379 – 408

Alloway, BJ and DC Ayres, 1993. *Chemical Principles of Environmental Pollution*, Blackie Academic & Professional/Chapman & Hall, London, UK

Al-Saleh, IA and L. Coate, 1995. Lead exposure in Saudi Arabia from the use of traditional cosmetics and medical remedies. *Environmental Geochemistry and Health*, 17: 29 – 31

Andelman, JB, and DW Underhill, 1988. *Health Effects From Hazardous Waste Sites*, Lewis Publishers, Chelsea, Michigan

Anderson, PS and AI Yuhas, 1996. Improving risk management by characterizing reality: a benefit of probabilistic risk assessment, *Human and Ecological Risk Assessment*, 2: 55 – 58

Asante-Duah, DK, 1996. *Managing Contaminated Sites: problem diagnosis and development of site restoration*, John Wiley & Sons, Chichester, England

Asante-Duah, DK, 1998. *Risk Assessment in Environmental Management: a guide for managing chemical contamination problems*, John Wiley & Sons, Chichester, England

ASTM, 1994. *Risk-based Corrective Action Guidance*, American Society for Testing and Materials (ASTM), Philadelphia, PA, USA

ASTM, 1995. *Standard Guide for Risk-based Corrective Action Applied at Petroleum-Release Sites*, American Society for Testing and Materials, ASTM (E1739-95), Philadelphia, PA

ASTM, 1997a. *ASTM Standards on Environmental Sampling*, 2nd edition, American Society for Testing and Materials, Philadelphia, PA, USA

ASTM, 1997b. *ASTM Standards Related to Environmental Site Characterization*, American Society for Testing & Materials, Philadelphia, PA, USA

ATSDR, 1999a. *Toxicological Profile for Lead*, Agency for Toxic Substances and Disease Registry (ATSDR), US Department of Health and Human Services, Atlanta, GA

ATSDR, 1999b. *Toxicological Profile for Mercury*, Agency for Toxic Substances and Disease Registry, US Dept. of Health and Human Service, Atlanta, GA

Barnes, DG and M. Dourson, 1988. Reference dose (RfD): description and use in health risk assessments, *Regulatory Toxicology and Pharmacology*, 8: 471 – 486

Barrie, LA, D. Gregor, B. Hargrave, R. Lake, D. Muir, R. Shearer, B. Tracey, and T. Biddleman, 1992. Arctic contaminants: sources, occurrence and pathways, *Science of the Total Environment*, 122: 1 – 74

Bartram, J. and R. Balance (eds.), 1996. *Water Quality Monitoring*, E & FN Spon/Chapman & Hall, London, UK

Batchelor, B., J. Valdés, and V. Araganth, 1998. Stochastic risk assessment of sites contaminated by hazardous wastes, *Journal of Environmental Engineering*, 124(4): 380 – 388

Bates, MN, A. Smith, and K. Cantor, 1995. Case-control study of bladder cancer and arsenic in drinking water, *American Journal of Epidemiology*, 141: 523 – 530

Bean, MC, 1988. Speaking of risk, *ASCE Civil Engr.*, 58(2): 59 – 61

Beck, LW, AW Maki, NR Artman and ER Wilson, 1981. Outline and criteria for evaluating the safety of new chemicals, *Regul. Toxicol. Pharmacol.*, 1: 19-58

Bentkover, JD, VT Covello, and J. Mumpower, 1986. *Benefits Assessment: The State of the Art*, Riedel Publ., Boston, MA

Berlow, PP, DJ Burton and JI Routh, 1982. *Introduction to the Chemistry of Life*, Saunders College Publishing, Philadelphia, PA, USA

Berne, RM and MN Levy, 1993. *Physiology*, 3rd edition, Mosby Year Book, St. Louis, MO

Bernhardsen, T., 1992. *Geographic Information Systems*, Viak IT, Arendal, Norway

Bertazzi, PA, L. Riboldi, A. Pesatori, L. Radice, and C. Zocchetti, 1987. Cancer mortality of capacitor manufacturing workers, *Am. J. Ind. Med.*, 11:165–176

Berthouex, PM and LC Brown, 1994. *Statistics for Environmental Engineers*, Lewis Publishers/CRC Press, Boca Raton, FL

Bertollini, R., MD Lebowitz, R. Saracci, and DA Savitz (eds.), 1996. *Environmental Epidemiology (Exposure and Disease)*, Lewis Publishers/CRC Press, Boca Raton, Florida

Bhatt, H.G., R.M. Sykes and T.L. Sweeney (ed), 1986. *Management of Toxic and Hazardous Wastes*, Lewis Publihers, Inc., Chelsea, MI

Binder, S., D. Sokal, and D. Maughan, 1986. Estimating the amount of soil ingested by young children through tracer elements, *Arch. Environ. Health*, 41(6): 341 – 345

Blumenthal, DS (ed.), 1985. *Introduction to Environmental Health*, Springer Publishing Co., New York

BMA (British Medical Association), 1991. *Hazardous Waste and Human Health*, Oxford University Press, Oxford, UK

Bogen, KT, 1990. *Uncertainty in Environmental Health Risk Assessment*, Garland Publishing, Inc., New York, NY

Bogen, KT, 1994. A note on compounded conservatism, *Risk Analysis*, 14: 379 – 381

Bogen, KT, 1995. Methods to approximate joint uncertainty and variability in risk, *Risk Analysis*, 15(3): 411 – 419

Boyce, CP, 1998. Comparison of approaches for developing distributions for carcinogenic slope factors, *Human and Ecological Risk Assessment*, 4(2): 527 – 577

Brooks, SM, *et al.*, 1995. *Environmental Medicine*, Mosby, Mosby-Year Book, Inc., St. Louis, Missouri

Brown, C., 1978. Statistical aspects of extrapolation of dichotomous dose response data, *J. National Cancer Inst.*, 60: 101 – 8

Brown, DP, 1987. Mortality of workers exposed to polychlorinated biphenyls – an update, *Arch. Environ. Health*, 42(6): 333 – 339

Brown, HS, 1986. A critical review of current approaches to determining "How Clean is Clean" at hazardous waste sites, *Hazardous Wastes and Hazardous Materials*, 3(3): 233-260

Brown, JF, Jr., 1994. Determination of PCB metabolic, excretion, and accumulation rates for use as indicators of biological response and relative risk, *Environ. Sci. Technol.*, 28(13): 2295 – 2305

Brown, HS, R. Guble, and S. Tatelbaum, 1988. Methodology for assessing hazards of contaminants to seafood, *Regul. Toxicol. Pharmacol.*, 8: 76 – 100

Brown, JF, Jr. and RE Wagner, 1990. PCB movement, dechlorination, and detoxication in the Acushnet Estuary, *Environ. Toxicol. Chem.*, 9: 1215 – 1233

Brum, G., L. McKane, and G. Karp. 1994. *Biology: Exploring Life*, 2nd Edition, John Wiley & Sons, New York, NY

BSI, 1988. *Draft for Development, DD175: 1988 Code of Practice for the Identification of Potentially Contaminated Land and its Investigation*, British Standards Institution (BSI), London, UK

Budd, P, J Montgomery, A Cox, *et al.*, 1998. The distribution of lead within ancient and modern human teeth: implications for long-term and historical exposure monitoring, *Science of the Total Environment*, 220(2-3): 121-136

Burmaster, DE. 1996. Benefits and costs of using probabilistic techniques in human health risk assessments – with emphasis on site-specific risk assessments, *Human and Ecological Risk Assessment*, 2: 35 – 43

Burmaster, DE and RH Harris, 1993. The magnitude of compounding conservatisms in Superfund risk assessments, *Risk Analysis*, 13: 131-134

Burmaster, DE and K. von Stackelberg, 1991. Using Monte Carlo simulations in public health risk assessments: estimating and presenting full distributions of risk, *Journal of Exposure Analysis and Environmental Epidemiology*, 1: 491 – 512

Cairney, T. (ed.), 1993. *Contaminated Land (Problems and Solutions)*, Blackie Academic & Professional, Glasgow/Chapman and Hall, London/Lewis Publishers, Boca Raton, FL

Cairney, T., 1995. *The Re-Use of Contaminated Land: A Handbook of Risk Assessment*, J. Wiley, Chichester, UK

Calabrese, EJ, 1984. *Principles of Animal Extrapolation*, John Wiley & Sons, New York, NY

Calabrese, EJ, R. Barnes, EJ Stanek III, H. Pastides, CE Gilbert, P. Veneman, X. Wang, A. Lasztity, and PT Kostecki, 1989. How much soil do young children ingest: an epidemiologic study, *Regulatory Toxicology and Pharmacology*, 10: 123 – 137

Calabrese, E.J. and P.T. Kostecki, 1992. *Risk Assessment and Environmental Fate Methodologies.* Lewis Publishers/CRC Press, Boca Raton, FL

Calabrese, EJ, EJ Stanek, and CE Gilbert, 1991. Evidence of soil-pica behavior and quantification of soil ingested, *Human Exposure and Toxicology,* 10: 245 – 249

Calabrese, EJ and EJ Stanek, 1992. Distinguishing outdoor soil ingestion from indoor dust ingestion in a soil pica child, *Regulatory Toxicology and Pharmacology,* 15: 83-85

Calabrese, EJ and EJ Stanek, 1995. Resolving intertracer inconsistencies in soil ingestion estimation, *Environmental Health Perspectives,* 103(5): 454-456

Calabrese, EJ, EJ Stanek, CE Gilbert, and RM Barnes, 1990. Preliminary adult soil ingestion estimates: results of a pilot study, *Regulatory Toxicology and Pharmacology,* 12: 88 – 95

Canter, LW, RC Knox, and DM Fairchild, 1988. *Ground Water Quality Protection,* Lewis Publishers, Inc., Chelsea, MI

CAPCOA, 1989. Air Toxics Assessment Manual. California Air Pollution Control Officers Association (CAPCOA), Draft Manual, August, 1987 (ammended, 1989), California

CAPCOA, 1990. Air Toxics "Hot Spots" Program. Risk Assessment Guidelines. California Air Pollution Control Officers Association (CAPCOA), California

Carrington, CD and PM Bolger, 1998. Uncertainty and risk assessment, *Human and Ecological Risk Assessment,* 4(2): 253 – 257

Carson, R., 1962. *Silent Spring,* Houghton Mifflin Co., New York, NY

Carson, R., 1994. *Silent Spring – with an Introduction by Vice President Al Gore,* Houghton Mifflin Co., New York, NY

Casarett, LJ and J. Doull, 1975. *Toxicology: The Basic Science of Poisons,* MacMillan Publishing Co., New York, NY

Cassidy, K., 1996. Approaches to the risk assessment and control of major industrial chemical and related hazards in the United Kingdom, *Int. J. Environment and Pollution,* 6(4-6): 361 – 387

CCME, 1991. Interim Canadian Environmental Quality Criteria for Contaminated Sites. Report CCME EPC-CS34, The National Contaminated Sites Remediation Program, Canadian Council of Ministers of the Environment (CCME), Winnipeg, Manitoba

CCME, 1993. *Guidance Manual on Sampling, Analysis, and Data Management for Contaminated Sites.* Volume I: Main Report (Report CCME EPC-NCS62E), and Volume II: Analytical Method Summaries (Report CCME EPC-NCS66E), Canadian Council of Ministers of the Environment, The National Contaminated Sites Remediation Program, Canadian Council of Ministers of the Environment (CCME), Winnipeg, Manitoba

CDHS, 1986. The California Site Mitigation Decision Tree Manual, California Department of Health Services (CDHS), Toxic Substances Control Division, Sacremento, California

CDHS, 1990. Scientific and Technical Standards for Hazardous Waste Sites, California Department of Health Services (CDHS), Toxic Substances Control Program, Technical Services Branch, Sacramento, California

CEQ, 1989. Risk analysis: A guide to principles and methods for analyzing health and environmental risks, Council on Environmental Quality. Washington, D.C. NTIS: PB89-137772.

Chen, JJ, DW Gaylor, and RL Kodell, 1990. Estimation of the joint risk from multiple-compound exposure based on single-compound experiments, *Risk Analysis,* 10: 285-290

Cheremisinoff, NP and ML Graffia, 1995. *Environmental and Health & Safety Management: A Guide to Compliance,* Noyes Publications, Park Ridge, New Jersey

Chouaniere, D., P. Wild, *et al.,* 2002. Neurobehavioral disturbances arising from occupational toluene exposure, *American Journal of Industrial Medicine,* 41(2): 77 – 88

Clausing, O., AB Brunekreef, and JH van Wijnen, 1987. A Method for Estimating Soil Ingestion by Children, *Int. Arch. Occup. Environ. Health,* 59(1): 73 – 82

Clark, M. 1996. *Transport Modeling for Environmental Engineers and Scientists,* J. Wiley, New York, NY

Clayton, CA, ED Pellizari, and JJ Quackenboss, 2002. National human exposure assessment survey: analysis of exposure pathways and routes for arsenic and lead in EPA Region 5, *Journal of Exposure Analysis and Environmental Epidemiology,* 12(1): 29 – 43

Cleek, RL and AL Bunge, 1993. A new method for estimating dermal absorption from chemical exposure, 1: general approach, *Pharm. Res.,* 10: 497-506

Clewell, HJ and ME Andersen, 1985. Risk assessment extrapolations and physiological modeling, *Toxicol. Ind. Health,* 1: 111 – 132

Cogliano, VJ, 1997. Plausible upper bounds: are their sums plausible?, *Risk Analysis,* 17(1): 77 – 84

Cohrssen, JJ and VT Covello, 1989. *Risk Analysis: A Guide to Principles and Methods for Analyzing Health and Environmental Risks,* National Technical Information Service (NTIS), US Dept. of Commerce, Springfield, VA

Colborn, T., FS Vom Saal, and AM Soto, 1993. Developmental effects of endocrine-disrupting chemicals in wild-life and humans, *Environmental Health Perspectives*, 101: 378 – 384

Colten, CE and PN Skinner, 1996. *The Road to Love Canal: Managing Industrial Waste before EPA*, Univ. of Texas Press, Austin, Texas

Conway, RA (ed.), 1982. *Environmental Risk Analysis of Chemicals*, Van Nostrand Reinhold Co., New York

Corn, M. (ed.), 1993. *Handbook of Hazardous Materials*, Academic Press, San Diego, CA

Cothern, CR (ed.), 1993. *Comparative Environmental Risk Assessment*, Lewis Publishers/CRC Press, Boca Raton, FL

Cothern, CR and NP Ross (eds.), 1994. *Environmental Statistics, Assessment, and Forecasting*, Lewis Publishers/CRC Press, Boca Raton, FL

Counter, SA, LH Buchanan, F. Ortega, and G. Laurell, 2002. Elevated blood mercury and neuro-otological observations in children of the Ecuadorian gold mines, *Journal of Toxicology and Environmental Health, Part A: Current Issues*, 65(2): 149 – 163

Couture, LA, Elwell, MR, and Birnbaum, LS, 1988. Dioxin-like effects observed in male rats following exposure to octachlorodibenzo-p-dioxin (OCDD) during a 13-week study, *Toxicol. Appl. Pharmacol.*, 93:3 1 – 46

Covello, VT, 1992. Trust and credibility in risk communication, *Health Environ. Dig.*, 6(1): 1-3

Covello, VT, 1993. Risk communication and occupational medicine, *J. Occup. Med.*, 35(1): 18-19

Covello, VT, *et al.*, 1987. *Uncertainty in Risk Assessment, Risk Management, and Decision Making*, Advances in Risk Analysis, Vol. 4, Plenum Press, New York

Covello, VT and F. Allen, 1988. *Seven Cardinal Rules of Risk Communication*, US EPA Office of Policy Analysis, Washington, DC

Covello VT, McCallum DB, and Pavolva MT, 1989. Principles and guidelines for improving risk communication, In: *Effective risk communication: the role and responsibility of government and nongovernment organizations*, New York: Plenum Press

Covello, VT, J. Menkes, and J. Mumpower (eds.), 1986. *Risk Evaluation and Management*, Contemporary Issues in Risk Analysis, Vol. 1, New York: Plenum Press

Covello, VT and MW Merkhofer, 1993. *Risk Assessment Methods: Approaches for Assessing Health and Environmental Risks*, Plenum Press, New York

Covello, VT and J. Mumpower, 1985. Risk analysis and risk management: an historical perspective, *Risk Analysis*, 5: 103-120

Cowherd, CM, GE Muleski, PJ Engelhart, and DA Gillette, 1985. Rapid Assessment of Exposure to Particulate Emissions From Surface Contamination Sites. Prepared for US EPA, Office of Health and Environmental Assessment, Washington, DC, EPA/600/8-85/002. Kansas City, MO: Midwest Research Institute

Cox, DC and PC Baybutt, 1981. Methods for uncertainty analysis: a comparative survey, *Risk Analysis*, 1(4): 251 – 258

Cox, LA and PF Ricci, 1992. Dealing with uncertainty – from health risk assessment to environmental decision making, *J. Energy Eng., ASCE*, 118(2): 77-94

Cox, SJ and NRS Tait, 1991. *Reliability, Safety & Risk Management: An Integrated Approach*, Butterworth-Heinemann, Oxford, England, UK

Crandall, RW and BL Lave (eds.), 1981. *The Scientific Basis of Risk Assessment*, Brookings Institution, Washington, DC

Cressie, NA, 1994. *Statistics for Spatial Data*, Revised Edition, J. Wiley & Sons, New York, NY

Crouch, EAC and R. Wilson, 1982. *Risk/Benefit Analysis*, Ballinger, Boston, Mass.

Crump, KS, 1981. An improved procedure for low-dose carcinogenic risk assessment from animal data, *J. Environ. Toxicol.*, 5: 339-46

Crump, KS, 1984. A new method for determining allowable daily intakes, *Fundamentals of Applied Toxicology*, 4: 854-871

Crump, KS and RB Howe, 1984. The multistage model with time-dependent dose pattern: applications of carcinogenic risk assessment, *Risk Analysis*, 4: 163-176

Csuros, M., 1994. *Environmental Sampling and Analysis for Technicians*, Lewis Publishers/CRC Press, Boca Raton, Florida

Cullen, AC, 1994. Measures of compounding conservatism in probabilistic risk assessment, *Risk Analysis*, 14(4): 389-393

D'Agostino, RB and MA Stephens, 1986. *Goodness-of-Fit Techniques*, Marcel Dekker, New York

Daugherty, J., 1998. *Assessment of Chemical Exposures – Calculation Methods for Environmental Professionals*, Lewis Publishers, Boca Raton, FL

Davey, B. and T. Halliday (eds.), 1994. *Human Biology and Health: An Evolutionary Approach*, Open University Press, Buckingham, UK

Davies, JC (ed.), 1996. *Comparing Environmental Risks: Tools for Setting Government Priorities*, Resources for the Future: Washington, DC

Davis, DL, HL Bradlow, M. Wolff, T. Wooddruff, DG Hoel, and H. Anton-Culver, 1993. Medical hypothesis: xenoestrogens as preventable causes of breast cancer, *Environmental Health Perspectives*, 101: 372-377

Decisioneering, Inc., 1996. *Crystal Ball, Version 4.0 User Manual*, Decisioneering, Inc., Denver, CO

Derelanko, MJ and MA Hollinger (eds.), 1995. *CRC Handbook of Toxicology*, CRC Press, Boca Raton, FL

de Serres, FJ and AD Bloom (eds.), 1996. *Ecotoxicity and Human Health (A Biological Approach to Environmental Remediation)*. Lewis Publishers/CRC Press, Boca Raton, FL

Dewailly, E., P. Ayotte, S. Bruneau, C. Laliberte, DCG Muir, and RJ Norstrom, 1993. Inuit exposure to organochlorines through the aquatic food chain in Arctic Quebec, *Environmental Health Perspectives*, 101(7): 618-620

Dewailly, E., P. Ayotte, C. Laliberte, J-P Webber, S. Gingras, and AJ Nantel, 1996. Polychlorinated biphenyl (PCB) and dichlorodiphenyl dichloroethylene (DDE) concentrations in the breast milk of women in Quebec, *American Journal of Public Health*, 86(9): 1241-1246

Dewailly, E., S. Dodin, R. Verreault, *et al.*, 1994. High organochlorine body burden in women with estrogen receptor-positive breast cancer, *J. National Cancer Inst.*, 86: 232-234

Dewailly, É., JJ Ryan, C. Laliberté, S. Bruneau, J-P. Weber, S. Gingras, and G. Carrier, 1994. Exposure of remote maritime populations to coplanar PCBs, *Environ. Health Perspect.*, 102(Suppl.1): 205–209

Dewailly, É., J-P. Weber, S. Gingras, and C. Laliberté, 1991. Coplanar PCBs in human milk in the province of Québec, Canada: are they more toxic than dioxin for breast-fed infants?, *Bull. Environ. Contam. Toxicol.*, 47: 491–498

Dienhart, CM, 1973. *Basic Human Anatomy and Physiology*, W.B. Saunders and Company, Philadelphia, PA

DoE (Department of the Environment), 1994. Sampling Strategies for Contaminated Land. CLR Report No.4, Department of the Environment, London, UK

DoE (Department of the Environment), 1995. *A Guide to Risk Assessment and Risk Management for Environmental Protection*. UK Department of the Environment, HMSO, London, UK

Donaldson, RM and RF Barreras, 1996. Intestinal absorption of trace quantities of chromium, *J. Lab. Clin. Med.*, 68: 484-493

Dourson, ML and SP Felter, 1997. Route-to-route extrapolation of the toxic potency of MTBE, *Risk Analysis*, 25: 43-57

Dourson, ML and FC Lu, 1995. Safety/Risk assessment of chemicals compared for different expert groups, *Biomedical and Environmental Sciences*, 8: 1-13

Dourson, ML and JF Stara, 1983. Regulatory history and experimental support of uncertainty (safety) factors, *Regulatory Toxicology and Pharmacology*, 3: 224-238

Dowdy, D., TE McKone, and DPH Hsieh, 1996. The use of molecular connectivity index for estimating biotransfer factors, *Environmental Science and Technology*, 30: 984-989

Driver, J., SR Baker, and D. McCallum, 2001. *Residential Exposure Assessment: A Sourcebook*, Kluwer Academic Publishers, Dordrecht, The Netherlands

Driver, JH, ME Ginevan, and GK Whitmyre, 1996. Estimation of dietary exposure to chemicals: a case study illustrating methods of distributional analyses for food consumption data, *Risk Analysis*, 16(6): 763-771.

Driver, JH, JJ Konz, and GK Whitmyre, 1989. Soil adherence to human skin, *Bull. Environ. Contam. Toxicol.*, 43: 814-820

DTSC, 1994a. *CalTOX, A Multimedia Exposure Model for Hazardous Waste Sites*, Office of Scientific Affairs, CalEPA/DTSC, Sacramento, CA

DTSC, 1994b. *Preliminary Endangerment Assessment Guidance Manual (A guidance manual for evaluating hazardous substance release sites)*, California Environmental Protection Agency, Department of Toxic Substances Control (DTSC), Sacramento, Calif.

DTSC/Cal-EPA, 1992. *Supplemental Guidance for Human Health Multimedia Risk Assessments of Hazardous Waste Sites and Permitted Facilities*, Department of Toxic Substances Control (DTSC), Cal-EPA, Sacramento, Calif.

Duff, RM and JC Kissel, 1996. Effect of soil loading on dermal absorption efficiency from contaminated soils, *J. Tox. And Environ. Health*, 48: 93-106

Earle, TC and G. Cvetkovich, 1997. Culture, cosmopolitanism, and risk management, *Risk Analysis*, 17(1): 55-65

Ellis, B. and JF Rees, 1995. Contaminated land remediation in the UK with reference to risk assessment: two case studies, *Journal of the Institute of Water and Environmental Management*, 9(1): 27-36

Erickson, BE, 2002. Analyzing the ignored environmental contaminants, *Environmental Science & Technology*, 30(7): 140A-145A

Eschenroeder, A., RJ Jaeger, JJ Ospital, and C. Doyle, 1986. Health risk assessment of human exposure to soil amended with sewage sludge contaminated with polychlorinated dibenzodioxins and dibenzofurans, *Vet. Hum. Toxicol.*, 28:356-442

Evans, JS and SJS Baird, 1998. Accounting for missing data in noncancer risk assessment, *Human and Ecological Risk Assessment*, 4(2): 291-317

Falck, F., A. Ricci, MS Wolff, J. Godbold, and P. Deckers, 1992. Pesticides and polychlorinated biphenyl residues in human breast lipids and their relation to breast cancer, *Archives of Environmental Health*, 47: 143-146

Farmer, A., 1997. *Managing Environmental Pollution*. Routledge, London, UK

Felter, SP and ML Dourson, 1997. Hexavalent chromium contaminated soils: options for risk assessment and risk management, *Reg. Tox. Pharmacol.*, 25: 43-59

Felter, S. and M. Dourson, 1998. The inexact science of risk assessment (and implications for risk management), *Human and Ecological Risk Assessment*, 4(2): 245-251

Fenner, K., C. Kooijman, M. Scheringer, and K. Hungerbuhler, 2002. Including transformation products into the risk assessment for chemicals: the case of nonylphenol ethoxylate usage in Switzerland, *Environmental Science & Technology*, 36(6): 1147-1154

Finkel, A., 1990. *Confronting Uncertainty in Risk Management*, Resources for the Future, Washington, DC

Finkel, AM, 1995. Toward less misleading comparisons of uncertain risks: the example of aflatoxin and alar, *Environmental Health Perspectives*, 103: 376-385

Finkel, AM and JS Evans, 1987. Evaluating the benefits of uncertainty reduction in environmental health risk management, *J. Air Pollut. Control Assoc.*. 37: 1164-71

Finkel, AM and D. Golding (eds.), 1994. *Worst Things First? (The Debate over Risk-Based National Environmental Priorities)*, Resources for the Future, Washington, DC

Finley, B. and DP Paustenbach, 1994. The benefits of probabilistic exposure assessment: three case studies involving contaminated air, water, and soil, *Risk Analysis*, 14:53-73

Finley, B., D. Proctor, *et al.*, 1994. Recommended distributions for exposure factors frequently used in health risk assessment, *Risk Analysis*, 14:533-553

Finley, B., PK Scott, and DA Mayhall, 1994. Development of a standard soil-to-skin adherence probability density function for use in Monte Carlo analyses of dermal exposures, *Risk Analysis*, 14:555-569

Fischhoff, B, S Lichtenstein, P Slovic, S Derby, and R. Keeney, 1981. *Acceptable Risk*, Cambridge Univ. Press, New York, NY

Fisher, A. and FR Johnson, 1989. Conventional wisdom on risk communication and evidence from a field experiment, *Risk Analysis*, 9(2): 209-213

Fitchko, J. 1989. Criteria for Contaminated Soil/Sediment Cleanup. Pudvan Publishing Co., Northbrook, IL.

Flegal, AR and DR Smith, 1992. Blood lead concentrations in preindustrial humans, *New England Journal of Medicine*, 326: 1293-4

Flegal, AR and DR Smith, 1995. Measurement of environmental lead contamination and human exposure, *Rev Environ Contam Toxicol*, 143: 1-45

FPC (Florida Petroleum Council), 1986. *Benzene in Florida Groundwater: An Assessment of the Significance to Human Health*. Florida Petroleum Council, Tallahassee, FL.

Francis, B., 1994. *Toxic Substances in the Environment*, Wiley-Interscience, New York

Freese, E., 1973. Thresholds in toxic, teratogenic, mutagenic, and carcinogenic effects. *Environ. Health Pers.*, 6, 171-178

Freeze RA and DB McWhorter, 1997. A Framework for Assessing Risk Reduction Due to DNAPL Mass Removal from Low-Permeability Soils, *Ground Water*, 35(1): 111-123

Freudenburg WR and SR Pastor, 1992. NIMBYs and LULUs: stalking the syndromes, *Journal of Social Issues*, 48(4): 39-61

Freund, J.E. and R.E. Walpole (eds.), 1987. *Mathematical Statistics*, Prentice-Hall, Englewood Cliffs, NJ

Frohse, F., M. Brodel, and L. Schlossberg, 1961. *Atlas of Human Anatomy*, Barnes & Noble Books/Harper & Row Publishers, New York

Furst, A., 1990. Yes, but is it a human carcinogen?, *Jour. of the American College of Toxicology*, 9:1-18

Gardels, MC and TJ Sorg. 1989. A laboratory study of the leaching of lead from water faucets. *Journal of the American Water Works Association*, 81(7): 101-113

Garrett, P., 1988. *How to Sample Groundwater and Soils*, National Water Well Association (NWWA), Dublin, OH

Gattrell, A. and M. Loytonen (eds.), 1998. *GIS and Health*, Taylor & Francis, London, UK

Gaylor, DW and JJ Chen, 1996. A simple upper limit for the sum of the risks of the components in a mixture, *Risk Analysis*, 16(3): 395-398

Gaylor, DW, and Kodell, RL, 1980. Linear interpolation algorithm for low dose risk assessment of toxic substances. *J. Environ. Pathol. Toxicol.* 4, 305-312

Gaylor, DW, and Shapiro, RE, 1979. Extrapolation and risk estimation for carcinogenesis. In *Advances in Modern Toxicology Volume 1, New Concepts in Safety Evaluation Part 2*, M.A. Mehlman, R.E. Shapiro, and H. Blumenthal (eds.) (pp. 65-85). New York: Hemisphere

Gaylor, DW and W. Slikker, 1990. Risk assessment for neurotoxic effects, *Neurotoxicology*, 11: 211-218

Gerber, GB, IA Leonard, and P. Jacquet, 1980. Toxicity, mutagenicity and teratogenicity of lead. Mutat Res 76:115-41

Gerity, TR and CJ Henry, 1990. *Principles of Route-to-Route Extrapolation for Risk Assessment*, Elsevier Publishing, Amsterdam

Gheorghe, AV and M. Nicolet-Monnier, 1995. *Integrated Regional Risk Assessment, Volume I & II*, Kluwer Academic Publishers, Dordrecht, The Netherlands

Gibbons, RD, 1994. *Statistical Methods for Groundwater Monitoring*, J. Wiley & Sons, New York, NY

Gibbs, LM, 1982. *Love Canal: My Story*. State Univ. New York Press, Albany, New York, NY

Gierthy, JF, KF Arcaro, and M. Floyd, 1995. Assessment and implications of PCB estrogenicity, *Organo-Halogen Compounds*, 25:419-423

Gilbert, RO, 1987. *Statistical Methods for Environmental Pollution Monitoring*, Van Nostrand-Reinhold, NY

Glasson, J., R. Therivel, and A. Chadwick (1994). Introduction to Environmental Impact Assessment. University College London, UCL Press, London, UK

Glickman, T S and M. Gough (eds.), 1990. *Readings in Risk*, Resources for the Future, Washington, DC

Glowa, JR, 1991. Dose-effect approaches to risk assessment, *Neurosci Biobehav Rev*, 15: 153-158

Gochfeld, M., 1997. Factors influencing susceptibility to metals, *Environmental Health Perspectives*, 105 (Suppl.4): 817-822

Goldman, M. 1996. Cancer risk of low-level exposure. *Science*, 271(5257): 1821-1822.

Goldstein, BD, 1989. The maximally exposed individual: an inappropriate basis for public health decision-making, *Environmental Forum*, 6: 13-16

Goldstein, BD, 1995. The who, what, when, where, and why of risk characterization. *Policy Studies Journal*, 23(1): 70-75

Goodchild, MF, BO Parks and LT Steyaert (eds.), 1993. Environmental Modeling with GIS. Oxford University Press, New York

Goodchild, MF, LT Steyaert, BO Parks, *et al.* (eds.), 1996. *GIS and Environmental Modeling: Progress and Research Issues*, GIS World Books, Fort Collins, Colorado

Goyer RA. 1996. Results of lead research: prenatal exposure and neurological consequences. *Environmental Health Perspectives*, 104:1050–4

Grandjean, P., RF White, and P. Weihe, 1996. Neurobehavioral epidemiology: application in risk assessment, *Environmental Health Perspectives*, 104 (Suppl.2): 397-400

Gratt, LB. 1996. Air Toxic Risk Assessment and Management: Public Health Risk from Normal Operations. Van Nostrand Reinhold, New York

Gregory, R. and H. Kunreuther, 1990. Successful siting incentives, *Civil Engineering*, 60, 4: 73-5. ASCE, NY

Grisham, J.W. (ed.), 1986. *Health Aspects of the Disposal of Waste Chemicals*. Pergamon Press, Oxford, England

Gulson BL, Jameson CW, Mahaffey KR, Mizon KJ, Patison N, Law AJ, *et al.*, 1998. Relationships of lead in breast milk to lead in blood, urine, and diet of the infant and mother, *Environmental Health Perspectives*, 106:667–74

Gulson BL, Mahaffey KR, Jameson CW, Vidal M, Law AJ, Mizon KJ, *et al.*, 1997. Dietary lead intakes for mother/child pairs and relevance to pharmacokinetic models, *Environmental Health Perspectives*, 105:1334–42

Gustavsson, P., C. Hogstedt, and C. Rappe, 1986. Short-term mortality and cancer incidence in capacitor manufacturing workers exposed to polychlorinated biphenyls (PCBs), *Am. J. Ind. Med.*, 10:341–344

Guyton, AC, 1968. *Textbook of Medical Physiology*, W.B. Saunders Company, Philadelphia, PA

Guyton, AC, 1971. *Basic Human Physiology: Normal Functions and Mechanisms of Disease*. W.B. Saunders Company, Philadelphia, PA

Guyton, AC, 1982. *Human Physiology and Mechanisms of Disease*, 3rd edition, W.B. Saunders Company, Philadelphia, PA

Guyton, AC, 1986. *Textbook of Medical Physiology*, 7th edition, W.B. Saunders Company, Philadelphia, PA

Haas, CN, 1997. Importance of distributional form in characterizing inputs to Monte Carlo risk assessments, *Risk Analysis*, 17(1):107-113

Hadley, PW and RM Sedman, 1990. A Health-Based Approach for Sampling Shallow Soils at Hazardous Waste Sites Using the AAL$_{soil\ contact}$ Criterion, *Environmental Health Perspectives*, 18: 203-207

Haimes, YY (ed.), 1981. *Risk/Benefit Analysis in Water Resources Planning and Management*, Plenum, NY

Haimes, Y.Y., L. Duan, and V. Tulsiani, 1990. Multiobjective Decision-Tree Analysis. *Risk Analysis*, 10(1): 111-129

Hallenbeck, WH, and Cunningham, KM, 1988. *Quantitative Risk Assessment for Environmental and Occupational Health*. 4th Printing. Chelsea, MI: Lewis Publishers, Inc.

Hamed, MM, 1999. Probabilistic sensitivity analysis of public health risk assessment from contaminated soil, *Journal of Soil Contamination*, 8(3): 285-306

Hamed, MM, 2000. Impact of Random Variables Probability Distribution on Public Health Risk Assessment from Contaminated Soil, *Journal of Soil Contamination*, 9(2): 99-117

Hamed, MM and PB Bedient. 1997a. On the effect of probability distributions of input variables in public health risk assessment, *Risk Analysis*, 17(1):97-105

Hamed, MM and PB Bedient, 1997b. On the Performance of Computational Methods for the Assessment of Risk from Ground-Water Contamination, *Ground Water*, 35 (4): 638-646

Hammitt, J.K, 1995. Can More Information Increase Uncertainty? *Chance*, 8(3): 15-17, 36

Hammitt, JK and AI Shlyakhter, 1999. The Expected Value of Information and the Probability of Surprise, *Risk Analysis*, 19(1): 135-152

Hance, BJ, C. Chess, and PM Sandman, 1990. Industry Risk Communication Manual: Improving Dialogue with Communities. Lewis Publishers, Boca Raton, FL

Hanley, N. and CL Spash, 1995. Cost-benefit Analysis and the Environment, Second reprint, Edward Elgar Publishing Limited, UK

Hansen, P.E. and S.E. Jorgensen (eds.), 1991. Introduction to Environmental Management. Developments in Environmental Modelling, 18, Elsevier, Amsterdam, The Netherlands

Hansson, S-O., 1989. Dimensions of Risk, *Risk Analysis*, 9(1): 107-112

Hansson, S-O., 1996a. Decision Making under Great Uncertainty, *Philosophy of the Social Sciences*, 26(3): 369-386

Hansson, S-O., 1996b. What is philosophy of risk? *Theoria*, 62: 169-186

Harper, N., K. Connor, M. Steinberg, and S. Safe, 1995. Immunosuppressive activity of polychlorinated biphenyl mixtures and congeners: nonadditive (antagonistic) interactions, *Fund. Appl. Toxicol.*, 27: 131–139

Harrad, S., 2000. *Persistent Organic Pollutants: Environmental Behavior and Pathways of Human Exposure*, Kluwer Academic Publishers, Dordrecht, The Netherlands

Harrison, RM and DPH Laxen (eds.), 1981. *Lead Pollution, Causes and Control.* Chapman & Hall, London, UK

Hathaway, GJ, NH Proctor, JP Hughes, and ML Fischman, 1991. *Proctor and Hughes' Chemical Hazards of the Workplace, 3rd ed.,* Van Nostrand Reinhold, New York

Hattis, D. and DE Burmaster, 1994. Assessment of variability and uncertainty distributions for practical risk analyses, *Risk Analysis*, 14(5): 713-729

Hawley, JK, 1985. Assessment of health risks from exposure to contaminated soil, *Risk Analysis*, 5, 4, pp. 289-302

Hayes, AW (ed.), 1982. *Principles and Methods of Toxicology.* New York,: Raven Press

Hayes, AW (ed.), 1994. *Principles and Methods of Toxicology,* 3rd ed., Raven Press, New York

Helton, JC, 1993. Risk, uncertainty in risk, and the EPA release limits for radioactive waste disposal, *Nuclear Technology*, 101:18-39

Hemond, HF and EJ Fechner, 1994. *Chemical Fate and Transport in the Environment.* Academic Press, San Diego, CA

Henderson, M. (1987). *Living with Risk: The Choices, The Decisions,* The British Medical Association Guide, John Wiley and Sons, New York

Henke, KR, V. Kuhnel, DJ Stepan, *et al.* 1993. Critical Review of Mercury Contamination Issues Relevant to Manometers at Natural Gas Industry Sites. Gas Research Institute

Hertwich, E.G., W.S. Pease, and T.E. McKone. 1998. Evaluating toxic impact assessment methods: what works best? *Environmental Science & Technology*, 32(5): 138A-144A

Hernández-Avila M, Smith D, Meneses F, Sanin LH, Hu H. 1998. The influence of bone and blood lead on plasma lead levels in environmentally exposed adults. *Environmental Health Perspectives*, 106:473–7.

Hertz, DB and H. Thomas, 1983. *Risk Analysis and Its Applications,* John Wiley & Sons, New York

Hilts SR, Bock SE, Oke TL, Yates CL, Copes RA. 1998. Effect of interventions on children's blood lead levels. *Environmental Health Perspectives*, 106:79–83

Hilts, SR, 1996. A co-operative approach to risk management in an active lead/zinc smelter community, *Environmental Geochemistry and Health*, 18: 17 – 24

Hipel, KW, 1988. Nonparametric Approaches to Environmental Impact Assessment. *Water Resources Bull., AWRA*, 24(3): 487-491

Hoddinott, KB (ed.), 1992. *Superfund Risk Assessment in Soil Contamination Studies,* American Society for Testing and Materials, ASTM Publication STP 1158, Philadelphia, PA

Hoel, D.G., Gaylor, D.W., Kirschstein, R.L., Saffiotti, U., and Schneiderman, M.A., 1975. Estimation of risks of irreversible, delayed toxicity. *J. Toxicol. Environ. Health*, 1, 133-151

Hoffmann, FO and JS Hammonds, 1992. *An Introductory Guide to Uncertainty Analysis in Environmental and Health Risk Assessment*, Environmental Sciences Division, Oak Ridge National Lab., TN, ESD Publication 3920. Prepared for the US Dept. of Energy, Washington, DC

Hogan, MD, 1983. Extrapolation of animal carcinogenicity data: Limitations and pitfalls, *Environ. Health Pers.*, 47, 333-337

Holmes, G., BR Singh and L. Theodore, 1993. *Handbook of Environmental Management and Technology*, J. Wiley & Sons, New York, NY

Holmes, Jr., KK, JH Shirai, KY Richter, and JC Kissel, 1999. Field measurement of dermal soil loadings in occupational and recreational activities, *Journal of Environmental Research*, 80(2): 148-157

Homburger, F, JA Hayes, and EW Pelikan (eds.), 1983. *A Guide to General Toxicology*, Karger Publishers, New York

HRI (Hampshire Research Institute), 1995. Risk*Assistant for Windows, The Hampshire Research Institute, Inc., Alexandria, VA

Hrudey, SE, W. Chen, and CG Rousseaux. 1996. Bioavailability in Environmental Risk Assessment. Lewis Publishers/CRC Press, Boca Raton, Florida

HSE (Health and Safety Executive), 1989a. *Risk Criteria for Land-Use Planning in the Vicinity of Major Industrial Hazards*. HMSO, London, UK

HSE (Health and Safety Executive), 1989b. *Quantified Risk Assessment – Its Input to Decision Making*. HMSO, London, UK

Huckle, K.R, 1991. Risk assessment--regulatory need or nightmare. Shell publications, Shell Center, London, England. (8pp)

Huff, JE, 1993. Chemicals and cancer in humans: first evidence in experimental animals, *Environ Health Perspect.*, 100: 201-210

Hughes, WW, 1996. *Essentials of Environmental Toxicology: the effects of environmentally hazardous substances on human health*, Taylor & Francis Publishers, Washington, DC

Hwang, ST and JW Falco, 1986. Estimation of Multimedia Exposures Related to Hazardous Waste Facilities, in, *Pollutants in a Multimedia Environment*, Y. Cohen (ed.), Plenum Press, New York

IARC (International Agency for Research on Cancer), [1972-1985]. *IARC Monographs on the evaluation of the carcinogenic risk of chemicals to man, (Multi-volume work)*, International Agency for Research on Cancer (IARC), World Health Organization, Geneva, Switzerland

IARC, 1982. *IARC Monographs on the Evaluation of the Carcinogenic Risk of Chemicals to Humans: Chemicals, Industrial Processes and Industries Associated with Cancer in Humans*, Supplement 4, International Agency for Research on Cancer (IARC), Lyon, France

IARC, 1984. *IARC Monographs on the Evaluation of the Carcinogenic Risk of Chemicals to Humans, Vol. 33*, International Agency for Research on Cancer (IARC), Lyon, France

IARC, 1987. *IARC Monographs on the Evaluation of Carcinogenic Risks of Chemicals to Humans: Overall Evaluations of Carcinogenicity, Supplement 7*, International Agency for Research on Cancer (IARC), Lyon, France

IARC, 1988. IARC Monographs on Evaluation of Carcinogenic Risks to Humans, Vol. 43, International Agency for Research on Cancer (IARC), Lyon, France

ICRCL, 1987. *Guidance on the Assessment and Redevelopment of Contaminated Land, ICRCL 59/83*, 2nd edition, Interdepartmental Committee on the Redevelopment of Contaminated Land (ICRCL), Department of the Environment, Central Directorate on Environmental Protection, London, UK

ICRP, 1975 – [current]. *Publication Series of the International Commission on Radiological Protection (ICRP)*, Pergamon Press, Oxford, UK

Illing, HP, 1999. Are societal judgments being incorporated into the uncertainty factors used in toxicological risk assessment, *Regulatory Toxicology and Pharmacology*, 29(3): 300-308

Iman, RL and JC Helton, 1988. An investigation of uncertainty and sensitivity analysis techniques for computer models, *Risk Analysis*, 8:71-90

Jain, RK, LV Urban, GS Stacey, and HE Balbach, 1993. *Environmental Assessment*, McGraw-Hill, Inc., New York, NY

Jarabek, AM, MG Menache, JH Overton, ML Dourson, and FJ Miller, 1990. The U.S. Environmental Protection Agency's inhalation RfD methodology: risk assessment in air toxics, *Toxicol Ind Health*, 6: 279-301

Jasonoff, S, 1993. Bridging the two cultures of risk analysis, *Risk Analysis*, 13: 122-128

Jo, WK, CP Weisel, and PJ Lioy, 1990a. Chloroform exposure and the health risk associated with multiple uses of chlorinated tap water, *Risk Analysis*, 10: 581 – 585

Jo, WK, CP Weisel, and PJ Lioy, 1990b. Routes of chloroform exposure and body burden from showering with chlorinated tap water, *Risk Analysis,* 10: 575-580

Johnson, B.B. and Covello, V.T. 1987. *Social and Cultural Construction of Risk: Essays on Risk Selection and Perception.* Norwell, MA: Kluwer Academic Publishers

Johnson, PC and RA Ettinger, 1991. A heuristic model for predicting the intrusion rate of contaminant vapours into buildings, *Environ. Sci. Technol.,* 25(8): 1445-1452

Johnson BL, Jones DE, 1992. ATSDR's activities and views on exposure assessment. Journal of Exposure Analysis and Environmental Epidemiology, Suppl. 1:1-17

Johnson, DL, K. McDade, and D. Griffith, 1996. Seasonal variation in paediatric blood lead levels in Syracuse, NY, USA, *Environmental Geochemistry and Health,* 18: 81-88

Jolley, R.L. and R.G.M. Wang (eds.), 1993. *Effective and Safe Waste Management: Interfacing Sciences and Engineering With Monitoring and Risk Analysis,* Lewis Publishers, Boca Raton, FL

Jones, RB, 1995. Risk-Based Management: A Reliability-Centered Approach, Gulf Publishing Co., Houston, TX

Jury, WA, WJ Farmer, and WF Spencer, 1984. Behavior assessment model for trace organics in soil: II. Chemical classification and parameter sensitivity, *J. Environ. Qual.,* 13(4): 567-572

Kaplan, A., 1964. *The Conduct of Enquiry: Methodology for Behavioral Science.* Chandler Publishing Co., San Francisco, Calif.

Kasperson, RE and PJM Stallen (eds.), 1991. *Communicating Risks to the Public: International Perspectives,* Kluwer Academic Press, Boston, MA

Kastenberg, WE and HC Yeh, 1993. Assessing public exposure to pesticides - contaminated ground water. *J. Ground Water,* Vol.31, No.5: 746-752

Kasting, GB and PJ Robinson, 1993. Can we assign an upper limit to skin permeability?, *Pharm. Res.,* 10: 930-939

Kates, RW, 1978. *Risk Assessment of Environmental Hazard,* SCOPE Report 8, J.Wiley & Sons, New York

Katz, M. and D. Thornton, 1997. *Environmental Management Tools on the Internet: Accessing the World of Environmental Information,* St. Lucie Press, Delray Beach, FL

Keeney, RL 1984. Ethics, Decision Analysis, and Public Risk, *Risk Analysis, 4, pp.117 - 129,* 1984

Keeney, RL, 1990. Mortality Risks Induced by Economic Expenditures, *Risk Analysis,* 10, 1: 147-59

Keeney, RD and H. Raiffa. 1976. *Decisions with Multiple Objectives: Preferences and Value Tradeoffs,* John Wiley & Sons, NY

Keith, LH, (ed.), 1988. *Principles of Environmental Sampling,* American Chemical Society (ACS), Washington, DC

Keith, LH, 1991. *Environmental Sampling and Analysis--A Practical Guide,* Lewis Publishers, Boca Raton, Florida

Keith, LH (ed.), 1992. *Compilation of E.P.A.'s Sampling and Analysis Methods,* Lewis Publishers/CRC Press, Boca Raton, Florida

Kim R, Hu H, Rotnitzky A, Bellinger D, Needleman H. 1995. A longitudinal study of chronic lead exposure and physical growth in Boston children, *Environ Health Perspect.,* 103:952–7.

Kimmel, CA and DW Gaylor, 1988. Issues in qualitative and quantitative risk analysis for developmental toxicology, *Risk Analysis,* 8:15-20

Kissel, J., KY Richter, and RA Fenske, 1996. Field measurement of dermal soil loading attributed to various activities: implications for exposure assessment, *Risk Analysis,* 16(1): 115-125

Kissel, J., JH Shirai, KY Richter, and RA Fenske, 1998. Investigation of dermal contact with soil in controlled trials, *J. Soil Contamination,* 7: 737-752

Kitchin, KT (ed.), 1999. *Carcinogenicity Testing, Predicting, and Interpreting Chemical Effects,* Marcel Dekker, New York

Klaassen, CD, Amdur, MO, and Doull, J. (eds.), 1986. *Casarett and Doull's Toxicology: The Basic Science of Poisons.* 3rd edition. New York: Macmillan Publishing Company

Klaassen, CD, Amdur, MO, and Doull, J. (eds.), 1996. *Casarett and Doull's Toxicology: The Basic Science of Poisons,* 5th edition. McGraw-Hill, New York

Kleindorfer, PR and HC Kunreuther (ed.), 1987. *Insuring and Managing Hazardous Risks: From Seveso to Bhopal and Beyond.* Springer-Verlag, Berlin, Germany

Kletz, T. 1994. *Learning from Accidents,* 2nd ed., Butterwort-Heinemann, Oxford, England.

Kocher, DC and FO Hoffman, 1991. Regulating environmental carcinogens: where do we draw the line?, *Environ. Sci. Technol.,* 25:1986-1989

Kodell, RL and JJ Chen, 1994. Reducing conservatism in risk estimation for mixtures of carcinogens, *Risk Analysis,* 14: 327-332

Koenig, JQ, 1999. *Health Effects of Ambient Air Pollution: How Safe is the Air we Breathe?,* Kluwer Academic Publishers, Dordrecht, The Netherlands

Kolluru, R. (ed.), 1994. *Environmental Strategies Handbook: A Guide to Effective Policies and Practices,* McGraw-Hill, NY

Kolluru, RV, SM Bartell, RM Pitblado, and RS Stricoff (eds.), 1996. *Risk Assessment and Management Handbook (for Environmental, Health, and Safety Professionals),* McGraw-Hill, New York, NY

Kreith, F. (ed.), 1994. *Handbook of Solid Waste Management,* McGraw-Hill, New York

Krewski, D., C. Brown and D. Murdoch, 1984. Determining safe levels of exposure: safety factors for mathematical models, *Fundam. Appl. Toxicol.,* 4: S383-S394

Krewski, D. and J. van Ryzin, 1981. Dose response models for quantal response toxicity data. In *Statistics and Related Topics,* M. Csorgo, D.A. Dawson, J.N.K. Rao, A.K.E. Saleh, (eds.) (pp. 201-231). New York: North Holland

Kunkel DB. 1986. The toxic emergency, *Emerg. Med.,* 18(Mar): 207–17

Kunreuther, H. and MV Rajeev Gowda (ed.), 1990. *Integrating Insurance and Risk Management for Hazardous Wastes,* Kluwer Academic Publishers, Boston, Mass.

Kunreuther, H. and P. Slovic, 1996. Science, Values, and Risk, *Annals of the American Academy of Political and Social Science,* 545: 116–25

Kuo, HW, TF Chiang, Il Lai, *et al.,* 1998. Estimates of cancer risk from chloroform exposure during showering in Taiwan, *Sci. Total Environ.,* 218: 1 – 7

LaGoy, PK, 1987. Estimated Soil Ingestion Rates for Use in Risk Assessment, *Risk Analysis,* 7 (3): 355-359

LaGoy, PK, 1994. *Risk Assessment, Principles and Applications for Hazardous Waste and Related Sites,* Noyes Data Corp., Park Ridge, NJ

LaGoy, PK and CO Schulz, 1993. Background Sampling: An Example of the Need for Reasonableness in Risk Assessment, *Risk Analysis,* 13(5): 483-484

Laib, R.J.; Rose, N.; Brunn, H., 1991. Hepatocarcinogenicity of PCB congeners, *Toxicol. Environ. Chem.,* 34:19–22

Laird, FN, 1989. The decline of deference: the political context of risk communication, *Risk Analysis,* 9(2): 543-550

Larsen, RJ and ML Marx, 1985. *An Introduction to Probability and its Applications,* Prentice-Hall, Englewood Cliffs, NJ

Lave, LB (ed.), 1982. *Quantitative Risk Assessment in Regulation,* The Brooking Institute, Washington, DC

Lave, LB and AC Upton (eds.), 1987. *Toxic Chemicals, Health, and the Environment,* The John Hopkins Univ. Press, Baltimore, MD

Layton, DW, 1993. Metabolically consistent breathing rates for use in dose assessments, *Health Physics,* 64(1): 23-36

Lee, BM, SD Yoo, and S. Kim, 2002. A proposed methodology of cancer risk assessment modeling using biomarkers, *Journal of Toxicology and Environmental Health, Part A: Current Issues,* 65(5-6): 341 – 354

Lee, RC, JR Fricke, WE Wright, and W. Haerer, 1995. Development of a probabilistic blood lead prediction model, *Environmental Geochemistry and Health,* 17:169-181

Lee, RC and JC Kissel, 1995. Probabilistic prediction of exposures to arsenic contaminated residential soil, *Environmental Geochemistry and Health,* 17:159-168

Lee, RG, WC Becker, and DW Collins. 1989. Lead at the tap: sources and control. *Journal of the American Water Works Association,* 81(7): 52-62

Lee, YW, MF Dahab, and I. Bogardi, 1995. Nitrate-risk assessment using fuzzy-set approach. *Journal of Environmental Engineering,* 121(3): 245-256

Leidel, N. and K. A. Busch, 1985. Statistical Design and Data Analysis Requirements. In, Patty's Industrial Hygiene and Toxicology, Vol. IIIa, 2nd ed., John Wiley & Sons, NY

Leiss, W., 1989. *Prospects and Problems in Risk Communication.* Institute for Risk Research. Waterloo, Ontario, Canada: University of Waterloo Press

Leiss, W. and C. Chociolko, 1994. *Risk and Responsibility,* McGill-Queen's Univ. Press, Montreal, Quebec

Lepow, ML, L. Bruckman, M. Gillette, S. Markowitz, R. Robino, J. Kapish, 1975. Investigations into Sources of Lead in the Environment of Urban Children, *Environmental Research,* 10:415-426

Lepow, ML, M. Bruckman, L. Robino, S. Markowitz, M. Gillette, J. Kapish, 1974: "Role of Airborne Lead in Increased Body Burden of Lead in Hartford Children," Environmental Health Perspectives 6:99-101

Lesage, S. and R.E. Jackson (ed.s), 1992. *Groundwater Contamination and Analysis at Hazardous Waste Sites,* Marcel Dekker, Inc., New York, NY

Levallois, P., M. Lavoie, L. Goulet, AJ Nantel, and S. Gingra, 1991. Blood lead levels in children and pregnant women living near a lead-reclamation plant. *Canadian Medical Association Journal,* 144(7): 877-885

Levesque, B., P. Ayotte, R. Tardif, *et al.,* 2002. Cancer risk associated with household exposure to chloroform, *Journal of Toxicology and Environmental Health, Part A: Current Issues,* 65(7): 489 – 502

Levine, AG, 1982. *Love Canal: Science, Politics, and People.* Lexington Books, Lexington, Mass.

Levine, DG and AC Upton (eds.), 1992. *Management of Hazardous Agents, Volumes 1 & 2,* Praeger, Westport, CT

Lifson, MW, 1972. *Decision and Risk Analysis for Practicing Engineers,* Barnes and Noble, Cahners Bks, Boston, MA

Lind, NC, JS Nathwani, and E. Siddall, 1991. *Managing Risks in the Public Interest,* Institute for Risk Research, Univ. of Waterloo, Waterloo, Ontario

Lippmann, M. (ed.), 1992. *Environmental Toxicants: Human Exposures and Their Health Effects.* Van Nostrand Reinhold, New York

Liptak, JF and G. Lombardo, 1996. The development of chemical-specific, risk-based soil cleanup guidelines results in timely and cost-effective remediation, *J. Soil Contamination,* 5(1):83-94

Little, JC, JM Daisey, and WW Nazaroff, 1992. Transport of subsurface contaminants in buildings: an exposure pathway for volatile organics, *Environmental Science and Technology,* 26(11)

Lockhart, WL, R. Wagemann, B. Tracey, D. Sutherland, and DJ Thomas, 1992. Presence and implications of chemical contaminants in the freshwaters of the Canadian Arctic, *Science of the Total Environment,* 122: 165-243

Long, F. A. and G. E. Schweitzer (eds.), 1982. *Risk Assessment at Hazardous Waste Sites.* American Chemical Society, Washington, DC

Louvar, J.F. and B.D. Louvar, 1998. *Health and Environmental Risk Analysis: Fundamentals with Applications.* Prentice-Hall, Upper Saddle River, New Jersey

Lowrance, WW, 1976. *Of Acceptable Risk: Science and the Determination of Safety,* William Kaufman, Inc., Los Altos, Calif.

Lu, FC, 1985a. *Basic Toxicology,* Washington, DC: Hemisphere

Lu, FC 1985b, Safety Assessments of Chemicals with Threshold Effects, *Regul. Toxicol. Pharmacol.,* 5:121-32

Lu, FC, 1988. Acceptable Daily Intake: Inception, Evolution, and Application, *Regul. Toxicol. Pharmacol.,* 8: 45-60

Lu, FC, 1996. *Basic Toxicology: Fundamentals, Target Organs, and Risk Assessment,* 3rd ed., Taylor & Francis, Washington, DC

Lum M., 1991. Benefits to conducting midcourse reviews, in: A. Fisher, M. Pavolva, and V. Covello, (eds.), *Evaluation and Effective Risk Communications Workshop Proceedings,* Pub. No. EPA/600/9-90/054, US Environmental Protection Agency, Washington, DC

Lundgren, R., 1994. *Risk Communication: A Handbook for Communicating Environmental, Safety, and Health Risks,* Battelle Press, Columbus, OH

Lurakis MF, Pitone JM, 1984. Occupational lead exposure, acute intoxication, and chronic nephropathy: report of a case and review of the literature, *J. Am Osteopath Assoc.,* 83:361–6

Lyman, WJ, WF Reehl and DH Rosenblatt, 1990. *Handbook of Chemical Property Estimation Methods: Environmental Behavior of Organic Compounds,* American Chemical Society, Washington, DC

Maas, RP and SC Patch, 1990. 'Dynamics of lead contamination of residential tapwater: implications for effective control.' Technical Report #90-004, Ashville, N.C.: University of North Carolina-Ashville Environmental Quality Institute

Macintosh, DL, GW Suter II, and FO Hoffman, 1994. Uses of probabilistic exposure models in ecological risk assessments of contaminated sites, *Risk Analysis,* 14:405-419

Mackay, D. and P.J Leinonen, 1975. Rate of Evaporation of Low-Solubility Contaminants from Water Bodies, *Environ. Sci. Technol.,* 9: 1178-1180

Mackay, D. and ATK Yeun, 1983. Mass Transfer Coefficient Correlations for Volatilization of Organic Solutes from Water, *Environ. Sci. Technol.,* 17: 211-217

Mahaffey KR, 1995. Nutrition and lead: strategies for public health, *Environ Health Perspect.,* 103(suppl 5):191–6

Marcus WL. 1986. Lead health effects in drinking water, *Toxicol. Ind. Health,* 2:363–400

Markowitz ME, Rosen JF, 1991. Need for the lead mobilization test in children with lead poisoning, *J. Pediatr.,* 119(2):305–10

Martin, WF, JM Lippitt, and TG Prothero, 1992. *Hazardous Waste Handbook for Health and Safety.* 2nd Edition. Butterworth-Heinemann, London, UK

Massaro, EJ, (ed.), 1998. *Handbook of Human Toxicology,* CRC Press, Boca Raton, FL

Massmann, J., and RA Freeze, 1987. Groundwater Contamination from Waste Management Sites: The Interaction Between Risk-Based Engineering Design and Regulatory Policy 1.Methodology 2. Results. *Water Resources Research,* 23(2): 351-380

Masters, GM, 1998. *Introduction to Environmental Engineering and Science,* 2nd edition, Prentice Hall, Upper Saddle River, New Jersey

Maughan, JT, 1993. *Ecological Assessment of Hazardous Waste Sites,* Van Nostrand Reinhold, NY

Maxwell, RM, SD Pelmulder, SD, AFB Tompson, and WE Kastenberg, 1998. On the development of a new methodology for groundwater driven health risk assessment, *Water Resources Research,* 34(4): 833-847

McColl, RS (ed.), 1987. *Environmental Health Risks: Assessment and Management,* Institute for Risk Research, University of Waterloo Press, Waterloo, Ontario, Canada

McKone, TE, 1987. Human exposure to volatile organic compounds in household tap water: the indoor inhalation pathway. *Environmental Science and Technology,* 21(12): 1194-1201

McKone, TE, 1989. Household exposure models, *Toxicology Letters,* 49: 321 - 339

McKone, TE, 1993. Linking a PBPK model for chloroform with measured breath concentrations in showers: implications for dermal exposure models, *Journal of Exposure Analysis and Environmental Epidemiology,* 3: 339-365

McKone, TE, 1994. Uncertainty and variability in human exposures to soil contaminants through home-grown food: a Monte Carlo assessment, *Risk Analysis,* 14:449-463

McKone, TE and KT Bogen, 1991. Predicting the uncertainties in risk assessment, *Environ. Sci. Technol.,* 25:1674-1681

McKone, TE and JI Daniels, 1991. Estimating human exposure through multiple pathways from air, water, and soil, *Regul. Toxicol. Pharmacol.,* 13: 36-61

McKone, TE and RA Howd, 1992. Estimating dermal uptake of nonionic organic chemicals from water and soil: part 1, unified fugacity-based models for risk assessments, *Risk Analysis,* 12: 543-557

McKone, TE and JP Knezovich, 1991. The transfer of trichloroethylene (TCE) from a shower to indoor air: experimental measurements and their implications, *Journal of Air and Waste management,* 41: 832-837

McTernan, WF and E. Kaplan (eds.), 1990. Risk Assessment for Groundwater Pollution Control, ASCE Monograph, American Society of Civil Engineers, New York, NY

Menzie-Cura & Associates, 1996. *An Assessment of the Risk Assessment Paradigm for Ecological Risk Assessment.* Report Prepared for the Presidential/Congressional Commission on Risk Assessment and Risk Management, Washington, DC

Merck, 1989. *The Merck Index: An Encyclopedia of Chemicals, Drugs and Biologicals,* Eleventh (Centennial) Edition, Rockway, NJ: Merck & Co., Inc.

Meyer, CR, 1983. Liver Dysfunction in Residents Exposed to Leachate from a Toxic Waste Dump. *Environ. Health Persp.,* 48:9-13

Meyer, PB, RH Williams, and KR Yount, 1995. *Contaminated Land: Reclamation, Redevelopment and Reuse in the United States and the European Union,* Edward Elgar Publishing Ltd., Aldershot, UK

Mielke HW, Dugas D, Mielke PW Jr, Smith KS, Smith SL, Gonzales CR. 1997. Associations between soil lead and childhood blood lead in urban New Orleans and rural Lafourche Parish of Louisiana, *Environ Health Perspect.,* 105:950–4

Miller, I. and JE Freund, 1985. *Probability and Statistics for Engineers, 3rd Edition,* Prentice-Hall, Englewood Cliffs, NJ

Millette, JR and SM Hays, 1994. *Settled Asbsestos Dust Sampling and Analysis,* Lewis Publishers/CRC Press, Boca Raton, FL

Millner, GC, RC James, and AC Nye, 1992. Human health-based soil cleanup guidelines for diesel fuel No.2, *J. Soil Contamination,* 1(2): 103-157

Mitchell, P. and D. Barr, 1995. The nature and significance of public exposure to arsenic: a review of its relevance to South West England, *Environmental Geochemistry and Health,* 17: 57 – 82

Mitchell JW (ed.), 1987. Occupational medicine forum: lead toxicity and reproduction, *J. Occup. Med.,* 29:397–9

Moeller, DW, 1997. *Environmental Health,* Revised Edition. Harvard University Press, Cambridge, MA

Moore MR, Goldberg A, Yeung-Laiwah AC, 1987. Lead effects on the hemebiosynthetic pathway, *Ann NY Acad Sci.,* 514:191–202

Morgan, MG and M. Henrion, 1991. *Uncertainty: A Guide to Dealing with Uncertainty in Quantitative Risk and Policy Analysis,* Oxford Univ. Press, Cambridge, UK

Morgan, MG and L. Lave, 1990. Ethical considerations in risk communication practice and research, *Risk Analysis,* 10(3): 355-358

Muir DCG, R. Wagemann, BT Hargrave, DJ Thomas, DB Peakall, and RJ Norstrom, 1992. Arctic marine ecosystem contamination, *Science of the Total Environment,* 122: 75-134

Mulkey, LA, 1984. Multimedia fate and transport models: an overview, *Journal of Toxicology-Clinical Toxicology,* 21(1-2): 65-95

Munro, IC, and Krewski, DR, 1981. Risk assessment and regulatory decision making, *Food Cosmet. Toxicol.,* 19, 549-560

Mushak, P. and AF Crocetti, 1995. Risk and revisionism in arsenic cancer risk assessment, *Environmental Health Perspectives,* 103: 684-688

NATO/CCMS, 1988a. *Pilot Study on International Information Exchange on Dioxins and Related Compounds*, North Atlantic Treaty Organization, Committee on the Challenges of Modern Society, Report 176, August 1988

NATO/CCMS, 1988b. *Scientific Basis for the Development of International Toxicity Equivalency Factor (I-TEF), Method of Risk Assessment for Complex Mixtures of Dioxins and Related Compounds*, North Atlantic Treaty Organization, Committee on the Challenges of Modern Society, Report No. 178, December 1988

NCI (National Cancer Institute), 1992. Making health communication programs work: a planner's guide, NIH Publication No. 92(1493): 64-65, National Cancer Institute, Washington, DC

Needleman, HL and CA Gatsonis, 1990. Low-level lead exposure and the IQ of children: a meta-analysis of modern studies, *Journal of the American Medical Association*, 263(5): 673-678

Needleman HL, Gunnoe C, Leviton A, Reed R, Peresie H, Maher C, et al. 1979. Deficits in psychologic and classroom performance of children with elevated dentine lead levels, *N. Engl. J. Med.*, 300:689–95

Neely, WB, 1980. *Chemicals in the Environment (Distribution, Transport, Fate, Analysis)*, Marcel Dekker, Inc., NY

Neely, WB, 1994. *Introduction to Chemical Exposure and Risk Assessment*, Lewis Publishers/CRC Press, Boca Raton, FL

Neubacher, FP, 1988. *Policy Recommendations for the Prevention of Hazardous Waste*, Elsevier Science Publishers, Amsterdam, The Netherlands

Ng, KL and DM Hamby, 1997. Fundamentals for establishing a risk communication program, *Health Physics*, 73(3): 473-482

Norrman, J., 2001. *Decision Analysis under Risk and Uncertainty at Contaminated Sites – A Literature Review*, SGI Varia 501, Swedish Geotechnical Institute (SGI), Linkoping, Sweden

NRC (National Research Council), 1982. *Risk and Decision-Making: Perspective and Research*, NRC Committee on Risk and Decision-Making, National Academy Press: Washington, DC

NRC, 1983. *Risk Assessment in the Federal Government: Managing the Process*, National Research Council, Committee on the Institutional Means for Assessment of Risks to Public Health, National Academy Press: Washington, DC

NRC, 1989a, *Ground Water Models: Scientific and Regulatory Applications*, National Research Council (NRC), National Academy Press, Washington, DC

NRC, 1989b, *Improving Risk Communication*, National Research Council, Committee on Risk Perception and Communication, National Academy Press: Washington, DC

NRC, 1991a, *Environmental Epidemiology (Public Health and Hazardous Wastes)*, National Academy Press, Washington, DC

NRC, 1991b, *Frontiers in Assessing Human Exposure to Environmental Toxicants*, National Academy Press, Washington, DC

NRC, 1991c, *Human Exposure Assessment for Airborne Pollutants: Advances and Opportunities*, National Academy Press, Washington, DC

NRC, 1993a, *Ground Water Vulnerability Assessment: Predicting Relative Contamination Potential Under Conditions of Uncertainty*, National Academy Press, Washington, DC

NRC, 1993b, *Issues in Risk Assessment*, National Academy Press, Washington, DC

NRC, 1993c, *Pesticides in the Diets of Infants and Children*, National Academy Press, Washington, DC

NRC, 1994a, *Building Consensus through Risk Assessment and Risk Management*, National Academy Press, Washington, DC

NRC, 1994b, *Science and Judgment in Risk Assessment*, National Research Council, Committee on Risk Assessment of Hazardous Air Pollutants, National Academy Press: Washington, DC

NRC, 1996. *Understanding Risk: Informing Decisions in a Democratic Society*, National Academy Press, Washington, DC

NRC, 1999. *Hormonally Active Agents in the Environment*, National Academy Press, Washington, DC

NTP, 1984. National Toxicology Program Report of the ad hoc Panel on Chemical Carcinogenesis Testing and Evaluation, National Toxicology Program (NTP) Board of Scientific Counselors, US Government Printing Office, Washington, DC

NTP, 1991. Sixth Annual Report on Carcinogens, National Toxicology Program (NTP), US Department of Health and Human Services, Public Health Service, Washington, DC

OECD (Organization for Economic Cooperation and Development), 1986. Report of the OECD Workshop on Practical Approaches to the Assessment of Environmental Exposure, 14-18 April 1986, Vienna, Austria

OECD (Organization for Economic Cooperation and Development), 1989. *Compendium of Environmental Exposure Assessment Methods for Chemicals*, Environmental Monographs, No.27, OECD, Paris, France

OECD (Organization for Economic Cooperation and Development), 1993. *Occupational and Consumer Exposure Assessment*, OECD Environment Monograph 69, Organization for Economic Cooperation and Development (OECD), Paris, France

OECD (Organization for Economic Co-operation and Development), 1994. *Environmental Indicators: OECD Core Set*, Organization for Economic Cooperation and Development (OECD), Paris, France

O'Flaherty, EJ, 1998. A physiologically based kinetic model for lead in children and adults, *Environmental Health Perspectives*, 106(Suppl.6): 1495 – 1503

Olin, SS (ed.), 1999. *Exposure to Contaminants in Drinking Water: Estimating Uptake through the Skin and by Inhalation*, CRC Press, Boca Raton, FL

Onalaja AO and L. Claudio, 2000. Genetic susceptibility to lead poisoning, *Environ Health Perspect.*, 108 (Suppl 1): 23–8

OSA, 1992. *Supplemental Guidance for Human Health Multimedia Risk Assessments of Hazardous Waste Sites and Permitted Facilities*, Office of Scientific Affairs (OSA), Cal-EPA, DTSC, Sacramento, CA

OSHA, 1980. Identification, Classification, and Regulation of Potential Occupational Carcinogens, Occupational Safety and Health Administration (OSHA), US Department of Labor, *Federal Register*, 45[13]: 5001-5296

OSTP (Office of Science and Technology Policy), 1985. Chemical Carcinogens: A Review of the Science and Its Associated Principles. Federal Register 50: 10372 - 442

OTA, 1993. *Researching Health Risks*, Office of Technology Assessment (OTA), US Congress, US Government Printing Office, Washington, DC

Ott, WR, 1995. *Environmental Statistics and Data Analysis*, Lewis Publishers/CRC Press, Boca Raton, FL

Page, GW and M. Greenberg, 1982. Maximum Contaminant Levels for Toxic Substances in Water: A Statistical Approach, *Water Resources Bulletin* 18.6 (December) 955-962

Parkhurst, DF, 1998. Arithmetic versus geometric means for environmental concentration data, *Environmental Science & Technology*, 32(3): 92A-98A

Patch, SC, RP Maas, and JP Pope, 1998. Lead leaching from faucet fixtures under residential conditions. *Environmental Health*, 61(3): 18-21

Patnaik, P., 1992. *A Comprehensive Guide to the Hazardous Properties of Chemical Substances*, Van Nostrand Reinhold, New York

Patrick, DR (ed.), 1996. *Toxic Air Pollution Handbook*, Van Nostrand Reinhold Co., New York

Patton, DE, 1993. The ABCs of risk assessment, *EPA Journal*, 19:10-15

Paustenbach, DJ (ed.), 1988. *The Risk Assessment of Environmental Hazards: A Textbook of Case Studies*, John Wiley & Sons, New York

Paustenbach, DJ, GM Bruce, and P. Chrostowski. 1997. 'Current views on the oral bioavailability of inorganic mercury in soil: implications for health risk assessments.' Risk Analysis, 17(5): 533-544

Peck, Dennis L.(ed.), 1989. *Psychosocial Effects of Hazardous Toxic Waste Disposal on Communities*, Charles C. Thomas Publishers, Springfield, Illinois

Pedersen, J., 1989. *Public Perception of Risk Associated with the Siting of Hazardous Waste Treatment Facilities*, European Foundation for the Improvement of Living and Working Conditions, Dublin, Eire

Petak, W.J, and A.A Atkisson, 1982. *Natural Hazard Risk Assessment and Public Policy: Anticipating the Unexpected*, Springer-Verlag, NY

Peters, RG, VT Covello, DB McCallum, 1997. The determinants of trust and credibility in environmental risk communication: an empirical study, *Risk Analysis*, 17(1):43-54

Petts, J., T. Cairney, and M. Smith. 1997. *Risk-Based Contaminated Land Investigation and Assessment*, J. Wiley, Chichester, England, UK

Piomelli S, Rosen JF, Chisolm JJ Jr, Graef JW, 1984. Management of childhood lead poisoning. J Pediatr 105(4):523–32

Pirkle, JL, WR Harlan, JR Landis, and J. Schwartz, 1985. The relationship between blood lead levels and blood pressure and its cardiovascular risk implications, *American Journal of Epidemiology*, 121: 246-258

Pirkle JL, Kaufmann RB, Brody DJ, Hickman T, Gunter EW, Paschal DC. 1998. Exposure of the U.S. population to lead, 1991–1994. Environ Health Perspect 106:745–50

Pollard, SJ, R. Yearsley, *et al.*, 2002. Current directions in the practice of environmental risk assessment in the United Kingdom, *Environmental Science & Technology*, 36(4): 530-538

Porcella, DB, 1994. 'Mercury in the Environment: Geochemistry.' In CJ Watras and JW Huckabee (eds.). *Mercury Pollution: Integration and Synthesis*, CRC Press, Boca Raton, FL. (pp.3-19)

Power, M. and LS McCarty, 1996. Probabilistic risk assessment: betting on its future., *Human and Ecological Risk Assessment*, 2:30-34

Power, M. and LS McCarty, 1998. A comparative analysis of environmental risk assessment/risk management frameworks. *Environmental Science and Technology*, 32(9): 224A-231A

Prager, JC, 1995. *Environmental Contaminant Reference Databook*, Van Nostrand Reinhold, New York
Price, PS, CL Curry, *et al*, 1996. Monte Carlo modeling of time-dependent exposures using a microexposure event approach, *Risk Analysis*, 16(3): 339-348
Prosser, CL and FA Brown, 1961. *Comparative Animal Physiology*, 2nd edition, WB Saunders Co., Philadelphia, PA
Pugh, DM and JV Tarazona, 1998. *Regulation for Chemical Safety in Europe: Analysis, Comment and Criticism*, Kluwer Academic Publishers, Dordrecht, The Netherlands
Putnam, RD, 1986. Review of toxicology of inorganic lead, *Am. Ind. Hyg. Assoc. J.*, 47:700–3

Rabinowitz, M., 1998. Historical perspective on lead biokinetic models, *Environmental Health Perspectives*, 106(Suppl.6): 1461 – 1465
Raloff, J., 1996. Tap water's toxic touch and vapors. *Science News*, 149(6): 84
Ramamoorthy, S. and E. Baddaloo. 1991. Evaluation of Environmental Data for Regulatory and Impact Assessment. Studies in Environmental Science 41, Elsevier Science Publishers B.V., Amsterdam, The Netherlands
Rappaport, S.M. and J. Selvin, 1987. A Method for Evaluating the Mean Exposure from a LogNormal Distribution, *J. Amer. Ind. Hyg. Assoc.*, 48:374-379
Rappe, C, Buser, HR, and Bosshardt, H-P, 1979. Dioxins, Dibenzofurans and Other Polyhalogenated Aromatics: Production, Use, Formation, and Destruction, *Ann NY Acad Sci.*, 320:1-18
Reeve, RN, 1994. *Environmental Analysis*, J. Wiley, NY
Regan MJ and Desvousges WH. 1990. Communicating environmental risks: a guide to practical evaluations. Washington, DC: U.S. Environmental Protection Agency, Pub. no. 230-01-91-001, pgs. 2-3.
Renn, O., 1999. A Model for an Analytic-Deliberative Process in Risk Management, *Environmental Science & Technology*, 33(18): 3049-3055
Renn, O., 1992. Risk communication: towards a rational dialogue with the public, *Journal of Hazardous Materials*, 29(3): 465-519
Ricci, PF (ed.), 1985. *Principles of Health Risk Assessment*, Prentice-Hall, Englewood Cliffs, NJ
Ricci, PF and MD Rowe (ed.), 1985. *Health and Environmental Risk Assessment*, Pergamon Press, NY
Rice DC. 1996. Behavioral effects of lead: commonalities between experimental and epidemiologic data, *Environ Health Perspect.*, 104(suppl 2):337–351
Richardson, ML (ed.), 1986. *Toxic Hazard Assessment of Chemicals*, Royal Society of Chemistry, London, England, UK
Richardson, ML (ed.), 1990. *Risk Assessment of Chemicals in the Environment*, Royal Society of Chemistry, Cambridge, UK
Richardson, ML (ed.), 1992. *Risk Management of Chemicals*, Royal Society of Chemicals, Cambridge, UK
Richardson, GM, 1996. Deterministic versus probabilistic risk assessment: strengths and weaknesses in a regulatory context, *Human and Ecological Risk Assessment*, 2:44-54
Richardson, SD, JE Simmons, and G. Rice, 2002. Disinfection byproducts: the next generation, *Environmental Science & Technology*, 36(9): 198A – 205A
Rodier PM. 1995. Developing brain as a target of toxicity, *Environ Health Perspect.*, 103(suppl 6):73–6
Rodricks, J. and Taylor, MR, 1983. Application of risk assessment to food safety decision making, *Regulatory Toxicol. Pharmacol.*, 3: 275-307
Romieu I, Carreon T, Lopez L, Palazuelos E, Rios C, Manuel Y, *et al.*, 1995. Environmental urban lead exposure and blood lead levels in children of Mexico City, *Environ Health Perspect.*, 103:1036–40
Romieu I, Lacasana M, McConnell R., 1997. Lead exposure in Latin America and the Caribbean, *Environ. Health Perspect.*, 105:398–405
Romieu I, Palazuelos E, Hernandez Avila M, Rios C, Muñoz I, Jimenez C, et al. 1994. Sources of lead exposure in Mexico City. Environ Health Perspect 102:384–9
Rosén, L. and HE LeGrand, 1997. An Outline of a Guidance Framework for Assessing Hydrogeological Risks at Early Stages, *Ground Water*, 35(2): 195-204
Rowe, WD, 1977. *An Anatomy of Risk*, John Wiley & Sons, New York
Rowe, WD, 1983. *Evaluation Methods for Environmental Standards*, CRC Press, Inc., Boca Raton, Florida
Rowe, WD, 1994. Understanding uncertainty, *Risk Analysis*, 14(5): 743-750
Rowland, AJ and P. Cooper, 1983. *Environment and Health*, Edward Arnold, England
Royston, P., 1995. A remark on algorithm AS 181: the W-test for normality, *Applied Statistics*, 44: 547-551
Ruckelshaus, WD, 1985. Risk, Science, and Democracy, *Issues in Science & Technology*, Spring 1985 (page 19-38)
Ryan, PB, 1991. An overview of human exposure modeling, *J. Exposure Anal. and Environ. Epidemiol.*, 1(4): 453-474

Sachs, L., 1984. *Applied Statistics – A Handbook of Techniques*, Springer-Verlag, New York

Safe, S., 1994. Polychlorinated biphenyls (PCBs): environmental impact, biochemical and toxic responses, and implications for risk assessment, *Crit. Rev. Toxicol.*, 24(2):87–149

Safe, SH, 1998. Hazard and risk assessment of chemical mixtures using the toxic equivalency factor approach, *Environmental Health Perspectives*, 106 (Suppl.4): 1051-1058

Saleh, MA, JN Blancato, and CH Nauman (eds.), 1994. *Biomarkers of Human Exposure to Pesticides*, ACS Symposium Series, American Chemical Society (ACS), Washington, DC

Saltzman, BE, 1997. Health risk assessment of fluctuating concentrations using lognormal models, *Journal of the Air & Waste Management Association*, 47: 1152-1160

Samiullah, Y., 1990. Prediction of the Environmental Fate of Chemicals. Elsevier Applied Science (in association with BP), London, UK

Samuels, ER and JC Meranger, 1984. Preliminary studies on the leaching of some trace metals from kitchen faucets. *Water Research*, 18(1): 75-80

Sandman, PM, 1993. *Responding to Community Outrage: Strategies for Effective Risk Communication*, American Industrial Hygiene Association, Fairfax, VA

Sax, NI, 1979. *Dangerous Properties of Industrial Materials*, 5th ed. New York: Van Nostrand Reinhold

Sax, NI and RJ Lewis, Sr., 1987. *Hawley's Condensed Chemical Dictionary*, Van Nostrad Reinhold Co., New York, NY

Saxena, J. and F. Fisher (eds.), 1981. *Hazard Assessment of Chemicals*, Academic Press, NY

Sayetta, RB, 1986. Pica: an overview, *American Family Physician*, 33(5): 181-185

Scanlon, VC and T. Sanders. 1995. *Essentials of Anatomy and Physiology*, 2nd ed. F.A. Davis, New York.

Scelfo GM, Flegal AR. 2000. Lead in calcium supplements. Environ Health Perspect 108:309–13

Schecter, A., 1994. *Dioxins and Health*, Kluwer Academic Publishers, Dordrecht, The Netherlands

Schecter, A., Startin, J, Wright, C, Kelly, M, Papke, O, Lis, A, Ball, M, and Olson, JR, 1994. Congener-specific Levels of Dioxins and Dibenzofurans in U.S. Food and Estimated Daily Dioxin Equivalent Intake, *Environ. Health Persp.*, 102:962-966

Schnoor, JL, 1996. Environmental Modeling: Fate and Transport of Pollutants in Water, Air and Soil. J. Wiley, NY

Schock, MR and CH Neff, 1988. Trace metal contamination from brass faucets. *Journal of the American Water Works Association*, 80(11): 47-56

Schrader-Frechette, KS. 1991. *Risk and Rationality*, University of California Press, Berkeley, Calif.

Schulz, TW and S. Griffin, 1999. Estimating risk assessment exposure point concentrations when data are not Normal or Lognormal, *Risk Analysis*, 19(4): 577-584

Schwartz, J. 1991. Lead, blood pressure, and cardiovascular disease in men and women. *Environmental Health Perspectives*, 91: 71-76

Schwartz, J., 1994. Low-level lead exposure and children's IQ: a meta-analysis and search for a threshold. *Environmental Research*, 65: 42-55

Schwartz J. and D. Otto, 1987. Blood lead, hearing thresholds, and neurobehavioral development in children and youth, *Arch Environ Health*, 42:153–9

Schwing, RC and WA Albers, Jr. (eds.), 1980. *Societal Risk Assessment: How Safe is Safe Enough*, Plenum Press, NY

Sedman, R.M. 1989. The Development of Applied Action Levels for Soil Contact: A Scenario for the Exposure of Humans to Soil in a Residential Setting. *Environmental Health Perspectives*, Vol. 79, pp.291-313

Sedman, R. and RS Mahmood, 1994. Soil ingestion by children and adults reconsidered using the results of recent tracer studies, *Air and Waste*, 44: 141-144

Seip HM and AB Heiberg (ed.s). 1989. *Risk Management of Chemicals in the Environment*, NATO, Challenges of Modern Society, Vol. 12. Plenum Press, New York, NY

Shalat, S.L.; True, L.D.; Fleming, L.E.; Pace, P.E. (1989) Kidney cancer in utility workers exposed to polychlorinated biphenyls (PCBs), *Br. J. Ind. Med.*, 46(11): 823–824

Shapiro, SS and MB Wilk, 1965. An analysis of variance test for normality (complete samples), *Biometrika*, 52: 591-611

Sharp, V.F. 1979. Statistics for the Social Sciences. Little, Brown & Co., Boston, MA

Sheppard, SC, 1995. Parameter values to model the soil ingestion pathway, *Environmental Monitoring and Assessment*, 34: 27-44

Shere, ME, 1995. The Myth of Meaningful Environmental Risk Assessment. *Harvard Environmental Law Review*, 19(2): 409-492

Shields PG, Harris CC, 1991. Molecular epidemiology and the genetics of environmental cancer. JAMA 266:681-687

Silbergeld, EK, 1991. Lead in bone: implications for toxicology during pregnancy and lactation, *Environmental Health Perspectives*, 91: 63 – 70

Silberhorn, EM; Glauert, H.; Robertson, LW, 1990. Carcinogenicity of polyhalogenated biphenyls: PCBs and PBBs, *Crit. Rev. Toxicol.*, 20(6): 439 – 496

Silk, JC and MB Kent (eds.). 1995. *Hazard Communication Compliance Manual*, Society for Chemical Hazard Communication/BNA Books, Washington, DC

Singhroy VH, DD Nebert, and AI Johnson (eds.), 1996. *Remote Sensing and GIS for Site Characterization: Applications and Standards*, ASTM Publication No. STP 1279, ASTM, Philadelphia, PA

Sinks, T.; Steele, G.; Smith, A.B.; Watkins, K.; Shults, R.A., 1992. Mortality among workers exposed to polychlorinated biphenyls, *Am. J. Epidemiol.*, 136(4): 389–398

Sitnig, M., 1985. Handbook of Toxic and Hazardous Chemicals and Carcinogens. Noyes Data Corp., Park Ridge, NJ

Sittig, M., 1994. *World-Wide Limits for Toxic and Hazardous Chemicals in Air, Water and Soil.* Noyes Publications, Park Ridge, New Jersey

Slob, W., 1994. 'Uncertainty analysis in multiplicative models.' Risk Analysis, 14: 571-576

Slovic, P., 1993. 'Perceived risk, trust, and democracy.' Risk Analysis, 13(6):675-682

Slovic, P., 1997. Public perception of risk. *Journal of Environmental Health*, 59(9): 22-24

Smith, AH, 1987. Infant exposure assessment for breast milk dioxins and furans derived from waste incineration emissions. *Risk Analysis*, 7(3): 347–353

Smith, RL, 1994. Use of Monte Carlo simulation for human exposure assessment at a Superfund site, *Risk Analysis*, 14(4): 433-439

Smith, RP, 1992. A Primer of Environmental Toxicology. Lea & Febiger, Philadelphia, PA

Smith, Jr., TT, 1996. Regulatory Reform in the USA and Europe, *Journal of Environmental Law*, 8(2): 257-282

Smith AE, PB Ryan, and JS Evans, 1992. The effect of neglecting correlations when propagating uncertainty and estimating population distribution of risk, *Risk Analysis*, 12:457-474

Smith, AH, S. Sciortino, H. Goeden, and CC Wright, 1996. Consideration of background exposures in the management of hazardous waste sites: a new approach to risk assessment, *Risk Analysis*, 16(5): 619-625

Splitstone, DE, 1991. How clean is clean .... statistically?, *Pollution Engineering*, 23: 90-96

Stanek, EJ and EJ Calabrese, 1990. A guide to interpreting soil ingestion studies, *Regulatory Toxicology and Pharmacology*, 13:263-292

Stanek, EJ and EJ Calabrese, 1995a. Daily estimates of soil ingestion in children, *Environmental Health Perspectives*, 103(3): 276-285

Stanek, EJ and EJ Calabrese, 1995b. Soil ingestion estimates for use in site evaluations based on the best tracer method, *Human and Ecological Risk Assessment*, 1: 133-156

Starr, C., R. Rudman, and C. Whipple, 1976. Philosophical basis for risk analysis, *Annual Review of Energy*, 1: 629-662

Starr C., and C.Whipple. 1980. Risks of Risk Decisions. *Science 208, 1114*

Steele, G.; Stehr-Green, P.; Welty, E. (1986) Estimates of the biologic half-life of polychlorinated biphenyls in human serum, *New Engl. J. Med.*, 314(14):926–927

Stenesh, J. , 1989. *Dictionary of Biochemistry and Molecular Biology*, J. Wiley, NY

Stepan, DJ, RH Fraley, and DS Charlton, 1995. *Remediation of Mercury-Contaminated Soils: Development and Testing of Technologies*, Gas Research Institute, USA

Suter, GW, 1993. *Ecological Risk Assessment*, Lewis Publishers, Boca Raton, FL

Suter II, GW, BW Cornaby, *et al.*, 1995. An approach for balancing health and ecological risks at hazardous waste sites, *Risk Analysis*, 15: 221 – 231

Swann, R.L., and A. Eschenroeder (eds.), 1983. *Fate of Chemicals in the Environment*, ACS Symposium Series 225, Amer. Chem. Society, Washington, DC

Talbot, EO and GF Craun (eds.), 1995. *Introduction to Environmental Epidemiology*, Lewis Publishers/CRC Press, Boca Raton, Florida

Talmage, SS and BT Walton, 1993. Food Chain Transfer and Potential Renal Toxicity to Small Mammals at a Contaminated Terrestrial Field Site, *Ecotoxicology*, 2: 243-256

Tan, WY, 1991. *Stochastic Models of Carcinogenesis*, Statistics: Textbooks and Monographs, Volume 116, Marcel Dekker, New York

Tardiff, RG and JV Rodricks (eds.), 1987. *Toxic Substances and Human Risk*, Plenum Press, New York

Taylor, AC, 1993. Using Objective and Subjective Information to Develop Distributions for Probabilistic Exposure Assessment, *Journal of Exposure Analysis and Environmental Epidemiology*, 3(3): 285-298

Tellez-Rojo, MM, M. Hernandez-Avila, *et al.*, 2002. Impact of breastfeeding on the mobilization of lead from bone, *American Journal of Epidemiology*, 155(5): 420-428

Tessier, A. and PGC Campbell, 1987. Partitioning of Trace Metals in Sediments: Relationships with Bioavailability, *Hydrobiologia, 149*: 43-52

Theiss, JC, 1983. The ranking of chemicals for carcinogenic potency, *Regulatory Toxicol. Pharmacol.*, 3, 320-328

Thibodeaux, LJ, 1979. *Chemodynamics: Environmental Movement of Chemicals in Air, Water and Soil*, J. Wiley, NY

Thibodeaux, LJ, 1996. *Environmental Chemodynamics, 2nd ed.*, J. Wiley, NY

Thibodeaux, LJ and ST Hwang, 1982. Landfarming of Petroleum Wastes--Modeling the Air Emission Problem. *Environmental Progress*, February, 1: 42-46

Thomas, DJ, B. Tracey, H. Marshall, and RJ Norstrom, 1992. Arctic terrestrial ecosystem contamination. *Science of the Total Environment*, 122: 135-164

Thompson, SK, 1992. *Sampling*, J. Wiley, NY

Thompson, KM and DE Burmaster, 1991. Parametric distributions for soil ingestion in children, *Risk Analysis*, 11: 339-342

Thompson, KM, DE Burmaster, and AC Crouch. 1992. 'Monte Carlo techniques for quantitative uncertainty analysis in public health risk assessments.' Risk Analysis, 12:53-63

Timbrell, JA, 1995. *Introduction to Toxicology*, 2nd ed., Taylor & Francis, London, UK

Todd AC, Wetmur JC, Moline JM, Godbold JH, Levin SM, Landrigan PJ, 1996. Unraveling the chronic toxicity of lead: an essential priority for environmental health, *Environ Health Perspect.*, 104(suppl 1): 141–6

Tohn E, Dixon S, Rupp R, Clark S. 2000. A pilot study examining changes in dust lead loading on walls and ceilings after lead hazard control interventions, *Environ Health Perspectives*, 108:453–6

Tomatis, L., J. Huff, I. Hertz-Picciotto, D. Dandler, J. Bucher, P. Boffetta, O. Axelson, A. Blair, J. Taylor, L. Stayner, and JC Barrett, 1997. Avoided and avoidable risks of cancer, *Carcinogenesis*, 18(1): 97-105

Travis, CC and AD Arms, 1988. Bioconcentration of organics in beef, milk, and vegetation, *Environmental Science and Technol.*, 22: 271 - 274

Travis, CC, SA Richter, EAC Crouch, R Wilson, and ED Klema, 1987. Cancer risk management, *Environ. Sci. Technol.*, 21:415-420

Tsuji, JS and KM Serl, 1996. Current uses of the EPA lead model to assess health risk and action levels for soil. *Environmental Geochemistry and Health*, 18:25-33

Turnberg, WL, 1996. *Biohazardous Waste: Risk Assessment, Policy, and Management*, J. Wiley, NY

USDHS, 1989. *Public Health Service, Fifth Annual Report on Carcinogens, Summary*, US Department of Health and Human Services (USDHS), Washington, DC

USEPA (US Environmental Protection Agency), 1984a. *Approaches to risk assessment for multiple chemical exposures*, EPA-600/9-84-008, Environmental Criteria and Assessment Office, US Environmental Protection Agency (USEPA), Cincinnati, OH

USEPA, 1984b. Proposed Guidelines for Carcinogen, Mutagenicity, and Developmental Toxicant Risk Assessment, *Federal Register*, 49, 46294-46331

USEPA, 1984c. *Risk Assessment and Management: Framework for Decision Making*, EPA 600/9-85-002, US Environmental Protection Agency (USEPA), Washington, DC

USEPA, 1985. *Principles of Risk Assessment: A Nontechnical Review*, Office of Policy Analysis, US Environmental Protection Agency (USEPA), Washington, DC

USEPA, 1986a. *Guidelines for the Health Risk Assessment of Chemical Mixtures*, EPA/630/R-98/002, Risk Assessment Forum, Office of Research and Development, US Environmental Protection Agency (USEPA), Washington, DC

USEPA, 1986b. Guidelines for Carcinogen Risk Assessment. *Federal Register*, 51(185): 33992-34003, CFR 2984, September 24, 1986

USEPA, 1986c. *Guidelines for Mutagenicity Risk Assessment*, EPA/630/R-98/003, Risk Assessment Forum, Office of Research and Development, USEPA, Washington, DC

USEPA, 1986d. *Methods for Assessing Exposure to Chemical Substances, Volume 8: Methods for Assessing Environmental Pathways of Food Contamination*. EPA/560/5-85-008, Exposure Evaluation Division, Office of Toxic Substances, Washington, DC

USEPA, 1986e. *Superfund Public Health Evaluation Manual*, EPA/540/1-86/060, Office of Emergency and Remedial Response, Washington, DC

USEPA, 1987. *Handbook for Conducting Endangerment Assessments*, USEPA, Research Triangle Park, NC

USEPA, 1988. *Superfund Exposure Assessment Manual*, EPA/540/1-88/001, OSWER Directive 9285.5-1, USEPA, Office of Remedial Response, Washington, DC

USEPA, 1989a. *Exposure Factors Handbook*, EPA/600/8-89/043, Office of Health and Environmental Assessment, Washington, DC

USEPA, 1989b. *Interim Methods for Development of Inhalation Reference Doses*, EPA/600/8-88/066F. Office of Health and Environmental Assessment, Washington, DC

USEPA, 1989c. *Interim Procedures for Estimating Risks Associated with Exposures to Mixtures of Chlorinated dibenzo-p-dioxins and -dibenzofurans (CDDs and CDFs)*, EPA/625/3-89/016, Risk Assessment Forum, Washington, DC

USEPA, 1989d. *Methods for Evaluating the Attainment of Cleanup Standards. Volume I: Soils and Solid Media*, EPA/230/2-89/042. Office of Policy, Planning and Evaluation, Washington, DC

USEPA, 1989e. *Risk Assessment Guidance for Superfund: Volume I--Human Health Evaluation Manual (Part A)*, EPA/540/1-89/002, Office of Emergency and Remedial Response, Washington, DC

USEPA, 1989f. *Risk Assessment Guidance for Superfund. Volume II--Environmental Evaluation Manual.* EPA/540/1-89/001. Office of Emergency and Remedial Response, Washington, DC

USEPA, 1990a. *Environmental Asbestos Assessment Manual, Superfund Method for the Determination of Asbestos in Ambient Air, Part 1: Method (EPA/540/2-90/005a) & Part 2: Technical Background Document (EPA/540/2-90/005b)*, US Environmental Protection Agency (USEPA), Washington, DC

USEPA, 1990b. *Guidance for Data Useability in Risk Assessment*, Interim Final, Washington, DC: Office of Emergency and Remedial Response. EPA/540/G-90/008

USEPA, 1991a. *Guidelines for Developmental Toxicity Risk Assessment*, EPA/600/FR-91/001, Risk Assessment Forum, Office of Research and Development, USEPA, Washington, DC

USEPA, 1991b. *Risk Assessment Guidance for Superfund, Volume 1: Human Health Evaluation Manual (Part B, Development of Risk-Based Preliminary Remediation Goals)*. EPA/540/R-92/003. Office of Emergency and Remedial Response, Washington, DC

USEPA, 1991c. *Risk Assessment Guidance for Superfund, Volume 1: Human Health Evaluation Manual (Part C, Risk Evaluation of Remedial Alternatives)*. EPA/540/R-92/004. Office of Emergency and Remedial Response, Washington, DC

USEPA, 1991d. *Risk Assessment Guidance for Superfund, Volume I: Human Health Evaluation Manual. Supplemental Guidance, "Standard Default Exposure Factors" (Interim Final)*, March, 1991, Washington, DC: Office of Emergency and Remedial Response. OSWER Directive: 9285.6-03

US EPA, 1992a. *Dermal Exposure Assessment: Principles and Applications*, EPA/600/8-91/011B, Office of Health and Environmental Assessment, US EPA, Washington, DC.

USEPA, 1992b. *Guidance for Data Useability in Risk Assessment (Parts A & B)*, Publication No. 9285.7-09A&B, Office of Emergency and Remedial Response, USEPA, Washington, DC

USEPA, 1992c. *Guidelines for Exposure Assessment*, EPA/600/Z-92/001, Risk Assessment Forum, Office of Research and Development, Office of Health and Environmental Assessment, USEPA, Washington, DC

USEPA, 1992d. *Supplemental Guidance to RAGS: Calculating the Concentration Term*, Publication No. 9285.7-08I, Office of Emergency and Remedial Response, USEPA, Washington, DC

USEPA, 1993. *Supplemental Guidance to RAGS: Estimating Risk from Groundwater Contamination*, Office of Solid Waste and Emergency Response, USEPA, Washington, DC

USEPA, 1994a. *Estimating Exposures to Dioxin-like Compounds*, EPA/600/6-88/005Cb, Office of Research and Development, USEPA, Washington, DC

USEPA, 1994b. *Estimating Radiogenic Cancer Risks*, EPA 402-R-93-076, USEPA, Washington, DC

USEPA, 1994c. *Guidance for the Data Quality Objectives Process*. EPA/600/R-96/055. Office of Research and Development, Washington, DC

USEPA, 1994d. *Guidance Manual for the Integrated Exposure Uptake Biokinetic Model for Lead in Children*, EPA/540/R-93/081, Office of Emergency and Remedial Response, US Environmental Protection Agency (USEPA), Washington, DC

US EPA, 1994e. *Radiation Site Cleanup Regulations: Technical Support Document for the Development of Radionuclide Cleanup Levels for Soil*, EPA 402-R-96-011A, September 1994

USEPA, 1995a. *A Guide to the Biosolids Risk Assessments for the EPA Part 503 Rule*, EPA/832-B-93-005, Office of Wastewater Management, USEPA, Washington, DC

USEPA, 1995b. *Guidance for Risk Characterization*, Science Policy Council, USEPA, Washington, DC

USEPA, 1995c. *The Use of the Benchmark Dose Approach in Health Risk Assessment*, EPA/630/R-94/007, Risk Assessment Forum, Office of Research and Development, US Environmental Protection Agency (USEPA), Washington, DC

USEPA, 1995-96. *Guidance for Assessing Chemical Contaminant Data for Use in Fish Advisories, Volumes I thru IV*, Office of Water, USEPA, Washington, DC

USEPA, 1996a. *Guidelines for Reproductive Toxicity Risk Assessment*, EPA/630-96/009, USEPA, Washington, DC

USEPA, 1996b. *Interim Approach to Assessing Risks Associated with Adult Exposures to Lead in Soil*, US Environmental Protection Agency (USEPA), Washington, DC

USEPA, 1996c. *PCBs: Cancer Dose-Response Assessment and Application to Environmental Mixtures*, EPA/600/P-96/001F, September 1996, National Center for Environmental Assessment, Office of Research and Development, U.S. Environmental Protection Agency, Washington, DC

USEPA, 1996d. *Soil Screening Guidance: Technical Background Document*, EPA/540/R-95/128. Office of Emergency and Remedial Response, Washington, DC

USEPA, 1996e. *Soil Screening Guidance: User's Guide*, EPA/540/R-96/018. Office of Emergency and Remedial Response, Washington, DC

US EPA, 1997a. *Establishment of Cleanup Levels for CERCLA Sites with Radioactive Contamination*, OSWER Directive 9200.4-18, August 1997

USEPA, 1997b. *Estimating Radiogenic Cancer Risks*, USEPA, Washington, DC

US EPA, 1997c. *Exposure Factors Handbook, Volumes I through III*, EPA/600/0-95-002Fa, Fb, Fc, Office of Research and Development, USEPA, Washington, DC

USEPA, 1997d. *Guiding Principles for Monte Carlo Analysis*, EPA/630/R-97/001, Office of Research and Development, USEPA, Washington, DC

USEPA, 1997e. *Policy for Use of Probabilistic Analysis in Risk Assessment*, EPA/630/R-97/001, Office of Research and Development, USEPA, Washington, DC

USEPA, 1997f. *The Lognormal Distribution in Environmental Applications*, EPA/600/r-97/006, Office of Research and Development/Office of Solid Waste and Emergency Response, USEPA, Washington, DC

USEPA, 1997g. *Users Guide for the Johnson and Ettinger (1991) Model for Subsurface Vapor Intrusion into Buildings*, Office of Emergency and Remedial Response, USEPA, Washington, DC

USEPA, 1998. *Guidelines for Neurotoxicity Risk Assessment*, EPA/630/R-95/001F, Risk Assessment Forum, Office of Research and Development, USEPA, Washington, DC

USEPA, 1998. *Human Health Risk Assessment Protocol for Hazardous Waste Combustion Facilities, Volumes One thru Three*, EPA530-D-98-001A/-001B/-001C, Office of Solid Waste and Emergency Response, USEPA, Washington, DC

USEPA, 1998. *Supplemental Guidance to RAGS: The Use of Probability Analysis in Risk Assessment (Part E), Draft Guidance*, Office of Solid Waste and Emergency Response, USEPA, Washington, DC

USEPA, 1999a. *Air Quality Criteria for Particulate Matter, Volumes I thru III*, EPA/600/P-99/002a, /002b, /002c, Office of Research and Development, USEPA, Washington, DC

USEPA, 1999b. *Cancer Risk Coefficients for Environmental Exposure to Radionuclides*, Federal Guidance Report No. 13, EPA 402-R-99-001, Office of Radiation and Indoor Air, USEPA, Washington, DC

USEPA, 1999c. *Estimating Radiogenic Cancer Risks, Addendum: Uncertainty Analysis*, EPA 402-R-99-003, USEPA, Washington, DC

USEPA, 1999d. *Report of the Workshop on Selecting Input Distributions for Probabilistic Assessments*, EPA/630/R-98/004, Risk Assessment Forum, Office of Research and Development, USEPA, Washington, DC

USEPA, 2002. *Risk Assessment Guidance for Superfund, Volume I: Human Health Evaluation Manual (Part E, Supplemental Guidance for Dermal Risk Assessment)*, EPA/540/R/99/005, Office of Emergency and Remedial Response, USEPA, Washington, DC

van Emden, HF and DB Peakall, 1996. *Beyond Silent Spring*, Kluwer Academic Publishers, Dordrecht, The Netherlands

van Leeuwen, CJ and JLM Hermens (eds.), 1995. *Risk Assessment of Chemicals: An Introduction*, Kluwer Academic Publishers, Dordrecht, The Netherlands

van Veen, MP, 1996. A general model for exposure and uptake from consumer products, *Risk Analysis*, 16(3): 331-338

van Wijnen, JH, P. Clausing, and B. Brunekreef, 1990. Estimated soil ingestion by children, *Environmental Research*, 51:147-162

Vaughan, E., 1995. The significance of socioeconomic and ethnic diversity for the risk communication process, *Risk Analysis*, 15(2): 169-180

Vesley, D., 1999. *Human Health and the Environment: A Turn of the Century Perspective*, Kluwer Academic Publishers, Dordrecht, The Netherlands

Vermeire, TG, P. van der Poel, R. van de Laar, and H. Roelfzema, 1993. Estimation of consumer exposure to chemicals: application of simple models, *Science of the Total Environment*, 135:155-176

Weisman, J. 1996. AMS adds realism to chemical risk assessment. *Science*, 271(5247): 286-287.

Weinstein, ND, K. Kolb, and BD Goldstein, 1996. Using time intervals between expected events to communicate risk magnitudes, *Risk Analysis*, 16(3): 305-308

Weinstein, ND and PM Sandman, 1993. Some criteria for evaluating risk messages, *Risk Analysis*, 13(1): 103-114

Whipple, C. , 1987. *De Minimis Risk. Contemporary Issues in Risk Analysis, Vol.2*, Plenum Press, NY

WHO (World Health Organization), 1990. *Public Health Impact of Pesticides Used in Agriculture*, World Health Organization and United Nations Environment Programme, WHO, Geneva, Switzerland

Wilkes, CR, MJ Small, CI Davidson, and JB Andelman, 1996. Modeling the effects of water usage and co-behavior on inhalation exposures to contaminant volatilized from household water, *Journal of Exposure and Environmental Epidemiology,* 6: 393 – 412

Williams, PL and JL Burson (eds.), 1985. *Industrial Toxicology,* Van Nostrand Reinhold, New York

Willis, MC, 1996. *Medical Terminology: The Language of Health Care,* Williams & Wilkins, Baltimore, MD

Wilson, R. and EAC Crouch, 1987. Risk assessment and comparisons: an introduction, *Science,* 236:267–270

Wilson, JD, 1996. 'Threshold for carcinogens: a review of the relevant science and its implications for regulatory policy.' Discussion Paper No. 96-21, Resources for the Future, Washington, DC

Wilson, JD, 1997. 'So carcinogens have thresholds: how do we decide what exposure levels should be considered safe?, *Risk Analysis,* 17(1): 1 – 3

Winter, CK, 1992. Dietary pesticide risk assessment, *Review Environ. Contam. Toxicol.,* 127: 23 – 67

Wolff, MS, PG Toniolo, EW Lee, M. Rivera, and N. Dubin, 1993. Blood levels of organochlorine residues and risk of breast cancer, *J. National Cancer Inst.,* 85: 648 – 653

Wonnacott, TH and RJ Wonnacott, 1972. *Introductory Statistics, 2nd Ed.,* John Wiley & Sons, NY

Woodside, G., 1993. *Hazardous Materials and Hazardous Waste Management: A Technical Guide,* John Wiley & Sons, New York, NY

Whyte, AV, and I. Burton (eds.), 1980. *Environmental Risk Assessment, SCOPE Report 15,* J.Wiley & Sons, New York, NY

Yu, R. and CP Weisel, 1998. Measurement of benzene in human breath associated with an environmental exposure, *J. Exposure Analysis and Environmental Epidemiology,* 6(3): 261 – 277

Zakrzewski, SF, 1991. *Principles of Environmental Toxicology,* American Chemical Society, Washington, DC

Zakrzewski, SF, 1997. *Principles of Environmental Toxicology,* 2nd edition, ACS Monograph 190, American Chemical Society, Washington, DC

Zartarian, VG and JO Leckie, 1998. Dermal exposure: the missing link, *Environmental Science & Technology,* 32(5): 134A – 137A

Zemba, SG, LC Green, EAC Crouch, and RR Lester, 1996. Quantitative risk assessment of stack emissions from municipal waste combusters, *J. Hazardous Materials,* 47: 229 – 275

Ziegler, J., 1993. Toxicity tests in animals: extrapolating to human risks, *Environmental Health Perspectives,* 101: 396-406

Appendix **A**

GLOSSARY OF
SELECTED TERMS AND DEFINITIONS

Absorption: Generally used to refer to the uptake of a chemical by a cell or an organism following exposure through the skin, lungs, and/or gastrointestinal tract. *Systemic absorption* – refers to the flow of chemicals into the bloodstream. In general, chemicals can be absorbed through the skin into the bloodstream and then transported to other organs; chemicals can also be absorbed into the bloodstream after breathing or oral intake.

Absorption barrier: Any of the exchange barriers of the human body that allow differential diffusion of various substances across a boundary; examples of absorption barriers are the skin, the lung tissue, and the gastrointestinal tract wall.

Absorption fraction: Refers to the percent or fraction of a chemical in contact with an organism that becomes absorbed into the receptor – i.e., the relative amount of a substance at the exchange barrier that actually penetrates into the body of an organism. Typically, this is reported as the unitless fraction of the applied dose or as the percent absorbed – e.g., relative amount of a substance on the skin that penetrates through the epidermis into the body.

Acceptable daily intake (ADI): An estimate of the maximum amount of a chemical (in mg/kg body weight/day) to which a potential receptor can be exposed to on a daily basis over an extended period of time (usually a lifetime) without suffering a deleterious effect, or without anticipating an adverse effect.

Acceptable risk: A risk level generally deemed by society to be acceptable or tolerable.

Action level (AL): The limit of a chemical in selected media of concern above which there are potential adverse health and/or environmental effects. This represents the environmental chemical concentration above which some corrective action (e.g., monitoring or remedial action) is typically required by regulation.

Acute: Of short-term duration – i.e., occurring over a short time, usually a few minutes or hours. *Acute exposure* – refers to a single large exposure or dose to a chemical, generally occurring over a short period (usually lasting <24 to 96 hours), in relation to the lifespan of the exposed organism. An acute exposure can result in short-term or long-term health effects. *Acute effect* – takes place a short time (up to 1 year) after exposure. *Acute toxicity* – refers to the development of symptoms of poisoning or the occurrence of adverse health effects after exposure to a single dose or multiple doses of a chemical within a short period of time. It represents the sudden onset of adverse health effects that are of short duration – generally resulting in cellular changes that are reversible.

Additivity (of chemical effects): A pharmacologic or toxicologic interaction in which the combined effect of two or more chemicals is approximately equal to the sum of the effect of each chemical acting alone.

Administered dose: The mass of substance administered to an organism, and that is in contact with an exchange boundary (e.g., gastrointestinal tract) per unit body weight per unit time (e.g., mg/kg-day). It actually is a measure of exposure – since it does *not* account for absorption. (See also, *applied dose.*)

Adsorption: The removal of contaminants from a fluid stream by concentrating the constituents onto a solid material. It consists of the physical process of attracting and holding molecules of other chemical substances on the surface of a solid, usually by the formation of chemical bonds. A substance is considered *adsorbed* if the concentration in the boundary region of a solid (e.g., soil) particle is greater than in the interior of the contiguous phase.

Adverse effect: A biochemical change, functional impairment, or pathologic lesion that affects the performance of the whole organism, or reduces an organism's ability to respond to a future environmental challenge.

Aerosol: A suspension of liquid or solid particles in air.

Aliphatic compounds: Organic compounds in which the carbon atoms exist as either straight or branched chains; examples include pentane, hexane, and octane.

Ambient: Pertaining to surrounding conditions or area. *Ambient medium* – one of the basic categories of material surrounding or contacting an organism (e.g., outdoor air, indoor air, water, or soil) through which chemicals or pollutants can move and reach the organism.

Analyte: A chemical component of a sample that is to be determined or measured; for example, if the *analyte of interest* in an environmental sample is mercury, then the laboratory testing or analysis will determine the amount of mercury in the sample. The *analytical method* defines the sample preparation and instrumentation procedures or steps that must be performed to estimate the quantity of analyte in a sample.

Antagonism (or, antagonistic chemical effect): A pharmacologic or toxicologic interaction in which the combined effect of two chemicals is less than the sum of the effect of each chemical acting alone. This phenomenon is the result of interference or inhibition of the effects of one chemical substance by the action of other chemicals – and reflects the counteracting effect of one chemical on another, thus diminishing their additive effects.

Anthropogenic: Caused or influenced by human activities or actions.

Applied dose: The amount of a substance in contact with the primary absorption boundaries of an organism (e.g., skin, lung, gastrointestinal tract), and that is available for absorption. This actually is a measure of exposure – since it does *not* take absorption into account. (See also, *administered dose.*)

Arithmetic mean (also, average): A statistical measure of central tendency for data from a normal distribution – defined, for a set of *n* values, by the sum of the values divided by *n*,

$$X_m = \frac{\sum\limits_{i=1}^{n} X_i}{n}$$

Aromatic compounds: Organic compounds that contain carbon molecular ring structures (i.e., a benzene ring); examples include benzene, toluene, ethylbenzene, xylenes (BTEX).

Attenuation: Any decrease in the amount or concentration of a pollutant in an environmental matrix as it moves in time and space. It represents the reduction or removal of contaminant constituents by a combination of physical, chemical, and/or biological factors acting upon the contaminated 'parent' media.

Attributable risk (also, *incremental risk*): The difference between risk of exhibiting a certain adverse effect in the presence of a toxic substance and that risk to be expected in the absence of the substance. (See also, *excess lifetime risk.*)

Average concentration: A mathematical average of chemical concentration(s) from more than one sample – typically represented by the arithmetic mean or the geometric mean for environmental samples.

Average daily dose (ADD): The average dose calculated for the duration of receptor exposure, defined by:

$$\text{ADD (mg/kg-day)} = \frac{[chemical\,concentration]\times[contact\,rate]}{[body\,weight]}$$

This is used to estimate risks for chronic non-carcinogenic effects of environmental chemicals.

Averaging time: The time period over which a function (e.g., human exposure concentration of a chemical) is measured – yielding a time-weighted value.

Background threshold level: The normal or typical average ambient environmental concentration of a chemical constituent. Two types of background levels may exist for chemical substances – namely, naturally-occurring concentrations and elevated anthropogenic levels resulting from non-site-related human activities. *Anthropogenic background levels* – refer to concentrations of chemicals that are present in the environment due to human-made, non-site sources (e.g., lead depositions from automobile exhaust and 'neighboring' industry). *Naturally occurring background levels* – refer to ambient concentrations of chemicals that are present in the environment and have not been influenced by human activities (e.g., natural formations of aluminum, arsenic, and manganese).

Benchmark dose (BMD) (or, *benchmark concentration, BMC):* A statistical lower confidence limit on the dose that produces a predetermined change in response rate of an adverse effect (called the benchmark response or BMR) compared to background.

Benchmark response (BMR): An adverse effect, used to define a benchmark dose from which an RfD (or RfC) can be developed. The change in response rate over background of the BMR is usually in the range of 5-10%, which is the limit of responses typically observed in well-conducted animal experiments.

Benchmark risk: A threshold level of risk, typically prescribed by regulations, above which corrective measures will almost certainly have to be implemented to mitigate the risks.

Bioaccessibility: A term used in describing an event that relates to the absorption process upon exposure of an organism – and generally refers to the fraction of the administered substance that becomes solubilized in the gastrointestinal fluid. For the most part, solubility is a prerequisite of absorption, although small amounts of some chemicals in particulate or suspended/emulsified form may be absorbed by pinocytosis. Moreover, it is not simply the fraction dissolved that determines bioavailability, but also the rate of dissolution, which has physiological and geochemical influences. In and of itself, bioaccessibility is not a direct measure of the movement of a substance across a biological membrane (i.e., absorption or bioavailability). Indeed, the relationship of bioaccessibility to bioavailability is ancillary and the former need not be known in order to measure the latter. However, bioaccessibility (i.e., solubility) may serve as a surrogate for bioavailability if certain conditions are met.

Bioaccumulation: The progressive increase in amount of a chemical in an organism or part of an organism that occurs because the rate of intake exceeds the organism's ability to remove the substance from the body. This represents the retention and concentration of a chemical by an organism – that results from a build-up of the chemical in the organism as a consequence of the organism taking in more of the chemical than it can rid of in the same length of time, and therefore stores the chemical in its tissue, etc. (See also, *bioconcentration.*)

Bioassay: Measuring the effect(s) of environmental exposures by intentional exposure of living organisms to a chemical. It consists of tests used to evaluate the relative potency of a chemical by comparing its effects on a living organism with the effect of a standard preparation on the same type of organism.

Bioavailability: A measure of the degree to which a dose of a chemical substance becomes physiologically available to the body target tissues after being administered or upon exposure. It refers to the fraction of the total amount of material in contact with a body portal-of-entry (viz., lung, gut, skin) that actually enters the blood – and this depends on the absorption, distribution, metabolism and excretion rates. *Absolute bioavailability* – refers to the fraction or percentage of a compound that is ingested, inhaled, or applied on the skin surface that actually is absorbed and reaches the systemic circulation. That is, this is the amount of the substance entering the blood via a particular route of exposure (e.g., gastrointestinal) divided by the total amount administered (e.g., soil lead ingested). *Relative bioavailability* – refers to a measure of the extent of absorption among two or more forms of the same chemical (e.g., lead carbonate vs. lead acetate), different vehicles (e.g., food, soil, water, etc.), or different doses. In the context of environmental risk assessment, relative bioavailability is the ratio of the absorbed fraction from the exposure medium in the risk assessment (e.g., food or soil) to the absorbed fraction from the dosing medium used in the critical toxicity study. It is indexed by measuring the bioavailability of a particular substance relative to the bioavailability of a standardized reference material, such as soluble lead acetate.

Bioconcentration: The accumulation of a chemical substance in tissues of organisms (such as fish) to levels greater than levels in the surrounding media (such as water) for the organism's habitat; this is often used synonymously with bioaccumulation. *Bioconcentration factor (BCF)* – is the ratio of the concentration of a chemical substance in an organism, at equilibrium to the concentration of the substance in the surrounding environmental medium. It is a measure of the amount of selected chemical substances that accumulate in humans or in biota. (See also, *bioaccumulation.*)

Biologically-based dose response model: A predictive tool used to estimate potential human health risks by describing and quantifying the key steps in the cellular, tissue, and organism responses as a result of chemical exposure.

Biological uptake: The transfer of hazardous substances from the environment to plants, animals, and humans. This may be evaluated through environmental measurements, such as measurement of the amount of the substance in an organ known to be susceptible to that substance. More commonly, *biological dose measurements* are used to determine whether exposure has occurred. The presence of a chemical compound, or its metabolite, in human biologic specimens (such as blood, hair, or urine) is used to confirm exposure – and this can be an independent variable in evaluating the relationship between the exposure and any observed adverse health effects.

Biomagnification: The serial accumulation of a chemical by organisms in the food chain – with higher concentrations occurring at each successive trophic level.

Biomarker of exposure: Exogenous chemicals, their metabolites, or products of interactions between a xenobiotic chemical and some target molecule or cell that is measured in a compartment within an organism to verify suspected exposures or degree of known exposures.

Biomedical testing: Biological testing of persons to evaluate a qualitative or quantitative change in a physiologic function that may be predictive of health impairment resulting from exposure to hazardous substance(s).

Body burden: The total amount of a particular chemical substance stored in the body (usually in fatty tissue, blood, and/or bone) at a particular time – especially relating to a potentially toxic chemical in the body that follows from exposure. Some chemicals build up in the body because they are stored in fat or bone, or are eliminated very slowly – e.g., the amount of metals such as lead in the bone; the amount of lipophilic compounds such as PCBs in the adipose tissue; etc. Indeed, body burdens can be the result of both long-term and short-term storage.

Cancer: Refers to the development of a malignant tumor or abnormal formation of tissue. It is a disease characterized by malignant, uncontrolled invasive growth of body tissue cells. *Tumor* – an uncontrolled growth of tissue cells forming an abnormal mass. *Benign tumor* – a tumor that does not spread to a secondary localization, but may impair normal biological function through obstruction or may progress to malignancy later. *Malignant tumor* – is an abnormal growth of tissue that can invade adjacent or distant tissues. *Neoplasm* – an abnormal growth of tissues, that may be benign or malignant. This relates to a genetically altered, relatively autonomous growth of tissue; it is composed of abnormal cells, the growth of which is more rapid than that of other tissues and is not coordinated with the growth of other tissues.

Cancer slope factor (CSF) (also, *slope factor, SF, cancer potency factor, CPF, or cancer potency slope, CPS*): Health effect information factor commonly used to evaluate health hazard potentials for carcinogens. It is a plausible upper-bound estimate of the probability of a response per unit intake of a chemical over a lifetime – represented by the slope of the dose-response curve in the low-dose region. This parameter is used to estimate an upper-bound probability for an individual to develop cancer as a result of a lifetime of exposure to a particular level of a carcinogen. Generally, cancer slope factors are available from databases such as US EPA's Integrated Risk Information System (IRIS).

Carcinogen: A cancer-producing chemical or substance. It represents any substance that is capable of inducing a cancer response in living organisms. *Co-carcinogen* – an agent that is not carcinogenic on its own, but enhances the activity of another agent that is carcinogenic when administered together with the carcinogen. *Complete carcinogen* – chemicals that are capable of inducing tumors in animals or humans without supplemental exposure to other agents; the term 'complete' refers to the three stages of carcinogenesis (namely: initiation, promotion, and progression) that need to be present in order to induce a cancer.

Carcinogenesis: The process by which normal tissue becomes cancerous – i.e., the production of cancer, most likely via a series of steps – viz., initiation, promotion, and progression. The carcinogenic event modifies the genome and/or other molecular control mechanisms of the target cells, giving rise to a population of altered cells. *Initiator* – a chemical/substance or agent capable of starting but not necessarily completing the process of producing an abnormal uncontrolled growth of tissue, usually by altering a cell's genetic material. Initiated cells may or may not be transformed into tumors. *Initiation* – refers to the first stage of carcinogenesis, and consists of the subtle alteration of DNA or proteins within target cells by carcinogens, which renders the cell capable of becoming cancerous. *Promoter* – a chemical that, when administered after an initiator has been given, promotes the change of an initiated cell – culminating in a cancer. This represents an agent that is not carcinogenic in itself, but when administered after an initiator of carcinogenesis, serves to dramatically potentiate the effect of a low dose of a carcinogen – by stimulating the clonal expansion of the initiated cell to produce a neoplasm. *Promotion* – the second hypothesized stage in a multistage process of cancer development, consisting of the conversion of initiated cells into tumorigenic cells; this occurs when initiated cells are acted upon by promoting agents to give rise to cancer.

Carcinogenic: Capable of causing, and tending to produce or incite cancer in living organisms. That is, a substance able to produce malignant tumor growth.

Carcinogenicity: The power, ability, or tendency of a chemical, physical, or biological agent to produce cancerous tissues from normal tissue – in order to cause cancer in a living organism.

Case-control study: A retrospective epidemiologic study in which individuals with the disease under study (cases) are compared with individuals without the disease (controls) in order to contrast the extent of exposure in the diseased group with the extent of exposure in the controls. The study consists of an investigation in which select cases with a specific diagnosis (such as cancer) are compared to individuals from the same or related population(s) without that specific diagnosis. In chemical exposure problems, this type of (retrospective) epidemiologic study looks back in time at the exposure history of individuals who have the health effects (cases) and at a group who do not (controls), in order to ascertain whether they differ in the proportion exposed to the chemical(s) under investigation.

Cell: The basic units of structure and function in a living organism. It consists of the complex assemblages of atoms, molecules, and complex molecules.

Central nervous system (CNS): The part of the nervous system that includes the brain and the spinal cord, and their connecting nerves.

Chronic: Of long-term duration – i.e., occurring over a long period of time (usually more than 1 year). *Chronic daily intake (CDI)* – refers to the receptor exposure, expressed in mg/kg-day, averaged over a long period of time. *Chronic effect* –

refers to an effect that is manifest after some time has elapsed from an initial exposure to a substance. *Chronic exposure* – refers to the long-term, low-level exposure to chemicals, i.e., the repeated exposure or doses to a chemical over a long period of time (usually lasting six months to a lifetime). It may cause latent damage that does not appear until a later period in time. *Chronic toxicity* – refers to the occurrence of symptoms, diseases, or other adverse health effects that develop and persist over time, following exposure to a single dose or multiple doses of a chemical delivered over a relatively long period of time. This represents the adverse health effects that are of a long and continuous duration – generally resulting in cellular changes that are irreversible. Chronic toxicity usually consists of a prolonged health effect that may not become evident until many years after exposure.

Cleanup: Actions taken to abate a situation involving the release or threat of release of contaminants that could potentially affect human health and/or the environment. This typically involves a process to remove or attenuate contamination levels, in order to restore the impacted media to an 'acceptable' or usable condition. *Cleanup level* – refers to the contaminant concentration goal of a remedial action, i.e., the concentration of media contaminant level to be attained through a remedial action.

Cluster investigation: A review of an unusual numbers (real or perceived) of health events (e.g., reports of cancer) grouped together in time and location. Cluster investigations are designed to confirm case reports; determine whether the reported cases represent an unusual disease occurrence; and, if possible, explore possible causes and environmental factors that are producing the cases.

*Cohort study (*or, *Prospective study):* An epidemiologic study comparing those with an exposure of interest to those without the exposure. It involves observing subjects in differently exposed groups and comparing the incidence of symptoms; these two cohorts are then followed over time to determine the differences in the rates of disease between the exposure subjects. The *relative risk* (or *risk ratio*) – defined as the rate of disease among the exposed divided by the rate of the disease among the unexposed – provides a relative measure of the difference in risk between the exposed and unexposed populations in a cohort study. A relative risk of two means that the exposed group has twice the disease risk as the unexposed group.

Community health investigation: Medical or epidemiologic evaluation of a descriptive health information about individual persons or a population group that is used to evaluate and determine observed health concerns, and to assess the likelihood that such prevailing conditions may be linked to exposure to hazardous substances.

Compliance: To conduct or implement in accordance with stipulated legislative or regulatory requirements.

Confidence interval (CI): A statistical parameter used to specify a range, and the probability that an uncertain quantity falls within this range. *Confidence limits* – the upper and lower boundary values of a range of statistical probability numbers that define the CI. *95 percent confidence limits (95% CL)* – refers to the limits of the range of values within which a single estimation will be included 95% of the time. For large samples sizes (i.e., $n > 30$),

$$95\% CL = X_m \pm \frac{1.96 s}{\sqrt{n}}$$

where CL is the confidence level, and s is the estimate of the standard deviation of the mean (X_m). For a limited number of samples $(n \leq 30)$, a confidence limit or confidence interval may be estimated from,

$$95\% CL = X_m \pm \frac{ts}{\sqrt{n}}$$

where t is the value of the Student t-distribution (refer to standard textbooks of statistics) for the desired confidence level and degrees of freedom, $(n\text{-}1)$.

Confounder (or, confounding factor): A condition or variable that may be a factor in producing the same response as the agent under study. This association between the exposure of interest and the confounder (a true risk factor for disease) may make it falsely appear that the exposure of interest is associated with disease. The effects of such factors may be discerned through careful design and analysis.

Consequence: The impacts resulting from the response associated with specified exposures, or loading or stress conditions.

Conservative assumption: Used in exposure and risk assessment, this expression refers to the selection of assumptions (when real-time data are absent) that are unlikely to lead to under-estimation of exposure or risk. Conservative assumptions are those which tend to maximize estimates of exposure or dose – such as choosing a value near the high end of the concentration or intake rate range. (See also, *worst case.*)

Contact rate: Amount of an exposure or environmental medium (e.g., air, groundwater, surface water, soil, cosmetics, etc.) contacted per unit time or per event (e.g., liters of groundwater ingested or milligrams of soil ingested per day).

Contaminant (or, pollutant): Any substance or material that enters a system (the environment, human body, food, etc.) where it is not normally found – as, e.g., any undesirable substance that is not naturally-occurring and therefore not normally found in the environmental media of concern. This typically consists of any potentially harmful physical, chemical, biological, or radiological agent occurring in the environment, in consumer products, or at the workplace as a result of human activities. Such materials can potentially have adverse impacts upon exposure to an organism and/or could adversely impact public health and the environment simply by their presence in the ambient setting. *Contaminant release* – refers to the ability of a contaminant to enter into other environmental media/matrices (e.g., air, water or soil) from its source(s) of origin. *Contaminant migration* – refers to the movement of a contaminant from its source through other matrices/media such as air, water, or soil. *Contaminant migration pathway* – is the path taken by the contaminants as they travel from the contaminated source through various environmental media.

Control group (or, reference group): A group used as the baseline for comparison in epidemiologic or laboratory studies. This group is selected because it either lacks the disease of interest (i.e., there is absence of an adverse response) (case-control group), or lacks the exposure of concern (i.e., there is absence of exposure to agent) (cohort study).

Corrective action: Action taken to correct a problematic situation. A typical/common example involves the remediation of chemical contamination in soil and groundwater.

Data quality objectives (DQOs): Qualitative and quantitative statements developed by analysts to specify the quality of data that, at a minimum, is needed and expected from a particular data collection activity (or hazard source characterization activity). This is determined based on the end use of the data to be collected.

Decision analysis: A process of systematic evaluation of alternative solutions to a problem where the decision is made under uncertainty. The approach is comprised of a conceptual and systematic procedure for analyzing complex sets of alternatives in a rational manner so as to improve the overall performance of a decision-making process.

Decision framework: Management tool designed to facilitate rational decision-making.

Degradation: The physical, chemical or biological breakdown of a complex compound into simpler compounds and byproducts.

Delayed toxicity: The development of disease states or symptoms a long-time (i.e., many months or years) after exposure to a toxicant.

de Minimus: A legal doctrine dealing with levels associated with insignificant versus significant issues relating to human exposures to chemicals that present very low risk. This represents the level below which one need not be concerned – and, therefore, is of no public health consequence.

Dermal absorption: The absorption of materials/substances through the skin. *Dermally absorbed dose* – the amount of the applied material (the dose) which becomes absorbed into the body.

Dermal adsorption: The process by which materials come into contact with the skin surface, but are then retained and adhered to the permeability barrier without being taken into the body.

Dermal exposure: Exposure of an organism or receptor through skin adsorption and possible absorption.

Dermatotoxicity: Adverse effects produced by toxicants contacting or entering the skin of an organism.

Detection limit (DL): The minimum concentration or weight of analyte that can be detected by a single measurement with a known confidence level. *Instrument detection limit (IDL)* – represents the lowest amount that can be distinguished from the normal 'noise' of an analytical instrument (i.e., the smallest amount of a chemical detectable by an analytical instrument under ideal conditions). *Method detection limit (MDL)* – represents the lowest amount that can be distinguished from the normal 'noise' of an analytical method (i.e., the smallest amount of a chemical detectable by a prescribed or specified method of analysis).

Developmental toxicity: Adverse effects on the developing organism that may result from exposure prior to conception (in either parent), during prenatal development, or post-natally until the time of sexual maturation. The major manifestations of developmental toxicity include death of the developing organism, structural abnormality, altered growth, and functional deficiency.

Diffusion: The migration of molecules, atoms, or ions from one fluid to another in a direction tending to equalize concentrations.

Digestive system: The organ system that is responsible for the conversion of ingested food into simple molecules that can be absorbed by the blood and lymph, and then used by cells. It is made up of the digestive tract and related accessory organs, such as the liver and pancreas.

Dispersion: The overall mass transport process resulting from both molecular diffusion (which always occurs if there is a concentration gradient in the system) and the mixing of the constituent due to turbulence and velocity gradients within the system.

DNA (Deoxyribonucleic acid): A nucleic acid molecule with the shape of a double helix that is present in chromosomes and that contains the genetic information. It is the repository of hereditary characteristics (genetic code).

Dose: A measure of the amount of a chemical substance received or taken in by potential receptors upon exposure – expressed as an amount of exposure (in mg) per unit body weight of the receptor (in kg). *Total dose* – is the sum of chemical doses received by an individual from multiple exposure sources in a given interval as a result of interaction with all exposure or environmental media that contain the chemical substances of concern. Units of dose and total dose (mass) are often converted to units of mass per volume of physiological fluid or mass of tissue. *Absorbed dose* (also called, *internal dose*) – is the amount of a chemical substance actually entering an exposed organism via the lungs (for inhalation exposures), the gastrointestinal tract (for ingestion exposures), and/or the skin (for dermal exposures). It represents the amount penetrating the exchange boundaries of the organism after contact – i.e., the amount of a substance penetrating across an absorption barrier (represented by the exchange boundaries such as the skin, lung, and gastrointestinal tract) of an organism, via either physical or biological processes. *Effective dose (ED_{10})* – the dose corresponding to a 10% increase in an adverse effect, relative to the control response. *Lower limit on effective dose (LED_{10})* – the 95% lower confidence limit of the dose of a chemical needed to produce an adverse effect in 10 percent of those exposed to the chemical, relative to control.

Dose-response: The quantitative relationship between the dose of a chemical substance and an effect caused by exposure to such substance. *Dose-response relationship* – refers to the relationship between a quantified exposure (dose), and the proportion of subjects demonstrating specific biological changes (response). *Dose-response curve* – a graphical representation of the relationship between the degree of exposure to a chemical substance and the observed or predicted biological effects or response. *Dose-response assessment* – consists of a determination of the relationship between the magnitude of an administered, applied, or internal dose and a specific biological response. Response can be expressed as measured or observed incidence, percent response within groups of subjects (or populations), or as the probability of occurrence within a population. *Dose-response evaluation* – refers to the process of quantitatively evaluating toxicity information, and then characterizing the relationship between the dose of a chemical administered or received and the incidence of adverse health effects in the exposed population.

Effect: The response produced due to a chemical contacting episode. *Local effect* – refers to the response that occurs at the site of first contact. *Systemic effect* – refers to the response that requires absorption and distribution of the chemical, and this tends to affect the receptor at sites farther away from the entry point(s).

Embryotoxicity: Any toxic effect on the *conceptus* as a result of prenatal exposure during the embryonic stages of development. These effects may include malformations and variations, altered growth, *in-utero* death, and altered postnatal function.

Endangerment assessment: A case-specific risk assessment of the actual or potential danger to human health and welfare, and also the environment, that is associated with the release of hazardous chemicals into various exposure or environmental media.

Endpoint (toxic): An observable or measurable biological or biochemical effect (e.g., metabolite concentration in a target tissue) used as an index of the impacts of a chemical on a cell, tissue, organ, organism, etc. This is usually referred to as *toxicological endpoint* or *physiological endpoint* in the context of chemical toxicity assessments.

Environmental fate: The 'destination' or 'destiny' of a chemical after release or escape from a given source into the environment, and following transport through various environmental compartments. For example, in a contaminated land situation, it may consist of the movement of a chemical through the environment by transport in air, water, sediment, and soil – culminating in exposures to living organisms. It represents the disposition of a material in the various environmental compartments (e.g., soil, sediment, water, air, biota) as result of transport, transformation, and degradation.

Environmental medium: A part of the environment for which reasonably distinct boundaries can be specified. Typical environmental media addressed in chemical risk assessments may include air, surface water, groundwater, soil, sediment, fruits, vegetables, meat, dairy, and fish – or indeed any other parts of the environment that could contain contaminants of concern.

Environmental toxicant: Agents present in the surroundings of an organism that are harmful to the health of such organisms.

Epidemiology: The study of the occurrence of disease, injury and other health effect patterns in human populations, as well as the causes and means of prevention or preventative strategies. An *epidemiological study* often compares two groups of people who are alike except for one factor – such as exposure to a chemical, or the presence of a health effect; the investigators endeavor to determine if any particular factor(s) is associated with the observed health effect(s). *Descriptive epidemiology* – consists of a study of the amounts and distributions of diseases within a population by person, place, and time.

Erythrocytes: Red blood cells.

Estimated exposure dose (EED): The measured or calculated dose to which humans are likely to be exposed – considering all sources and routes of exposure.

Event-tree analysis: A procedure, utilizing deductive logic, often used to evaluate a series of events that lead to an upset or accident scenario. It offers a systematic approach for analyzing the types of exposure scenarios that can result from a chemical exposure problem.

Excess (or incremental) lifetime risk: The additional or extra risk (above normal background rate) incurred over the lifetime of an individual as a result of exposure to a toxic substance. (See also, *attributable risk.*)

Exposure: The situation of receiving a dose of a substance, or coming in contact with a hazard. It represents the contact of an organism with a chemical, biological, or physical agent available at the exchange boundary (e.g., lungs, gut, skin) during a specified time period. *Exposure conditions* – refer to factors (such as location, time, etc.) that may have significant effects on an exposed population's response to a hazard situation. *Exposure duration* – refers to the length of time that a potential receptor is exposed to the hazards or contaminants of concern in a defined exposure scenario. *Exposure event* – refers to an incident of contact with a chemical or physical agent, usually defined by time (e.g., number of days or hours of contact). *Exposure frequency* – refers to the number of times (per year or per event) that a potential receptor would be exposed to contaminants of concern in a defined exposure scenario. *Exposure parameters (or factors)* – refer to the variables used in the calculation of intake (e.g., exposure duration, breathing rate, food ingestion rate, and average body weight); these may consist of standard factors that may be needed to calculate a potential receptor's exposure to toxic chemicals (in the environment). *Exposure point* – refers to a location of potential contact between an organism and a hazardous (viz., biological, chemical or physical) agent. *Exposure route* – refers to the avenue (such as inhalation, ingestion, and dermal contact) by which an organism contacts a chemical. It represents the way in which a potential receptor may contact a chemical substance; for example, drinking (ingestion) and bathing (skin contact) are two different routes of exposure to contaminants that may be found in water.

Exposure assessment: The qualitative or quantitative estimation, or the measurement, of the dose or amount of a chemical to which potential receptors have been exposed, or could potentially be exposed to. This process comprises of the determination of the magnitude, frequency, duration, route, and extent of exposure (to the chemicals or hazards of potential concern).

Exposure (point) concentration (EPC): The concentration of a chemical (in its transport or carrier medium) at the point of receptor contact.

Exposure investigation: The collection and analysis of site-specific information to determine if human populations have been exposed to hazardous substances. The site-specific information may include environmental sampling, exposure-dose reconstruction, biologic or biomedical testing, and evaluation of medical information. The information from an exposure investigation can be used to support a complete public health risk assessment and subsequent risk management programs.

Exposure pathway: The course a chemical, biological, or physical agent takes from a source to an exposed population or organism. It describes a unique mechanism by which an individual or population is exposed to chemical, biological, or physical agents at or originating from a contaminant release source.

Exposure scenario: A set of conditions or assumptions about hazard sources, exposure pathways, concentrations of chemicals, and potential receptors that aids in the evaluation and quantification of exposure in a given situation. *Potentially exposed* – refers to the situation where valid information, usually analytical environmental data, indicates the presence of chemical(s) of a public health concern in one or more environmental media contacting humans (e.g., air, drinking water, soil, food chain, surface water), and where there is evidence that some of the target populations have well-defined route(s) of exposure (e.g., drinking contaminated water, breathing contaminated air, contacting contaminated soil, or eating contaminated food) associated with them. Although actual exposure is generally not confirmed for a 'potentially exposed' receptor, this type of exposure scenario would typically have to be adequately evaluated during the exposure assessment.

Extrapolation: An estimate of response or quantity at a point outside the range of the experimental data, generally via the use of a mathematical model. This consists of the estimation of unknown numerical values of an empirical (measured) function by extending or projecting from known values/observations to points outside the range of data that were used to calibrate the function. In chemical exposure situations, this may comprise of the estimation of a measured response in a different species, or by a different route than that used in the experimental study of interest (i.e., species-to-species; route-to-route; acute-to-chronic; high-to-low dose; etc.). For instance, the quantitative risk estimates for carcinogens are generally low-dose extrapolations based on observations made at higher doses.

Frank effect level (FEL): A level of exposure or dose which produces irreversible adverse effects (such as irreversible functional impairment or mortality) and a statistically or biologically significant increase in frequency or severity between those exposed and those not exposed (i.e., the appropriate control).

Fugitive dust: Atmospheric dust arising from disturbances of particulate matter exposed to the air. Fugitive dust emissions typically consist of the release of chemicals from contaminated surface soil into the air, attached to dust particles.

Genetic toxicity (or, genotoxicity): An adverse event resulting in damage to genetic material; damage may occur in exposed individuals or may be expressed in subsequent generations. *Genotoxic –* is a broad term that usually refers to a chemical that has the ability to damage DNA or the chromosomes.

Geographic Information System (GIS): Computer-based tool used to capture, manipulate, process, and display spatial or geo-referenced data for solving complex resource, environmental, and social problems. GIS is indeed a rapidly developing technology for handling, analyzing, and modeling geographic information. It is not a source of information *per se*, but only a way to manipulate information. When the manipulation and presentation of data relates to geographic locations of interest, then our understanding of the real world is enhanced.

Geometric mean: A statistical measure of the central tendency for data from a positively skewed distribution (viz., lognormal), given by:

$$X_{gm} = [(X_1)(X_2)(X_3)...(X_n)]^{1/n}$$

or,

$$X_{gm} = anti\log\left\{\frac{\sum_{i=1}^{n}\ell n[X_i]}{n}\right\}$$

Hazard: That innate character which has the potential for creating adverse and/or undesirable consequences. It represents the inherent adverse effect that a chemical or other object poses – and defines the chance that a particular substance will have an adverse effect on human health or the environment under a particular set of circumstances that creates an exposure to that substance. Thus, hazard is simply a source of risk that does not necessarily imply potential for occurrence; a hazard produces risk only if an exposure pathway exists, and if exposures create the possibility of adverse consequences.

Hazard assessment: The process of determining whether exposure to an agent can cause an increase in the incidence of a particular adverse health effect (e.g., cancer, birth defects, etc.), and whether the adverse health effect is likely to occur in the target receptor populations potentially at risk. This involves gathering and evaluating data on types of injury or consequences that may be produced by a hazardous situation or substance.

Hazard identification: The systematic identification of potential accidents, upset conditions, etc. – consisting of a process of determining whether or not, for instance, a particular substance or chemical is causally linked to particular health effects. Thus, the process involves determining whether exposure to an agent or hazard can cause an increase in the incidence of a particular adverse response or health effect in receptors of interest.

Hazard quotient (HQ): The ratio of a single substance exposure level for a specified time period to the 'allowable' or 'acceptable' intake limit/level of that substance derived from a similar exposure period. For a particular chemical and mechanism of intake (e.g., oral, dermal, inhalation), the hazard quotient is defined by the ratio of the average daily dose (ADD) of the chemical to the reference dose (RfD) for that chemical; or the ratio of the exposure concentration to the reference concentration (RfC). A value of less than 1.0 indicates the risk of exposure is likely insignificant; a value greater than 1.0 indicates a potentially significant risk. *Hazard index (HI)* – is the sum of several hazard quotients (HQs) for multiple substances and/or multiple exposure pathways.

Hazardous substance: Any substance that can cause harm to human health or the environment whenever excessive exposure occurs. *Hazardous waste* – is that byproduct which has the potential to cause detrimental effects on human health and/or the environment if it is not managed in an efficient manner. Typically, this refers to wastes that are ignitable, explosive, corrosive, reactive, toxic, radioactive, pathological, or has some other property that produces substantial risk to life.

Heavy metals: Members of a group of metallic elements that are recognized as toxic, and are generally bioaccumulative. The term arises from the relatively high atomic weights of these elements.

Hematotoxicity: Adverse effects or diseases in the blood as produced by toxicants contacting or entering an organism. *Hematotoxins* – refer to agents that produce toxic symptoms or diseases in the blood of an organism.

Hemoglobin: The respiratory compound of red blood cells. It consists of the oxygen-carrying protein in red blood cells.

Hepatotoxicity: Adverse effects or diseases in the liver as produced by toxicants contacting or entering an organism. *Hepatotoxins* – refer to agents that produce toxic symptoms or diseases in the liver of an organism.

'Hot-spot': Term often used to denote zones where contaminants are present at much higher concentrations than the immediate surrounding areas. It tends to represent a relatively small area that is highly contaminated within a study area.

Human equivalent dose: A dose that, when administered to humans, produces effects comparable to that produced by a dose in experimental animals.

Human health risk: The likelihood (or probability) that a given exposure or series of exposures to a hazardous substance will cause adverse health impacts on individual receptors experiencing the exposures.

Hydrocarbon: Organic chemicals/compounds, such as benzene, that contain atoms of both hydrogen and carbon.

Hydrophilic: Having greater affinity for water – or 'water-loving'. Hydrophilic compounds tend to become dissolved in water.

Hydrophobic: Tending *not* to combine with water – or less affinity for water. Hydrophobic compounds tend to avoid dissolving in water and are more attracted to non-polar liquids (e.g., oils) or solids.

Incidence: The number of new cases of a disease that develop within a specified population over a specified period of time. *Incidence rate* – is the ratio of new cases within a population to the total population at risk given a specified period of time.

Individual excess lifetime cancer risk: An upper-bound estimate of the increased cancer risk, expressed as a probability that an individual receptor could expect from exposure over a lifetime.

Ingestion exposure: An exposure type whereby chemical substances enter the body through the mouth, and into the gastrointestinal system. After ingestion, chemicals can be absorbed into the blood and distributed throughout the body.

Inhalation exposure: The intake of a substance by receptors through the respiratory tract system. Exposure may occur from inhaling contaminants – which become deposited in the lungs, taken into the blood, or both.

Initiating event: The specific trigger action that results in a risk being incurred.

Intake: The amount of material inhaled, ingested or dermally absorbed during a specified time period. It is a measure of exposure – expressed in mg/kg-day.

Integrated Risk Information System (IRIS): A US EPA database containing verified toxicity parameters (viz., reference doses [RfDs] and slope factors [SFs]), and also up-to-date health risk and EPA regulatory information for numerous chemicals. It does indeed serve as a very important source of toxicity information for health and environmental risk assessment. (See Appendix B.)

Interspecies: Between different species. *Interspecies dose conversion* – the process of extrapolating from animal doses to human equivalent doses.

Intraspecies: Within a particular species.

In vitro: Processes or reactions occurring in an artificial environment – outside of a living organism. For example, *in vitro* laboratory studies refer to studies conducted in a laboratory setup that do not use live animals (i.e., tests conducted outside the whole body in an artificially maintained environment) – as in a test tube, culture dish, or bottle.

In vivo: Processes or reactions occurring within a living organism. For example, *in vivo* laboratory studies refer to those tests conducted using live animals, or whole living body – i.e., tests conducted within the whole living body.

Latency period: A seemingly inactive period – such as the time between the initial induction of a health effect from first exposures to a chemical agent and the manifestation or detection of actual health effects of interest. (Often used to identify the period between exposure to a carcinogen and development of a tumor.)

LC$_{50}$ (Mean lethal concentration): The lowest concentration of a chemical in air or water that will be fatal to 50% of test organisms living in that media, under specified conditions.

LD$_{50}$ (Mean lethal dose): The single dose (ingested or dermally absorbed) that is required to kill 50% of a test animal group. Also represents the median lethal dose value.

Leukocytes: White blood cells.

Lifetime average daily dose (LADD): The exposure, expressed as mass of a substance contacted and absorbed per unit body weight per unit time, averaged over a lifetime. It is usually used to calculate carcinogenic risks – and takes into account the fact that, whereas carcinogenic risks are determined with an assumption of lifetime exposure, actual exposures may be for a shorter period of time. Indeed, the

LADD may be derived from the ADD – to reflect the difference between the length of the exposure period and the exposed person's lifetime, as follows:

$$LADD = ADD \times \frac{\text{Exposure period}}{\text{Lifetime}}$$

Lifetime exposure: The total amount of exposure to a substance or hazard that a potential receptor would be subjected to in a lifetime.

Lifetime risk: Risk that results from lifetime exposure to a chemical substance or hazard.

Linear dose-response: A pattern of frequency or severity of biological response that varies proportionately with the amount of dose of an agent. *Non-linear dose-response* – shows a pattern of frequency or severity of biological response that does not vary proportionately with the amount of dose of an agent. When mode of action information indicates that responses may not follow a linear pattern below the dose range of the observed data, non-linear methods for determining risk at low dose may be justified.

Lipophilic: The property of a chemical/substance to have a strong affinity for lipid, fats, or oils – i.e., being highly soluble in nonpolar organic solvents. Also, refers to a physicochemical property that describes a partitioning equilibrium of solute molecules between water and an immiscible organic solvent that favors the latter.

Lipophobic: The property of a chemical to be antagonistic to lipid – i.e., incapable of dissolving in or dispersing uniformly in fats, oils, or nonpolar organic solvents.

LOAEC (Lowest-observed-adverse-effect-concentration): The lowest concentration in an exposure medium in a study that is associated with an adverse effect on the test organisms.

LOAEL (Lowest-observed-adverse-effect level): The lowest dose or exposure level, expressed in mg/kg body weight/day, at which adverse effects are noted in the exposed population. It represents the chemical dose rate or exposure level causing statistically or biologically significant increases in frequency or severity of adverse effects between the exposed and control groups. $LOAEL_a$ – refers to the LOAEL values adjusted by dividing by one or more safety factors.

Local effect: A biological response occurring at the site of first contact between the toxic substance and the organism.

LOEL (Lowest-observed-effect-level): The lowest exposure or dose level to a substance at which effects are observed in the exposed population; the effects may or may not be serious. In a given study, it is the lowest dose or exposure level at which a statistically or biologically significant effect is observed in the exposed population compared with an appropriate unexposed control group.

Margin-of-exposure (MOE): Defined by the ratio of the no-observed-adverse-effect-level (NOAEL) to the estimated human exposure.

Matrix (or, medium): The predominant material comprising the environmental or exposure sample being investigated (e.g., food, cosmetics, soils, water, and air).

Maximum daily dose (MDD): The maximum dose calculated for the duration of receptor exposure – and used to estimate risks for subchronic or acute non-carcinogenic effects from chemical exposures.

Maximum Likelihood Estimate (MLE): Statistical method for estimating model parameters. It generally provides a mean or central tendency estimate, as opposed to a confidence limit on the estimate.

Minimal risk level (MRL): An estimate of daily human exposure to a substance that is likely to be without an appreciable risk of adverse (non-cancer) effects over a specified duration of exposure. *MRLs* are derived when reliable and sufficient data exist to identify the target organ(s) of effect or the most sensitive health effect(s) for a specific duration via a given route of exposure. *MRLs* are based on non-cancer health effects only – and can be derived for acute, intermediate, and chronic duration exposures by the inhalation and oral routes.

Mitigation: The process of reducing or alleviating a hazard or problem situation.

Modeling: Refers to the use of mathematical equations to simulate and predict real events and processes. A *model* – refers to a mathematical function with parameters that can be adjusted so that the function closely describes a set of empirical data. A *mechanistic model* – usually reflects observed or hypothesized biological or physical mechanisms, and has model parameters with real world interpretation. In contrast, *statistical or empirical models* selected for particular numerical properties are fitted to a given data – and model parameters in this case may or may not have real world interpretation. When data quality is otherwise equivalent, extrapolation from mechanistic models (e.g., biologically based dose-response models) often carries higher confidence than extrapolation using empirical models (e.g., logistic model – representing a dose-response model used for low-dose extrapolation).

Monitoring: Process involving the measurement of concentrations of chemicals in environmental media, or in tissues of human receptors and other biological organisms. *Biological monitoring* – consists of measuring chemicals in biological materials (e.g., blood, urine, breath, etc.) to determine whether chemical exposure in living organisms (e.g., humans, animals, or plants) has occurred.

Monte Carlo simulation: A process in which outcomes of events or variables are determined by selecting random numbers – subject to a defined probability law. The technique is used to obtain information about the propagation of uncertainty in mathematical simulation models. The *Monte Carlo technique* involves a repeated random sampling from the distribution of values for each of the parameters in a calculation (e.g., lifetime average daily dose), in order to derive a distribution of estimates (of exposures) in the population.

Morbidity: Illness or disease state. *Morbidity rate* – is the number of illnesses or cases of disease in a population.

Mortality: The number of individual deaths in a population.

Multi-hit models: Dose-response models that assume more than one exposure to a toxic material is necessary before effects are manifested.

Multi-stage models: Dose-response models that assume there are a given number of biological stages through which the carcinogenic agent must pass, without being deactivated, for cancer to occur. The multistage model consists of a mathematical function that is used to extrapolate the probability of cancer from animal bioassay data. A *linearized multistage model* – is a derivation of the multistage model for which the data are assumed to be linear at low doses. The linearized multistage procedure consists of a modification of the multistage model, used for estimating carcinogenic risk – and that incorporates a linear upper bound on extra risk for exposures below the experimental range. A *multistage Weibull model* – is a dose-response model for low-dose extrapolation that includes a term for decreased survival time associated with tumor incidence.

Mutagen: A substance that can cause an alteration in the structure of the DNA of an organism. *Mutagenic compounds* – have the ability to induce structural changes in genetic material.

Neuron: Nerve cells.

Neurotoxicity: Adverse effects or diseases in the nervous system as produced by toxicants contacting or entering an organism – i.e., the hazard effects that are poisonous to the nerve cells. *Neurotoxic* – having toxic effect on any aspect of the central or perpheral nervous system. *Neurotoxins* – refers to agents that have the ability to damage nervous tissues (i.e., produce toxic symptoms or diseases in the nervous system of an organism).

NOAEC (No-observed-adverse-effect-concentration): The highest concentration in an exposure medium in a study that is *not* associated with an adverse effect on the test organisms.

NOAEL (No-observed-adverse-effect level): The highest level at which a chemical causes no observable adverse effect in the species being tested or the exposed population. It represents chemical intakes or exposure levels at which there are no statistically or biologically significant increases in frequency or severity of adverse effects between the exposed and control groups – meaning statistically significant effects are observed at this level, but they are not considered to be adverse nor precursors to adverse effects. $NOAEL_a$ – refers to NOAEL values adjusted by dividing by one or more safety factors.

NOEL (No-observed-effect level): The highest level at which a chemical causes no observable changes in the species or exposed populations under investigation. In a given study, this represents the dose rate or exposure level of chemical at which there are no statistically or biologically significant increases in frequency or severity of any effects between the exposed and control groups.

Nonparametric statistics: Statistical techniques whose application is independent of the actual distribution of the underlying population from which the data were collected.

One-hit model: A dose-response mathematical model that assumes a single biological event can initiate a response. It is represented by a dose-response model of the form $P(d) = [1 - e^{-\lambda d}]$, where $P(d)$ is the probability of cancer from a lifetime continuous exposure at a dose rate, d, and λ is a constant. The one-hit model is based on the concept that a tumour can be induced after a single susceptible target or receptor has been exposed to a single effective unit dose of an agent. *Gamma (Multi-hit) model* – is a generalization of the one-hit model for low-dose extrapolation. It defines the probability, $P(d)$, that an individual will respond to lifetime, continuous exposure to dose, d, by the use of a gamma function.

Organ: A group of several tissue types that unite to form structures to perform a special function within an organism.

Particulate matter: Small/fine, discrete, solid or liquid particles/bodies, especially those suspended in a liquid or gaseous medium – such as dust, smoke, mist, fumes, or smog suspended in air or atmospheric emissions.

Partitioning: The separation or division of a substance into two or more compartments. This consists of a chemical equilibrium condition in which a chemical's concentration is apportioned between two different phases, according to the partition coefficient. *Partition coefficient* – is a term used to describe the relative amount of a substance partitioned between two different phases, such as a solid and a liquid or a liquid and a gas. It is the ratio of the chemical's concentration in one phase to its concentration in the other phase. For instance, a *blood-to-air partition coefficient* is the ratio of a chemical's concentration between blood and air when at equilibrium.

Pathway: Any specific route via which environmental chemicals or stressors take in order to travel away from the source in order to reach potential receptors or individuals.

PEL (Permissible exposure limit): A maximum (legally enforceable) allowable level for a chemical in workplace air.

Persistence: Attribute of a chemical substance which describes the length of time that such substance remains in a particular environmental compartment before it is physically removed, chemically modified, or biologically transformed.

pH: A measure of the acidity or alkalinity of a material or medium.

Pharmacokinetics: Study of changes in toxicant or substance characteristics (e.g., via absorption, distribution, metabolism/biotransformation, and excretion) in parts of the body of an organism over time.

Physiologically-based pharmacokinetic (PB-PK) models: Type of models that find usefulness especially in predicting specific tissue dose under a range of exposure conditions. Physiologically-based compartmental models are used in characterizing the pharmacokinetic behavior of a chemical; available data on blood flow rates, and on metabolic and other processes that the chemical undergoes within each compartment are used to construct a mass-balance framework for the PB-PK model.

Pica: The behavior in children and toddlers (usually under age 6 years) involving the intentional eating/mouthing of large quantities of dirt and other objects.

Plume (or contaminant plume): A zone containing predominantly dissolved (or vapor phase) and sorbed contaminants that usually originates from the contaminant or pollution source areas. It refers to an area of chemicals in a particular medium (such as air or groundwater), moving away from its source in a long band or column. A plume can be a column of smoke from a chimney or chemicals moving with groundwater. Common examples may consist of a body of contaminated groundwater or vapor originating from a specific source and spreading out due to influences of environmental factors such as local groundwater conditions or soil vapor flow patterns, or wind directions.

PM-10, PM_{10}: Particulate matter with physical/aerodynamic diameter <10 μm. It represents the respirable particulate emissions. *Aerodynamic diameter* – is the diameter of a particle with the same settling velocity as a spherical particle with unit density ($1g/cm^3$); this parameter, which depends on particle density, is often used to describe particle size. *Respirable fraction (of dust)* (also, *respirable particulate matter)* – is the fraction of dust particles that enter the respiratory system because of their size distribution; generally, the size of these particles correspond to aerodynamic diameter of ≤ 10 μm.

Pollution (or contamination): Refers to the release of a physical, chemical, or biological agent into an environment; this typically has the potential to impact human and/or ecological health.

Population-at-risk (PAR): A population group or subgroup that is more susceptible or sensitive to a hazard or chemical exposure than is the general population.

Population excess cancer burden: An upper-bound estimate of the increase in cancer cases in a population as a result of exposure to a carcinogen.

Potency: A measure of the relative toxicity of a chemical.

Potentiation (of chemical effects): The effect of a chemical that enhances the toxicity of another chemical.

ppb (parts per billion): An amount of substance in a billion parts of another material – also expressed by μg/kg or μg/liter, and equivalent to (1×10^{-9}). (NB: A billion is often used to represent a thousand millions, i.e., 10^{-9}, in some places, such as in the USA and France; whereas a billion represents a million millions, i.e., 10^{-12}, in some other places, such as in the UK and Germany.)

ppm (parts per million): An amount of substance in a million parts of another material – also expressed by mg/kg or mg/liter, and equivalent to (1×10^{-6}).

ppt (parts per trillion): An amount of substance in a trillion parts of another material – also expressed by ng/kg or ng/liter and equivalent to (1×10^{-12}). (NB: A trillion is often used to represent a million times a million or a thousand billions, i.e., 10^{-12}, in some places, such as in the USA and France; whereas a trillion represents a million billions, i.e., 10^{-18}, in some other places, such as in the UK and Germany.)

Prevalence: The proportion of disease cases that exist within a population at a specific point in time, relative to the number of individuals within that population at the same point in time.

Probability: The likelihood of an event occurring – numerically represented by a value between 0 and 1; a probability of 1 means an event is certain to happen, whereas a probability of 0 means an event is certain *not* to happen.

Probit model: A dose-response model that can be derived under the assumption that individual tolerance is a random variable following a lognormal distribution. A probit, or probability unit, is obtained by modifying the standard variate of the standardized normal distribution; this transformation can then be used in the analysis of dose-response data used in a risk characterization.

Proxy concentration: Assigned chemical concentration value for situations where sample data may not be available, or when it is impossible to quantify accurately.

Public health education: A program of activities to promote health and provide information and training about hazardous substances in the human environments that will result in the reduction of exposure, illness, and/or disease. This type of program may include diagnosis and treatment information for health care providers, as well as activities in communities to enable them to prevent or mitigate the health effects from exposure to hazardous substances in the human environments.

Public health risk management: Action designed to prevent exposures and/or to mitigate or prevent adverse health effects in populations experiencing chemical exposure problems. Public health mitigation actions can be identified from information developed in public health exposure and risk assessments, as well as from environmental and public health monitoring activities. These actions may be comprised of the removal or separation of individuals from exposure sources (for example, by providing an alternative water supply), conducting biologic indicators of exposure studies to assess exposure; and providing health education for health care providers and community members.

Pulmonotoxicity: Adverse effects or disease states produced by toxicants in the respiratory system of an organism.

Qualitative: Description of a situation without numerical specifications.

Quality assurance (QA): A system of activities designed to assure that the quality control system in a study or investigation is performing adequately. It consists of the management of information and data sets from an investigation, to ensure that they meet the data quality objectives. *Quality control (QC)* – a system of specific efforts designed to test and control the quality of data obtained in an investigation. This comprises of the management of activities involved in the collection and analysis of data to assure they meet the data quality objectives – and it represents the system of activities required to provide information as to whether the quality assurance system is performing adequately.

Quantitation limit (QL): The lowest level at which a chemical can be accurately and reproducibly quantitated. It usually is equal to the instrument detection limit (IDL) multiplied by a factor of 3 to 5, but varies for different chemicals and different samples.

Quantitative: Description of a situation that is presented in reasonably exact numerical terms.

Reasonable maximum exposure (RME): A concept that attempts to identify the highest exposure (and, therefore, the greatest risk) that could reasonably be expected to occur in a given population.

Receptor: Members of a potentially exposed population, such as persons or organisms that are potentially exposed to concentrations of a particular chemical compound of concern. *Sensitive receptor* – individual in a population who is particularly susceptible to health impacts due to exposure to a chemical substance.

Reference concentration (RfC): An estimate of a daily inhalation exposure to the human population (including sensitive subgroups) that is likely to be without an appreciable risk of deleterious non-cancer effects during a lifetime. It represents a concentration of a chemical substance in an environmental medium to which exposure can occur over a prolonged period without expected adverse effect; the medium in this case is usually air – with the concentration expressed in mg of chemical per m^3 of air.

Reference dose (RfD): The estimate of lifetime daily oral exposure of a non-carcinogenic substance for the general human population (including sensitive receptors) which appears to be without an appreciable risk of deleterious effects, consistent with the threshold concept. This constitutes the maximum amount of a chemical that the human body can absorb without experiencing chronic health effects, expressed in mg of chemical per kg body weight per day. Generally, RfDs are available from databases such as US EPA's Integrated Risk Information System (IRIS) – and serves as the toxicity value for a chemical in human health risk assessment used for evaluating the non-carcinogenic effects that could result from exposures to chemicals of concern. (See Appendix B.)

Regulatory standard: A general term used to describe legally-established values above which regulatory action will usually be required. *Regulatory limit* – refers to an estimated chemical concentration in specific media that is not likely to cause adverse health effects, given a standard daily intake rate and standard body weight. The regulatory limits are calculated from the scientific literature available on exposure and health effects. *Regulatory dose* – the daily exposure to the human population, as reflected in a final risk management decision.

Representative sample: A sample that is assumed *not* to be significantly different from the population of samples available.

Respiratory system: The organ system that functions to distribute air and gas exchange in an organism.

Response (toxic): The reaction of a body or organ to a chemical substance or other physical, chemical, or biological agent.

Risk: The probability or likelihood of an adverse consequence from a hazardous situation or hazard, or the potential for the realization of undesirable adverse consequences from impending events. In chemical exposure situations, it is generally used to provide a measure of the probability and severity of an adverse effect to health, property, or the environment under specific circumstances of exposure to a chemical agent or a mixture of chemicals. In quantitative probability terms, risk may be expressed in values ranging from zero (representing the certainty that harm will not occur) to one (representing the certainty that harm will occur). In risk assessment practice, the following represent examples of how risk is typically expressed: 1E-4 or 10^{-4} = a risk of 1/10,000; 1E-5 or 10^{-5} = 1/100,000; 1E-6 or 10^{-6} = 1/1,000,000; 1.3E-3 or 1.3×10^{-3} = a risk of 1.3/1,000=1/770; 8E-3 or 8×10^{-3} = a risk of 1/125; and 1.2E-5 or 1.2×10^{-5} = a risk of 1/83,000. *Individual risk* – refers to the probability that an individual person in a population will experience an adverse effect from exposures to hazards. It is used to define the frequency at which an individual may be expected to sustain a given level of harm from the realization of specified hazards. In general, this is identical to 'population' or 'societal' risk – unless if specific population subgroups can be identified that have different (i.e., higher or lower) risks. *Societal (or population) risk* – refers to the relationship between the frequency and the number of people suffering from a specified level of harm in a given population, as a result of the realization of specified hazards. *Relative risk* – refers to the ratio of incidence or risk among exposed individuals to incidence or risk among non-exposed individuals. *Residual risk* – refers to the risk of adverse consequences that remains after corrective actions have been implemented. *Cumulative risk* – refers to the total added risks from all sources and exposure routes that an individual or group is exposed to. *Aggregate risk* – refers to the sum total of individual increased risks of an adverse health effect in an exposed population.

Risk acceptability/acceptance: Refers to the willingness of an individual, group, or society to accept a specific level of risk in order to obtain some gain or benefit.

Risk appraisal: A review of whether existing or potential biologic receptors are presently, or may in the future, be at risk of adverse effects as a result of exposures to chemicals originating or found in the human environments.

Risk assessment: The determination of the kind and degree of hazard posed by an agent, the extent to which a particular group of receptors have been or may be exposed to the agent, and the present or potential future health risk that exists due to the agent. It comprises of a methodology that combines exposure assessment with health and environmental effect data to estimate risks to human or environmental target organisms that may result from exposure to various hazardous substances. In the context of human exposure to chemical substances, risk assessment involves the determination of potential adverse health effects from exposure to the chemicals of potential concern – including both quantitative and qualitative expressions of risk. Overall, the process of risk assessment involves four key steps, namely: hazard identification, dose-response assessment, exposure assessment, and risk characterization.

Risk-based concentration: A chemical concentration determined based on an evaluation of the compound's overall risk to health upon exposure.

Risk characterization: The estimation of the incidence and severity of the adverse effects likely to occur in a human population or ecological group due to actual or predicted exposure to a substance or hazard.

Risk communication: Activities carried out to ensure that messages and strategies designed to prevent exposure, adverse health effects, and diminished quality of life are effectively communicated to the public and/or stakeholders. As part of a broader prevention strategy, risk communication supports education efforts by promoting public awareness, increasing knowledge, and motivating individuals to take action to reduce their exposure to hazardous substances.

Risk control: The process of managing risks associated with a hazard situation. It may involve the implementation, enforcement, and re-evaluation of the effectiveness of corrective measures from time to time.

Risk decision: The complex public policy decision relating to the control of risks associated with hazardous situations.

Risk determination: An evaluation of the environmental and health impacts associated with chemical releases and/or exposures.

Risk estimation: The process of quantifying the probability and consequence values for a hazard situation. The process is used to determine the extent and probability of adverse effects of the hazards identified, and to produce a measure of the level of health, property, or environmental risks being assessed. A *risk estimate* is comprised of a description of the probability that a potential receptor exposed to a specified dose of a chemical will develop an adverse response.

Risk evaluation: The complex process of developing acceptable levels of risk to individuals or society. It is the stage at which values and judgments enter into the decision-making process.

Risk group: A real or hypothetical exposure group composed of the general or specific population groups.

Risk management: The steps and processes taken to reduce, abate, or eliminate the risk that has been revealed by a risk assessment. For chemical exposure situations, it consists of measures or actions taken to ensure that the level of risk to human health or the environment as a result of possible exposure to the chemicals of concern does not exceed the pre-established acceptable limit (e.g., 1E-06). The process focuses on decisions about whether an assessed risk is sufficiently high to present a public health concern, and also about the appropriate means for controlling the risks that are judged to be significant. The decision-making process involved takes account of political, social, economic, and engineering constraints, together with risk-related information, in order to develop, analyze, and compare management options and then select the appropriate managerial response to a potential hazard situation.

Risk perception: Refers to the magnitude of a risk as is perceived by an individual or population. It consists of a convolution of the measured risk together with the pre-conceptions of the observer.

Risk reduction: The action of lowering the probability of occurrence and/or the value of a risk consequence, thereby reducing the magnitude of the risk.

Risk-specific dose (RSD): An estimate of the daily dose of a carcinogen which, over a lifetime, will result in an incidence of cancer equal to a given specified (usually the acceptable) risk level.

Risk tolerability: A willingness to 'live with' a risk, and to keep it under review. 'Tolerances' refer to the extent to which different groups or individuals are prepared to tolerate identified risks.

Sample quantitation limit (SQL) (also called *practical quantitation limit, PQL):* The lowest level that can be reliably achieved within specified limits of precision and accuracy during routine laboratory operating conditions. It represents a detection limit that has been corrected for sample characteristics, sample preparation, and analytical adjustments such as dilution. Typically, the PQL or SQL will be about 5 to 10 times the chemical-specific detection limit.

Sampling and analysis plan (SAP): Documentation that consists of a quality assurance project plan (QAPP) and a field sampling plan (FSP). The *QAPP* – contains documentation of all relevant QA and QC programs for the case-specific project. The *FSP* – is a documentation that defines in detail, the sampling and data gathering activities to be used in the investigation of a potential environmental contamination or chemical exposure problem.

Sensitivity analysis: A method used to examine the operation of a system by measuring the deviation of its nominal behavior due to perturbations in the performance of its components from their nominal values. In risk assessment, this may involve an analysis of the relationship of individual factors (such as chemical concentration, population parameter, exposure parameter, and environmental medium) to variability in the resulting estimates of exposure and risk.

Skin adherence: The property of a material which causes it to be retained on the surface of the epidermis (i.e., adheres to the skin).

Skin (or dermal) permeability coefficient: Denoted by K_p (cm/hr), this is a flux value (normalized for concentration) that represents the rate at which a chemical penetrates the skin.

Solubility: A measure of the ability of a substance to dissolve in a fluid.

Sorption: The processes that remove solutes from the fluid phase and concentrate them on the solid phase of a medium.

Standard deviation: The most widely used statistical measure to describe the dispersion of a data set – defined for a set of n values as follows:

$$s = \sqrt{\frac{\sum\limits_{i=1}^{n}(X_i - X_m)^2}{(n-1)}}$$

where X_m is the arithmetic mean for the data set of n values. The higher the value of this descriptor, the broader is the dispersion of data set about the mean.

Stochasticity: Variability in parameters (or in models containing such parameters) that may be attributed to the inherent variability of the system under consideration.

Stressor (also, Agent): Any physical, chemical, or biological entity that can induce an adverse response in an organism. *Stressor-response profile* – summarizes the data on the effects of a stressor and the relationship of the data to the assessment endpoint.

Structure-activity relationship: Relationships of biological activity or toxicity of a chemical to its chemical structure or sub-structure.

Subchronic: Relating to intermediate duration, usually used to describe studies or exposure levels spanning 5 to 90 days duration. *Subchronic daily intake (SDI)* – refers to the exposure, expressed in mg/kg-day, averaged over a portion of a lifetime. *Subchronic exposure* – refers to the short-term, high-level exposure to chemicals, i.e., the maximum exposure or doses to a chemical over a portion (approximately 10%) of a lifetime of an organism.

Surrogate data: A substitute data or measurement on a given substance or agent that is used to estimate analogous or corresponding values of another substance/agent.

Synergism (of chemical effects): A pharmacologic or toxicologic interaction in which the combined effect of two or more chemicals is greater than the sum of the effect of each chemical acting alone. That is, the aspect of two or more agents interacting to produce an effect greater than the sum of the agents' individual effects. More generally, this represents the effects from a combination of two or more events, efforts, or substances that are greater than would be expected from adding the individual effects.

Systemic: Pertaining to or affecting the body as a whole or acting in a portion of the body other than the site of entry; generally used to refer to non-cancer effects. *Systemic effect* – relates to those effects that require absorption and distribution of the toxicant to a site distant from its original entry point, and at which distant point any effects are produced. Most chemical substances that produce systemic toxicity do not cause a similar degree of toxicity in all organs, but usually demonstrate major toxicity to one or two organs; these are referred to as the *target organs* of toxicity for that chemical. *Systemic toxicity* – relates to toxic effects as a result of absorption and distribution of a toxicant to a site distant from its entry point, at which distant point any effects are produced. Not all chemicals that produce systemic effects cause the same degree of toxicity in all organs.

Target organ: The biological organ(s) most adversely affected by exposure to a chemical substance. That is, the organ affected by a specific chemical in a particular species. *Target organ toxicity* – the adverse effects or disease states manifested in specific organs of the body of an organism.

Teratogenic: Structural developmental defects due to exposure to a chemical agent during formation of individual organs.

Threshold: The lowest dose or exposure of a chemical at which a specified measurable/deleterious effect is observed and below which such effect is not observed. *Threshold dose* – is the minimum exposure dose of a chemical that will evoke a stipulated toxicological response. *Toxicological threshold* – refers to the concentration at which a compound begins to exhibit toxic effects. *Threshold limit* – a chemical concentration above which adverse health and/or environmental effects may occur. *Threshold hypothesis* – refers to the assumption that no chemical injury occurs below a specified level of exposure or dose.

Threshold chemical (also, *nonzero threshold chemical*): Refers to a substance that is known or assumed to have no adverse effects below a certain dose. *Nonthreshold chemical* (also called, *zero threshold chemical*) – refers to a substance that is known, suspected, or assumed to potentially cause some adverse response or toxic effect at any dose above zero. Thus, any level of exposure is deemed to involve some risk – and this is usually used only in regard to carcinogenesis.

Thrombocytes (also, *Platelets*): The smallest cellular components of blood in an organism.

Tissue: A collection of cells that together perform a similar function within an organism.

Tolerance limit: The level or concentration of a chemical residue in media of concern above which adverse health effects are possible, and above which levels corrective action should therefore be undertaken.

Toxic: Harmful or deleterious with respect to the effects produced by exposure to a chemical substance. *Toxic substance* – refers to any material or mixture that is capable of causing an unreasonable threat or adverse effects to human health and/or the environment.

Toxicant: Any synthetic or natural chemical with an ability to produce adverse health effects.

Toxicity: The property of a substance to cause any adverse physiological effects (on living organisms). It represents the degree to which a chemical substance elicits a deleterious or adverse effect upon the biological system of an organism exposed to the substance over a designated time period. This indicates the harmful effects produced by a chemical substance – and reflects on the quality or degree of being poisonous or harmful to human or ecological receptors. *Delayed toxicity* – refers to the development of disease states or symptoms long (usually several months or years) after exposure to a toxicant. *Immediate toxicity* – refers to the rapid development or onset of disease states or symptoms following exposure to a toxicant.

Toxicity assessment: Evaluation of the toxicity of a chemical based on all available human and animal data. It consists of the characterization of the toxicological properties and effects of a chemical substance, with special emphasis on the establishment of dose-response characteristics.

Toxicity equivalency factors (TEFs): Toxicity parameters that are based on congener-specific data and the assumption that the toxicity of dioxin and dioxin-like compounds is mediated by the *Ah* receptor, and is additive. The TEF scheme compares the relative toxicity of individual dioxin-like compounds to that of TCDD (i.e., 2,3,7,8-Tetrachlorodibenzo-*p*-dioxin and related compounds), which is the known most toxic halogenated aromatic hydrocarbon in that family.

Toxicity equivalent (TEQ): Is defined as the product of the concentration, CI, of an individual 'dioxin-like compound' in a complex environmental mixture and the corresponding TCDD toxicity equivalency factor (TEF$_i$) for that compound. The total TEQs is the sum of the TEQs for each of the congeners in a given mixture, viz.,

$$Total\,TEQs = \sum_{i=1}^{n}\left(CI_i \times TEF_i\right)$$

Toxicodynamics: The study of the mechanisms by which toxicants produce their unique effects within an organism – i.e., the study of how a toxic chemical (or metabolites derived from it) interacts with specific molecular components of cellular processes in the body.

Toxicokinetics: The study of the time-dependent processes of toxicants in their interactions with living organisms – i.e., the study of how a toxic chemical is absorbed, distributed, metabolized, and excreted into and from the body. Thus, this includes the study of the absorption, distribution, storage, biotransformation, and elimination processes taking place within an organism. *Distribution* – refers to a toxicokinetic process that occurs after absorption, when toxicants enter the lymph or blood supply for transport to other regions of the body. *Storage* – the accumulation of toxicants or their metabolites in specific tissues of an organism or as bound to circulating plasma proteins in the organism. *Elimination* – refers to the toxicokinetic processes responsible for the removal of toxicants or their metabolites from the body.

Toxicological profile: A documentation about a specific substance in which scientific interpretation is provided from all known information on the substance; this also includes specifying the levels at which individuals or populations may be harmed if exposed. The toxicological profile may also identify significant gaps in knowledge on the substance, and serves to initiate further research, where needed.

Toxicology: The study of the adverse effects of chemical, biological, and physical agents on living organisms.

Uncertainty: The lack of confidence in the estimate of a variable's magnitude or probability of occurrence.

Uncertainty factor (UF) (also called, *safety factor*): In toxicological evaluations, this refers to a factor that is used to provide a margin of error when extrapolating from experimental animals to estimate human health risks. It represents one of several, generally 10-fold factors, used to operationally derive the reference dose (RfD) and reference concentration (RfC) from experimental data. UFs are intended to account for: (1) the variation in sensitivity among the members of the human population, i.e., inter-human or intraspecies variability; (2) the uncertainty in extrapolating animal data to humans, i.e., interspecies variability; (3) the uncertainty in extrapolating from data obtained in a study with less-than-lifetime exposure to lifetime exposure, i.e., extrapolating from subchronic to chronic exposure; (4) the uncertainty in extrapolating from a LOAEL rather than from a NOAEL; and (5) the uncertainty associated with extrapolation from animal data when the data base is incomplete. A *modifying factor (MF)* – serving as a companion factor to the UF, refers to a factor used in the derivation of a RfD or RfC. The magnitude of the MF reflects the scientific uncertainties of the study and database not explicitly treated with standard uncertainty factors (e.g., completeness of the overall database). A MF is greater than zero and less than or equal to 10 – with a typical default value of 1.

Unit cancer risk (UCR): The excess lifetime risk of cancer due to a continuous lifetime exposure/dose of one unit of carcinogenic chemical concentration (caused by one unit of exposure in the low exposure region). It is a measure of the probability of an individual developing cancer as a result of exposure to a specified unit ambient concentration.

Unit risk (UR): The upper-bound (plausible upper limit) estimate of the probability of contracting cancer as a result of constant/continuous exposure to an agent at a concentration of 1 $\mu g/L$ in water, or 1 $\mu g/m^3$ in air over the individual lifetime. The interpretation of unit risk would be as follows: if unit risk = 5.5 x 10^{-6} $\mu g/L$, 5.5 excess tumors are expected to develop per 1,000,000 people if exposed daily for a lifetime to 1 μg of the chemical in 1 liter of drinking water.

Upper-bound estimate: The estimate not likely to be lower than the true (risk) value. That is, an estimate of the plausible upper limit to the true value of the quantity.

Upper confidence limit, 95% (95% UCL): The upper limit on a normal distribution curve below which the observed mean of a data set will occur 95% of the time. This is also equivalent to stating that, there is at most a 5% chance of the true mean being greater than the observed value. (It is a value that equals or exceeds the true mean 95% of the time.) That is, assuming a random and normal distribution, this is the range of values below which a given value will fall 95% of the time. (See also, *confidence interval.*)

Volatile organic compound (VOC): Any organic compound that has a great tendency to vaporize, and is susceptible to atmospheric photochemical reactions. Such chemical volatilizes (evaporates) relatively easily when exposed to air. VOCs generally consist of substances containing carbon and different proportions of other elements such as hydrogen, oxygen, fluorine, chlorine, bromine, sulfur, or nitrogen – and these substances easily become vapors or gases. A significant number of the VOCs found in human environments are commonly used as solvents (e.g., paint thinners, lacquer thinner, degreasers, and dry cleaning fluids). In general, volatile compounds are amenable to analysis by the purge and trap techniques.

Volatilization: The transfer of a chemical from the liquid or solid into the gaseous phase. *Volatility* – is a measure of the tendency of a compound to vaporize or evaporate, usually from a liquid state.

Weight-/Strength-of-evidence for carcinogenicity: The extent to which the available biomedical and related data support the hypothesis that a substance causes cancer in humans.

Worst case: A semi-qualitative term that refers to the maximum possible exposure, dose, or risk to an exposed person or group, that could conceivably occur – regardless of whether or not this exposure, dose, or risk actually occurs or is observed in a specific population. Typically, this should refer to a hypothetical situation in which everything that can plausibly happen to maximize exposure, dose, or risk actually takes place. This worst case may indeed occur (or may even be observed) in a given population; however, since this is usually a very unlikely set of circumstances, in most cases, a worst-case estimate will be somewhat higher than occurs in a specific population. In most health risk assessments, a worst-case scenario is essentially a type of bounding estimate. (See also, *conservative assumption.*)

Xenobiotics: Substances that are foreign to an organism – i.e., substances that are not naturally produced within the organism. Also, substances not normally present in the environment – such as a pesticide or other environmental pollutant. Most xenobiotics are considered pollutants.

Appendix **B**

SELECTED TOOLS, CHEMICAL EXPOSURE MODELS, AND CHEMICAL DATABASES FOR PUBLIC HEALTH RISK POLICY DECISION-MAKING

A variety of decision-making tools and logistics, as well as computer databases and information libraries, may find several useful applications in public health risk assessment and environmental management programs designed to address chemical exposure problems. A limited number and select examples of such logistical tools, models, and database systems are presented below in this appendix. It must be emphasized here that, the list provided here is by no means complete and exhaustive – and neither does it cover the broad spectrum of what is available to the scientific communities and/or the general public.

B.1. Example Decision Support Tools and Logistical Computer Software

Oftentimes, a variety of scientific and analytical tools are employed to assist the decision-maker with various issues associated with the management of a chemical exposure problem. A select number of application tools (consisting of scientific models and software) appropriate for such purposes are highlighted below. This listing is by no means complete and exhaustive; several other similar logistical tools can indeed be used to support environmental and risk management programs, in order to arrive at informed decisions on chemical exposure problems. In fact, recent years have seen a proliferation of software systems for a variety of environmental chemical fate, chemical exposure, and risk management studies. Care must therefore be exercised in the choice of an appropriate tool for specific problems.

A primary resource to consider in order to obtain further information on the listed software (and indeed several other current ones) would be the on-line communication system of the *Internet* service – the most widely used international network communication service. Also, traditional libraries and directories of environment- and health-related professional groups/associations may provide the necessary up-to-date contacts.

B.1.1.   THE *IEUBK* MODEL

Lead (Pb) poisoning presents potentially significant risks to the health and welfare of children all over the world today. The Integrated Exposure Uptake Biokinetic (IEUBK) model for lead in children attempts to predict blood-lead concentrations (PbBs) for children exposed to Pb in their environment. Indeed, measured PbB concentration is not only an indication of exposure, but is a widely used index to discern potential future health problems.

The IEUBK model for lead in children is a menu-driven, user-friendly model designed to predict the probable PbB concentrations (via pharmacokinetic modeling) for children aged between six months and seven years who have been exposed to Pb in various environmental media (e.g., air, water, soil, dust, paint, diet and other sources). The model has the following four key functional components:

- *Exposure component* – compares Pb concentrations in environmental media with the amount of Pb entering a child's body.
- *Uptake component* – compares Pb intake into the lungs or digestive tract with the amount of Pb absorbed into the child's blood.
- *Biokinetic component* – shows the transfer of Pb between blood and other body tissues, or the elimination of Pb from the body altogether.
- *Probability distribution component* – shows a probability of a certain outcome (*e.g.*, a PbB concentration greater than 10 µgPb/dL in an exposed child based on the parameters used in the model).

It is noteworthy that, in the United States, the US EPA and the Centers for Disease Control and Prevention (CDC) have determined that childhood PbB concentrations at or above 10 micrograms of Pb per deciliter of blood (µgPb/dL) present risks to children's health. The IEUBK model calculates the probability of children's PbB concentrations exceeding 10 µgPb/dL (or other user-entered value). By varying the data entered into the model, the user can evaluate how changes in environmental conditions may affect PbB levels in exposed children.

The IEUBK model allows the user to input relevant absorption parameters, (e.g., the fraction of Pb absorbed from water) as well as rates for intake and exposure. Using these inputs, the IEUBK model then rapidly calculates and recalculates a complex set of equations to estimate the potential concentration of Pb in the blood for a hypothetical child or population of children (6 months to 7 years). Overall, the model is intended to:

- Estimate a typical child's long-term exposure to Pb in and around his/her residence;
- Provide an accurate estimate of the geometric average PbB concentration for a typical child aged six months to seven years;
- Provide a basis for estimating the risk of elevated PbB concentration for a hypothetical child;
- Predict likely changes in the risk of elevated PbB concentration from exposure to soil, dust, water, or air following concerted action to reduce such exposure;
- Provide assistance in determining target cleanup levels at specific residential sites for soil or dust containing high amounts of Pb; and

- Provide assistance in estimating PbB levels associated with the Pb concentration of soil or dust at undeveloped sites that may be developed at a later date.

A major advantage of the IEUBK model is the fact that it takes into consideration the several different media through which children can be exposed.

Further information on IEUBK may be obtained from the US EPA's Office of Emergency and Remedial Response, Washington, DC, USA (see also, USEPA, 1994: *Guidance Manual for the Integrated Exposure Uptake Biokinetic Model for Lead in Children*, NTIS #PB93-963510, OSWER #9285.7-15-1).

B.1.2.   *CATREG* SOFTWARE FOR CATEGORICAL REGRESSION ANALYSIS

CatReg is a computer program, written in the S-PLUS® (MathSoft, Inc.) programming language, to support the conduct of exposure-response analyses by toxicologists and health scientists. *CatReg* can be used to perform categorical regression analyses on toxicity data after effects have been assigned to ordinal severity categories (e.g., no effect, adverse effect, severe effect) and associated with up to two independent variables corresponding to the exposure conditions (e.g., concentration and duration) under which the effects occurred. *CatReg* calculates the probabilities of the different severity categories over the continuum of the variables describing exposure conditions. The categorization of observed responses allows expression of dichotomous, continuous, and descriptive data in terms of effect severity – and supports the analysis of data from single studies or a combination of similar studies.

There are many potential applications of the *CatReg* program in the analysis of health effects studies and other types of data. The software was developed to support toxicity assessment for acute inhalation exposures, but the US Environmental Protection Agency (EPA) encourages the broad application of this software.

The special features offered by *CatReg* include options for the following:

- Stratifying the analysis by user-specified covariates (e.g., species, sex, etc.)
- Choosing among several basic forms of the exposure-response curve;
- Using effects assigned to a range of severity categories, rather than a single category;
- Using cluster-correlated data;
- Incorporating user-specified weights;
- Using aggregate data; and
- Query-based exclusion of user-specified data (i.e., filtering) for sensitivity analysis.

CatReg reads data from ordinary text files in which data are separated by commas. A query-based interface guides the user through the modeling process. Simple commands provide model summary statistics, parameter estimates, diagnostics, and graphical displays.

CatReg requires a Windows95/98 environment and S-PLUS® Professional version 3 or higher. Knowledge of the S-PLUS® programming language is not required. The *CatReg* software is accompanied by: *CatReg Software Documentation* (USEPA, 2000: *CatReg Software Documentation*, EPA/600/R-98/053F, Office of Research and Development, National Center for Environmental Assessment, Research Triangle Park,

NC) – that provides a technical description of the statistical methods used by the program; and *CatReg Software User Manual* (USEPA, 2000: *CatReg Software User Manual*, EPA EPA/600/R-98/052F, Office of Research and Development, National Center for Environmental Assessment, Research Triangle Park, NC) – that provides instructions for installing the software, creating data input files, performing analyses, using hypothesis testing and diagnostics, and plotting graphics. The user manual, which contains illustrated examples, provides ideas on adapting *CatReg* for situation-specific applications.

B.1.3.   OTHER MISCELLANEOUS COMPUTER SOFTWARE

A variety of other decision support tools and computer software are available to facilitate various risk assessment and/or environmental management tasks – including the following:

- *LEADSPREAD*. LEADSPREAD, the DTSC Lead Risk Assessment Spreadsheet, is a tool for evaluating exposure and the potential for adverse health effects that could result from exposure to lead in the environment. Basically, it consists of a mathematical model for estimating blood lead concentrations as a result of contacts with lead-contaminated environmental media. The model can be used to determine blood levels associated with multiple pathway exposures to lead. A distributional approach is used with this model – allowing estimation of various percentiles of blood lead concentration associated with a given set of inputs.

  Overall, the LEADSPREAD model provides a computer spreadsheet methodology for evaluating exposure and the potential for adverse health effects resulting from multipathway exposure to inorganic lead via dietary intake, drinking water, soil and dust ingestion, inhalation, and dermal contact. Each pathway is represented by an equation relating incremental blood lead increase to a concentration in a medium, using contact rates and empirically determined ratios. The contributions via all pathways are added to arrive at an estimate of median blood lead concentration resulting from the multipathway exposure.

  Further information on LEADSPREAD may be obtained from the Office of Scientific Affairs, Department of Toxic Substances Control (DTSC), California EPA, Sacramento, California, USA.

- *RISK*ASSISTANT*. The RISK*ASSISTANT software is designed to assist the user in rapidly evaluating exposures and human health risks from chemicals in the environment at a particular site. It is designed to evaluate human health risks associated with *chronic* exposures to chemicals. The user need only provide measurements or estimates of the concentrations of chemicals in the air, surface water, groundwater, soil, sediment, and/or biota.

  RISK*ASSISTANT generates two types of risk estimate: (1) for potential cancer-causing chemicals, the increased probability of individuals getting cancer from a particular exposure; and (2) for potential non-cancer toxic effects, a comparison of the expected level of exposure to an exposure that is assumed to be essentially without risk. Indeed, the model provides an array of analytical tools, databases, and information-handling capabilities for human health risk assessment – and it has the ability to tailor exposure and risk assessments to local conditions. In fact, RISK*ASSISTANT uses standard approaches to generate estimates of exposure and risk using detailed, locally relevant information, and then test them against many alternative assumptions.

Further information on RISK*ASSISTANT may be obtained from the following: Hampshire Research Institute, Alexandria, Virginia; US EPA, Research Triangle Park, North Carolina; California EPA, Sacramento, California; New Jersey Department of Environmental Protection, Trenton, New Jersey; and Delaware Department of Natural Resources and Environmental Control, Dover, Delaware, USA.

- *RISKPRO.* RISKPRO is a complete software system designed to predict the environmental risks and effects of a wide range of human health-threatening situations. It consists of a multimedia/multipathway environmental pollution modeling system – providing for modeling tools to predict exposure from pollutants in the air, soil and water.

  RISKPRO is used to evaluate receptor exposures and risks from environmental contaminants. It graphically represents its results through maps, bar charts, wind-rose diagrams, isopleth diagrams, pie charts, and distributional charts. Its mapping capabilities can also allow the user to create custom maps showing data and locations of environmental contaminant plumes.

  Further information on RISKPRO may be obtained from General Sciences Corporation (GSC), Laurel, Maryland, USA.

- *TOXIC.* TOXIC is a microcomputer program that calculates the incremental risk to the hypothetical maximum exposed individual from hazardous waste incineration. It calculates exposure to each pollutant individually, using a specified dispersion coefficient (which is the ratio of pollutant concentration in air [in $\mu g/m^3$] to pollutant emission rate [in gm/s]).

  TOXIC is used in hazardous waste facility risk analysis. It is a flexible and convenient tool for performing inhalation risk assessments for hazardous waste incinerators – and produces point estimates of inhalation risks.

  Further information on TOXIC may be obtained from Rowe Research and Engineering Associates, Alexandria, Virginia, USA.

- *CRYSTAL BALL. Crystall Ball* is a user-friendly, graphical-oriented forecasting and risk analysis program that helps diminish the uncertainty associated with decision-making. The system employs standard spreadsheet models in its application.

  Through the use of Monte Carlo simulation techniques, *Crystal Ball* forecasts the entire range of results possible for a given situation. It also shows the confidence levels, in order that the analyst will know the likelihood for any specific event to take place. The tool further allows sensitivity evaluations to be carried out in a very effective manner – by the use of sensitivity charts; *Crystal Ball* calculates sensitivity by computing rank correlation coefficients between every assumption and every forecast cell during simulation. With an intuitive graphical interface, *Crystal Ball* gives users powerful capabilities to perform uncertainty analyses based on Monte Carlo simulations.

  Further information on Crystal Ball may be obtained from Decisioneering, Inc., Boulder, Colorado, USA.

- *CALTOX.* CalTOX is an innovative spreadsheet model that relates the concentration of a chemical in soil to the risk of an adverse health effect for a person living or working on or near the contaminated soil. The model computes site-specific health-based soil clean-up concentrations for specified target risk levels, and/or estimates human health risks for given soil concentrations at the site.

  The CalTOX spreadsheet contains a multimedia transport and transformation model that uses equations based on conservation of mass and chemical equilibrium. This model predicts the time-dependent concentrations of a chemical in the seven environmental compartments of air, water, three soil layers, sediment, and plants at a site. After partitioning the concentration of the chemical to these environmental compartments, CalTOX determines the chemical concentration in the exposure media of breathing zone air, drinking water, food, and soil that people inhale, ingest, and contact dermally. CalTOX then uses the standard equations (found in the US Environmental Protection Agency Risk Assessment Guidance for Superfund [USEPA, 1989]) to estimate exposure and risk.

  CalTOX has the capability of conducting Monte Carlo simulations with a spreadsheet add-in program. Used in this manner, CalTOX will produce a range of risks and/or health-based soil target clean-up levels that reflect the uncertainty/variability of the estimates.

  CalTOX was developed by the California EPA, and is available for free downloading from their website. In addition to the site-specific risk assessments, results from CalTOX can be exported to other programs (such as *Crystal Ball*) for Monte Carlo Simulation.

B.2.   Selected Databases and Information Library with Important Risk Information for Risk Assessment and Environmental Management

Several databases containing information on numerous chemical substances exist within the scientific community that may find extensive useful applications in the management of various types of chemical exposure problems. Example databases of general interest to chemical exposure assessments and risk management programs are presented below; this select list is of interest, especially because of their international appeal and/or their wealth of risk assessment support information. Indeed, the presentation is meant to demonstrate the overall wealth of scientific information that already exists – and that should generally be consulted to provide the relevant chemical exposure and risk assessment support information necessary for risk determination and/or risk management actions.

B.2.1.   THE INTERNATIONAL REGISTER OF POTENTIALLY TOXIC CHEMICALS (IRPTC) DATABASE

In 1972, the United Nations Conference on the Human Environment, held in Stockholm, recommended the setting up of an international registry of data on chemicals likely to enter and damage the environment. Subsequently, in 1974, the Governing Council of the United Nations Environment Programme (UNEP) decided to establish both a chemicals register and a global network for the exchange of information that the register would contain. The definition of the register's objectives was subsequently elaborated to address the following:

- Make data on chemicals readily available to those who need it
- Identify and draw attention to the major gaps in the available information and encourage research to fill those gaps
- Help identify the potential hazards of using chemicals and improve people awareness of such hazards
- Assemble information on existing policies for control and regulation of hazardous chemicals at national, regional and global levels.

In 1976, a central unit for the register – named the International Register of Potentially Toxic Chemicals (IRPTC) – was created in Geneva, Switzerland, with the main function of collecting, storing and disseminating data on chemicals, and also to operate a global network for information exchange. IRPTC network partners (the designation assigned to participants outside the central unit) consist of National Correspondents appointed by governments, national and international institutions, national academies of science, industrial research centers and specialized research institutions.

Chemicals examined by the IRPTC have been chosen from national and international priority lists. The selection criteria used include the quantity of production and use, the toxicity to humans and ecosystems, persistence in the environment, and the rate of accumulation in living organisms.

IRPTC stores information that would aid in the assessment of the risks and hazards posed by a chemical substance to human health and environment. The major types of information collected include that relating to the behavior of chemicals, and information on chemical regulations. Information on the behavior of chemicals is obtained from various sources such as national and international institutions, industries, universities, private databanks, libraries, academic institutions, scientific journals and United Nations bodies such as the International Programme on Chemical Safety (IPCS). Regulatory information on chemicals is largely contributed by IRPTC National Correspondents. Specific criteria are used in the selection of information for entry into the databases. Whenever possible, IRPTC uses data sources cited in the secondary literature produced by national and international panels of experts to maximize reliability and quality. The data are then extracted from the primary literature. Validation is performed prior to data entry and storage on a computer at the United Nations International Computing Centre (ICC).

General Types of Information in the IRPTC Databases
The complete IRPTC file structure consists of databases relating to the following key subject matter and areas of interest: Legal; Mammalian and Special Toxicity Studies; Chemobiokinetics and Effects on Organisms in the Environment; Environmental Fate Tests, and Environmental Fate and Pathways into the Environment; and Identifiers, Production, Processes and Waste.

The IRPTC *Legal* database contains national and international recommendations and legal mechanisms related to chemical substances control in environmental media such as air, water, wastes, soils, sediments, biota, foods, drugs, consumer products, etc. This organization allows for rapid access to the regulatory mechanisms of several nations and to international recommendations for safe handling and use of chemicals.

The *Mammalian Toxicity* database provides information on the toxic behavior of chemical substances in humans; toxicity studies on laboratory animals are included as a means of predicting potential human effects. The Special Toxicity databases contain information on particular effects of chemicals on mammals, such as mutagenicity and carcinogenicity, as well as data on non-mammalian species when relevant for the description of a particular effect.

The *Chemobiokinetics and Effects on Organisms in the Environment* databases provide data that will permit the reliable assessment of the hazard of chemicals present in the environment to man. The absorption, distribution, metabolism and excretion of drugs, chemicals and endogenous substances are described in the Chemobiokinetics databases. *The Effects on Organisms in the Environment* databases contain toxicological information regarding chemicals in relation to ecosystems and to aquatic and terrestrial organisms at various nutritional levels.

The *Environmental Fate Tests, and Environmental Fate and Pathways into the Environment* databases assess the risk presented by chemicals to the environment.

The *Identifiers, Production, Processes and Waste* databases contain miscellaneous information about chemicals, including physical and chemical properties; hazard classification for chemical production and trade statistics of chemicals on worldwide or regional basis; information on production methods; information on uses and quantities of use for chemicals; data on persistence of chemicals in various environmental compartments or media; information on the intake of chemicals by humans in different geographical areas; sampling methods for various media and species, as well as analytical protocols for obtaining reliable data; recommendable methods for the treatment and disposal of chemicals; etc.

The Role of IRPTC in Risk Assessment and Environmental Management
The IPRTC, with its carefully designed database structure, provides a sound model for national and regional data systems. More importantly, it brings consistency to information exchange procedures within the international community. The IPRTC is serving as an essential international tool for chemicals hazard assessment, as well as a mechanism for information exchange on several chemicals. The wealth of scientific information contained in the IRPTC can serve as an invaluable database for a variety of environmental and public health risk management programs.

Further information on the IRPTC may be obtained from the National Correspondent to the IRPTC and scientific bodies/institutions – such as a country's National Academy of Sciences. (Also, it is noteworthy that, following the successful implementation of the IRPTC databases, a number of countries created National Registers of Potentially Toxic Chemicals [NRPTCs] that is completely compatible with the IRPTC system.)

B.2.2.   THE INTEGRATED RISK INFORMATION SYSTEM (IRIS) DATABASE

The Integrated Risk Information System (IRIS), prepared and maintained by the Office of Health and Environmental Assessment of the United States Environmental Protection Agency (US EPA), is an electronic database containing health risk and regulatory information on several specific chemicals. It serves as an on-line database of chemical-specific risk information; it is also a primary source of EPA health hazard assessment and related information on several chemicals of environmental concern.

IRIS was originally developed for the US EPA staff – in response to a growing demand for consistent risk information on chemical substances for use in decision-making and regulatory activities. The information in IRIS is accessible to those without extensive training in toxicology, but with some rudimentary knowledge of health and related sciences.

The IRIS database provides information on how chemicals affect human health and is a primary source of EPA risk assessment information on chemicals of environmental and public health concern. It serves as a guide for the hazard identification and dose-response assessment steps of EPA risk assessments. More importantly, IRIS makes chemical-specific risk information readily available to those who must perform risk assessments – and also increases consistency in risk management decisions. The information in IRIS represents expert Agency consensus. In fact, this Agency-wide agreement on risk information is one of the most valuable aspects of IRIS. Chemicals are added to IRIS on a regular basis. Chemical file sections in the system are updated as new information is made available to the responsible review groups.

The primary types of health assessment information in IRIS are oral reference doses (RfDs) and inhalation reference concentrations (RfCs) for non-carcinogens; and oral and inhalation carcinogen assessments. Reference doses and concentrations are estimated human chemical exposures over a lifetime which are just below the expected threshold for adverse health effects. The carcinogen assessments include: a weight-of-evidence classification, oral and inhalation quantitative risk information, including slope factors, along with unit risks calculated from those slope factors. (A slope factor is the estimated lifetime cancer risk per unit of the chemical absorbed, assuming lifetime exposure.)

Overall, summary information in IRIS consists of three components: derivation of oral chronic RfD and inhalation chronic RfC, for non-cancer critical effects, cancer classification (and cancer hazard narrative for the more recent assessments) and quantitative cancer risk estimates. Indeed, the IRIS information has focused on the documentation of toxicity values (i.e., RfD, RfC, cancer unit risk and slope factor) and cancer classification. The bases for these numerical values and evaluative outcomes are provided in an abbreviated and succinct manner. Details for the scientific rationale can be found in supporting documents, and references for these assessment documents, and key studies are provided in the bibliography sections. Since 1997, IRIS summaries and accompanying support documents, including a summary and response to external peer review comments, have been publicly available in full text on the IRIS web site. The Internet site is now EPA's primary repository for IRIS. Together they comprise the 'IRIS assessment' for a given chemical substance. The information currently on IRIS represents the state-of-the-science and state-of-the-practice in risk assessment, as existed when each assessment was prepared.

General Types of Information in IRIS
The IRIS database consists of a collection of computer files covering several individual chemicals. To aid users in accessing and understanding the data in the IRIS chemical files, the following key supportive documentation is provided as an important component of the system:

- Alphabetical list of the chemical files in IRIS and list of chemicals by Chemical Abstracts Service (CAS) number.
- Background documents describing the rationales and methods used in arriving at the results shown in the chemical files.
- A user's guide that presents step-by-step procedures for using IRIS to retrieve chemical information.

- An example exercise in which the use of IRIS is demonstrated.
- Glossaries in which definitions are provided for the acronyms, abbreviations, and specialized risk assessment terms used in the chemical files and in the background documents.

The chemical files contain descriptive and numerical information on several subjects – including oral and inhalation reference doses (RfDs) for chronic non-carcinogenic health effects, and oral and inhalation cancer slope factors (SFs) and unit cancer risks (UCRs) for chronic exposures to carcinogens. It also contains supplementary data on acute health hazards and physical/chemical properties of the chemicals.

It is noteworthy that, because exposure assessment pertains to exposure at a particular place, IRIS cannot provide situation-specific information on exposure. However, IRIS can be used with an exposure assessment to characterize the risk of chemical exposure. This risk characterization can then be used to decide what actions to take to protect human health.

The Role of IRIS in Risk Assessment and Environmental Management
IRIS is a tool that provides hazard identification and dose-response assessment information, but does not provide situation- or problem-specific information on individual instances of exposure. It is a computerized library of current information that is updated periodically. Combined with specific exposure information, the data in IRIS can be used to characterize the public health risks of a chemical of potential concern under specific scenarios, which can then facilitate the development of effectual corrective action decisions designed to protect public health. The information in IRIS can indeed be used to develop corrective action and risk management decision for chemical exposure problems – such as achieved via the application of risk assessment and risk management procedures.

The IRIS database was created by the USEPA in 1986 (and made publicly available in 1988) as a mechanism for developing consistent consensus positions on potential health effects of chemical substances. Combined with site-specific or national exposure information, the summary health information in IRIS could then be used by risk assessors and others to evaluate potential public health risks from environmental chemicals.

Further information on, and access to the IRIS database may be obtained via the US EPA internet website. Alternatively, the following groups may be contacted: IRIS User Support, US EPA, Environmental Criteria and Assessment Office, Cincinnati, Ohio, USA; and National Library of Medicine [NLM], Bethesda, Maryland, USA.

B.2.3.   THE INTERNATIONAL TOXICITY ESTIMATES
          FOR RISKS (*ITER*) DATABASE

International Toxicity Estimates for Risk (ITER) is a database of human health risk values and supporting information. The ITER database consists of chemical files, with information from the US Environmental Protection Agency (EPA), the US Agency for Toxic Substances and Disease Registry (ATSDR), and Health Canada. It includes direct links to EPA's IRIS and to ATSDR's Toxicological Profiles for each chemical file – and also has the ability to print reports. The data in the ITER database are presented in a comparative fashion – allowing the user to view what conclusions each organization has reached; in addition, a brief explanation of differences is provided.

The values and text in the ITER database have been extracted from credible published documents and data systems of the original author organizations. The risk values are compiled into a consistent format, so that comparisons can be made readily by informed users. The necessary conversions are performed so that direct comparisons can be made – and the synopsis text is written so as to help the user better understand the similarities and differences between the values of the different organizations. At the end of each so-identified 'Level 3' summary in the ITER database, the user will find listed a source and/or link for further information about that particular assessment.

Independently-derived values, which have undergone external peer review at a TERA-sponsored peer-review meeting, are also listed in the ITER database; TERA (Toxicology Excellence for Risk Assessment) is a non-profit corporation dedicated to the best use of toxicity data for the development of risk values. It is noteworthy that TERA prevents conflicts of interest in part through its nonprofit status and policy of informed and neutral guidance. Consequently, TERA helps environmental, industry, and government groups find common ground through the application of good science to risk assessment. The motivation is, in fostering successful partnerships, improvements in the science and practice of risk assessment will follow.

As of the early part of the year 2002, ITER contained values from three organizations – namely, Health Canada, US EPA, and the US ATSDR. In the future, it is expected that the ITER database will include additional chemicals and health information from organizations such as the World Health Organization/International Programme on Chemical Safety (WHO/IPCS) and Rijksinstituut Voor Volksgezondheid en Miliouhygiene (RIVM) (National Institute of Public Health and the Environment) of the Netherlands. RIVM develops human-toxicological risk limits (i.e., maximum permissible risk levels, MPRs) for a variety of chemicals based on chemical assessments that are compiled in the framework of the Dutch government program on risks in relation to soil quality.

The Role of ITER in Risk Assessment and Environmental Management
ITER is a compilation of human health risk values for chemicals of environmental and/or public health concern from several health organizations worldwide. These values are developed for multiple purposes depending on the particular organization's function. They are principally used as guidance or regulatory levels against which human exposures from chemicals in the air, food, soil, and water can be compared. The information in the ITER database is useful to risk assessors and risk managers needing human health toxicity values to make risk-based decisions. ITER allows the user to compare a number of key organizations' values and to determine the best value to use for the human exposure situation being evaluated.

Further information on, and access to the ITER database may be obtained via the ITER and/or TERA internet websites.

B.2.4.   OTHER MISCELLANEOUS INFORMATION SOURCES

A variety of other information sources are available to facilitate various risk assessment and/or environmental management tasks – such as the following:

- *Health Effects Assessment Summary Tables (HEAST).* The HEAST is a comprehensive listing consisting almost entirely of provisional risk assessment information relative to oral and inhalation routes for various chemical compounds.

These entries in the HEAST are limited to analytes that have undergone review and have the concurrence of individual US EPA program offices, and each is supported by an Agency reference. This risk assessment information has not, however, had enough review to be recognized as high quality, Agency-wide consensus information. Chemicals listed as 'HEAST Table 2' (HEAST2) were derived from alternate methods that are not currently practiced by the RfD/RfC Work Group. 'HEAST Table 2' values consist primarily of inhalation RfC values determined from a methodology that does not follow the interim inhalation methods adopted by the US EPA, and RfC or RfD values based on route-to-route extrapolation with inadequate pharmacokinetic and toxicity data.

- *ATSDR Toxicological Profiles.* Information on the toxic effects of chemical exposure in humans and experimental animals is contained in the ATSDR Toxicological Profiles. These documents also contain dose-response information for different routes of exposure. When information is available, the Toxicological Profiles also contain a discussion of toxic interactive effects with other chemicals, as well as a description of potentially sensitive human populations.

  Further information on, and access to the ATSDR toxicological profiles may be obtained via the ATSDR internet website, or by contacting the Agency for Toxic Substances and Disease Registry (ATSDR), US Department of Health and Human Service, Atlanta, GA, USA.

- *The Chemical Substances Information Network (CSIN).* The Chemical Substances Information Network (CSIN) is not a database, but rather an interactive network system that links together a number of databases relating to several chemical substances. The CSIN accesses data on chemical nomenclature, composition, structure, properties, toxicological information, health and environmental effects, production and uses, regulations, etc.

  The CSIN and the databases it accesses are in the public domain; however, users may have to make independent arrangements with vendors of those databases in the network that needs to be used for specific assignments. Further information on CSIN may be obtained from CIS, Baltimore, Maryland, USA.

- *MMEDE (Multimedia-Modeling Environmental Database and Editor).* The Multimedia-Modeling Environmental Database and Editor (MMEDE) is a user-friendly database interface for physical, chemical, and toxicological parameters typically associated with environmental assessments. The parameters for evaluation of impacts of hazardous and radioactive materials can be viewed, estimated, modified, printed, deleted, and exported. The database parameters include: physical parameters, dose factors, toxicity factors, environmental transfer factors, environmental decay half times, and other parameters of relevance. A source-of-information citation is used for every parameter value.

  Further information on MMEDE may be obtained from Battelle – Pacific Northwest National Laboratory, Richland, Washington, USA.

Further listings may generally be available on the *internet* – which serves as a very important and contemporary international network communication system.

Appendix C

TOXICITY INDEX PARAMETERS
FOR SELECTED CHEMICALS

Carcinogenic and non-carcinogenic toxicity indices relevant to the estimation of human health risks – represented by the cancer slope factor (SF) and reference dose (RfD), respectively – are presented in Table C.1 for selected chemical constituents that may be encountered in the human environments. A more complete and up-to-date listing may be obtained from a variety of toxicological databases – such as the Integrated Risk Information System (IRIS), developed and maintained by the US EPA (see Appendix B.2.2.).

Table C.1. Toxicological parameters of selected chemicals

Chemical name	Toxicity index			
	Oral *SF* (mg/kg-day)$^{-1}$	Inhalation *SF* (mg/kg-day)$^{-1}$	Oral *RfD* (mg/kg-day)	Inhalation *RfD* (mg/kg-day)
Inorganic Chemicals				
Aluminum (Al)			1.00E+00	
Antimony (Sb)			4.00E-04	
Arsenic (As)	1.75E+00	1.20E+01	3.00E-04	5.00E+01
Barium (Ba)			7.00E-02	1.40E-04
Beryllium (Be)		8.40E+00	5.00E-03	
Cadmium (Cd)		1.50E+01	5.00E-04	
Chromium (Cr – total)			1.00E+00	
Chromium VI (Cr^{+6})^a		4.10E+01	3.00E-03	
Cobalt (Co)			2.90E-04	2.90E-04
Cyanide (CN) – free			2.00E-02	
Manganese (Mn)			1.40E-01	1.40E-05
Mercury (Hg)			3.00E-04	8.60E-05
Molybdenum (Mo)			5.00E-03	5.00E-03
Nickel (Ni)		9.10E-01	2.00E-02	
Selenium (Se)			5.00E-03	
Silver (Ag)			5.00E-03	
Thallium (Tl)			8.00E-05	
Vanadium (V)			7.00E-03	
Zinc (Zn)			3.00E-01	

(continues overleaf)

Chemical name	Toxicity index			
	Oral SF $(mg/kg\text{-}day)^{-1}$	Inhalation SF $(mg/kg\text{-}day)^{-1}$	Oral RfD $(mg/kg\text{-}day)$	Inhalation RfD $(mg/kg\text{-}day)$
Organic Compounds				
Acetone			1.00E-01	1.00E-01
Alachlor			1.00E-02	
Aldicarb			2.00E-04	
Anthracene			3.00E-01	
Atrazine			5.00E-03	
Benzene	2.90E-02	2.90E-02		
Benzo(*a*)anthracene	1.20E+00	3.90E-01		
Benzo(*a*)pyrene [BaP]	1.20E+01	3.90E+00		
Benzo(*b*)fluoranthene	1.20E+00	3.90E-01		
Benzo(*k*)fluoranthene	1.20E+00	3.90E-01		
Benzoic acid			4.00E+00	4.00E+00
Bis(2-ethylhexyl)phthalate	1.40E-02	1.40E-02	2.00E-02	2.20E-02
Bromodichloromethane	1.30E-01	1.30E-01	2.00E-02	2.00E-02
Bromoform	7.90E-03	3.90E-03	2.00E-02	2.00E-02
Carbon disulfide			1.00E-01	2.90E-03
Carbon tetrachloride	1.30E-01	1.30E-01	7.00E-04	
Chlordane	1.30E+00	1.30E+00	6.00E-05	
Chlorobenzene			2.00E-02	
Chloroform	3.10E-02	1.90E-02	1.00E-02	1.00E-02
2-Chlorophenol			5.00E-03	
Chrysene	1.20E-01	3.90E-02		
o-Cresol [2-Methylphenol]			5.00E-02	
m-Cresol [3-Methylphenol]			5.00E-02	
Cyclohexanone			5.00E+00	
1,4-Dibromobenzene			1.00E-02	
Dibromochloromethane	8.40E-02	8.40E-02	2.00E-02	2.00E-02
1,2-Dibromomethane [EDB]	8.50E+01	7.70E-01		
1,2-Dichlorobenzene			9.00E-02	
Dichlorodifluoromethane			2.00E-01	
p,p'-Dichlorodiphenyl- dichloroethane [DDD]	2.40E-01			
p,p'-Dichlorodiphenyl- dichloroethylene [DDE]	3.40E-01			
p,p'-Dichlorodiphenyl- trichloroethane [DDT]	3.40E-01	3.40E-01	5.00E-04	
1,1-Dichloroethane	5.70E-03	5.70E-03		
1,2-Dichloroethane	7.00E-02	7.00E-02		
1,1-Dichloroethene	6.00E-01	1.80E-01	9.00E-03	9.00E-03
cis-1,2-Dichloroethene			1.00E-02	1.00E-02
trans-1,2-Dichloroethene			2.00E-02	2.00E-02
2,4-Dichlorophenol			3.00E-03	
Dieldrin	1.60E+01	1.60E+01	5.00E-05	
Di(2-ethylhexyl)phthalate [DEHP]	1.40E-02		2.00E-02	2.20E-02
Diethyl phthalate			8.00E-01	
2,4-Dimethylphenol			2.00E-02	
2,6-Dimethylphenol			6.00E-04	
3,4-Dimethylphenol			1.00E-03	
m-Dinitrobenzene			1.00E-04	

Chemical name	Toxicity index			
	Oral SF $(mg/kg\text{-}day)^{-1}$	Inhalation SF $(mg/kg\text{-}day)^{-1}$	Oral RfD $(mg/kg\text{-}day)$	Inhalation RfD $(mg/kg\text{-}day)$
1,4-Dioxane	1.10E-02			
Endosulfan			5.00E-05	
Endrin			3.00E-04	
Ethylbenzene			1.00E-01	2.90E-01
Ethyl chloride				2.90E+00
Ethyl ether			2.00E-01	
Ethylene glycol			2.00E+00	
Fluoranthene			4.00E-02	4.00E-02
Fluorene			4.00E-02	
Formaldehyde		4.50E-02	2.00E-01	
Furan			1.00E-03	
Heptachlor	4.50E+00	4.50E+00	5.00E-04	
Hexachlorobenzene	1.60E+00	1.60E+00	8.00E-04	
Hexachlorodibenzo-p-dioxin [HxCDD]	6.20E+03	6.20E+03		
Hexachloroethane	1.40E-02	1.40E-02	1.00E-03	
n-Hexane				5.72E-02
Indeno(1,2,3-c,d)pyrene	1.20E+00	3.90E-01		
Isobutyl alcohol			3.00E-01	
Lindane [gamma-HCH]			3.40E-04	
Malathion			2.00E-02	
Methanol			5.00E-01	
Methyl mercury			3.00E-04	
Methyl parathion			2.50E-04	
Methylene chloride [Dichloromethane]	7.50E-03	1.65E-03	6.00E-02	
Methyl ethyl ketone [MEK]			6.00E-01	2.90E-01
Methyl isobutyl ketone [MIBK]			8.00E-02	2.30E-02
Mirex			2.00E-06	
Nitrobenzene			5.00E-04	
n-Nitroso-di-n-butylamine	5.40E+00	5.40E+00		
n-Nitroso-di-n-methylethylamine	2.20E+01			
n-Nitroso-di-n-propylamine	7.00E+00			
n-Nitrosodiethanolamine	2.80E+00			
n-Nitrosodiethylamine	1.50E+02	1.50E+02		
n-Nitrosodimethylamine	5.10E+01	5.10E+01		
n-Nitrosodiphenylamine	4.90E-03			
Pentachlorobenzene			8.00E-04	
Pentachlorophenol	1.80E-02	1.80E-02	3.00E-02	
Phenol			6.00E-01	
Polychlorinated biphenyls [PCBs][b]	4.00E-02 to 2.00E+00			
Pyrene			3.00E-02	3.00E-02
Styrene			2.00E-01	2.90E-01
1,2,4,5-Tetrachlorobenzene			3.00E-04	
1,1,1,2-Tetrachloroethane	2.60E-02	2.60E-02	3.00E-02	
1,1,2,2-Tetrachloroethane	2.70E-01	2.70E-01		

(continues overleaf)

Chemical name	Toxicity index			
	Oral SF $(mg/kg\text{-}day)^{-1}$	Inhalation SF $(mg/kg\text{-}day)^{-1}$	Oral RfD $(mg/kg\text{-}day)$	Inhalation RfD $(mg/kg\text{-}day)$
Tetrachloroethene	5.10E-02	2.10E-02	1.00E-02	1.00E-02
2,3,4,6-Tetrachlorophenol			3.00E-02	
Toluene			2.00E-01	1.40E+00
Toxaphene	1.10E+00	1.10E+00		
1,2,4-Trichlorobenzene			1.00E-02	
1,1,1-Trichloroethane			9.00E-02	2.90E-01
1,1,2-Trichloroethane	5.70E-02	5.60E-02	4.00E-03	4.00E-03
Trichloroeth[yl]ene	1.50E-02	1.00E-02	6.00 E-03	6.00E-03
1,1,2-Trichloro-1,2,2-trifluoroethane [CFC-113]			3.00E+01	
Trichlorofluoromethane			3.00E-01	
2,4,5-Trichlorophenol			1.00E-01	
2,4,6-Trichlorophenol	1.10E-02	1.10E-02		
1,1,2-Trichloropropane			5.00E-03	
1,2,3-Trichloropropane			6.00E-03	
Triethylamine				2.00E-03
1,3,5-Trinitrobenzene			5.00E-05	
2,4,6-Trinitrotoluene [TNT]	3.00E-02		5.00E-04	
o-Xylene			2.00E+00	2.00E-01
Xylenes (mixed)			2.00E+00	
Others				
Asbestos (*units of per fibers/mL*)^c		2.3E-01		
Hydrazine	3.00E+00	1.7E+01		
Hydrogen chloride				2.00E-03
Hydrogen cyanide			2.00E-02	
Hydrogen sulfide			3.00E-03	2.60E-04

Notes:

a – Inhalation unit risk = 8 x 10^{-6} mg/m^3 (for exposure to Cr^{+6} acid mists and dissolved aerosols); and inhalation unit risk = 1 x 10^{-4} mg/m^3 (for exposure to Cr^{+6} particulate matter).

b – Tiers of human potency and slope estimates exist for environmental mixtures of PCBs. For high risk and persistent PCB congeners or isomers, an upper-bound slope of 2E+00 per mg/kg-day and a central slope of 1E+00 per mg/kg-day may be used; for low risk and persistent PCBs, an upper-bound slope of 4E-01 per mg/kg-day and a central slope of 3E-01 per mg/kg-day may be used; and for the lowest risk and persistent PCBs, an upper-bound slope of 7E-02 per mg/kg-day and a central slope of 4E-02 per mg/kg-day may be used.

c – Note the different set of units applied; also, 2.3E-01 per fibers/mL ≡ 2.3E-07 per fibers/m^3. It is also noteworthy that, regulatory agencies (such as the California EPA) use a significantly more restrictive value of 1.9 per fibers/mL (≡ 1.9E-06 per fibers/m^3) as the inhalation SF for asbestos.

Appendix **D**

SELECTED UNITS OF MEASUREMENT
& NOTEWORTHY EXPRESSIONS

Some selected units of measurements and noteworthy expressions (of potential interest to the environmental and public health professional, analyst, or decision-maker) are provided below.

◆ **Mass/Weight Units**

g	gram(s)
ton (metric)	tonne $= 1 \times 10^6$ g
Mg	Megagram(s), metric ton(s) $= 10^6$ g
kg	kilogram(s) $= 10^3$ g
mg	milligram(s) $= 10^{-3}$ g
µg	microgram(s) $= 10^{-6}$ g
ng	nanogram(s) $= 10^{-9}$ g
pg	picogram(s) $= 10^{-12}$ g
mol	mole, molecular weight (mol. wt.) in grams

◆ **Volumetric Units**

cc or cm^3	cubic centimeter(s) ≈ 1 mL $= 10^{-3}$ L
mL	milliliter(s) $= 10^{-3}$ L
L	liter(s) $= 10^3$ cm^3
m^3	cubic meter(s) $= 10^3$ L

◆ **Environmental/Chemical Concentration Units**

ppm	parts per million
ppb	parts per billion
ppt	parts per trillion

These are used for expressing/specifying the relative masses of contaminant and medium. It is noteworthy that, because water is assigned a mass of 1 kilogram per liter, mass-to-mass and mass-to-volume measurements are interchangeable for this particular medium.

NB: (1) A billion is often used to represent a thousand millions, i.e., 10^9, in some places, such as in the USA and France; whereas a billion represents a million millions, i.e., 10^{12}, in some other places, such as in the UK and Germany. (2) A trillion is often used to represent a million times a million or a thousand billions, i.e., 10^{12}, in some places, such as in the USA and France; whereas a trillion represents a million billions, i.e., 10^{18}, in some other places, such as in the UK and Germany.

◆ **Concentration Equivalents**

1 ppm   ≡   mg/kg or mg/L ≡ 10^{-6}
1 ppb   ≡   μg/kg or μg/L ≡ 10^{-9}
1 ppt   ≡   ng/kg or ng/L ≡ 10^{-12}

▶ *Concentrations in soils or other solid media:*

mg/kg   mg chemical per kg weight of sampled
               medium
μg /kg   μg chemical per kg weight of sampled
               medium

▶ *Concentrations in water or other liquid media:*

mg/L   mg chemical per liter of total liquid volume

μg /L   μg chemical per liter of total liquid volume

▶ *Concentrations in air media:*

mg/m^3 mg chemical per m^3 of total fluid volume

$μg /m^3$ μg chemical per m^3 of total fluid volume

◆ To convert from ppm to mg/m^3, use the following conversion relationship:

$$[mg/m^3] = [ppm] \times \frac{[molecular\ weight\ of\ substance,\ in\ g/mol]}{24.45}$$

◆ To convert from ppm to $μg/m^3$, use the following conversion relationship:

$$[μg/m^3] = [ppm] \times [molecular\ weight\ of\ substance,\ in\ g/mol] \times 40.9$$

Note: The above conversion relationships assume standard temperature and pressure (STP), i.e., temperature of 25°C and barometric pressure of 760 mmHg (or 1 atm).

◆   **Units of Chemical Intake and Dose**

mg/kg-day = milligrams of chemical exposure per unit body weight
of exposed receptor per day

◆   **Typical Expressions Commonly Used in Risk Assessment
and Environmental Management Programs**

▶   *'Order of Magnitude'*
Reference to an 'order of magnitude' means a ten-fold difference or a
multiplicative factor of ten – i.e., the base parameter may vary by a factor of
10. Hence, 'two orders of magnitude' means a factor of about 100; 'three
orders of magnitude' implies a factor of about 1,000; etc. For example, 'three
orders of magnitude' may be used to describe the difference between 3 and
3,000 (= 3×10^3). The expression is often used in reference to the calculation
of environmental quantities or risk probabilities.

▶   *Exponentials denoted by 10^K*
Superscript refers to the number of times *10* that is multiplied by itself. For
example, $10^2 = 10 \times 10 = 100$; $10^3 = 10 \times 10 \times 10 = 1,000$; $10^6 = 10 \times 10 \times 10$
$\times 10 \times 10 \times 10 = 1,000,000$.

▶   *Exponentials denoted by 10^{-K}*
Negative superscript is equivalent to the reciprocal of the positive term, i.e.,
10^{-K} equals $1/10^K$. For example, $10^{-2} = 1/10^2 = 1/(10 \times 10) = 0.01$; $10^{-3} =$
$1/10^3 = 1/(10 \times 10 \times 10) = 0.001$; $10^{-6} = 1/10^6 = 1/(10 \times 10 \times 10 \times 10 \times 10 \times$
$10) = 0.000001$.

▶   *Exponentials denoted by $X.YZ\,E+K$*
Number after the *E* indicates the power to which *10* is raised, and then
multiplied by the preceding term (i.e., the number of times 10^K is multiplied
by preceding term, or $X.YZ \times 10^K$). For example, $1.00\text{E-}01 = 1.00 \times 10^{-1} =$
0.1; $1.23\text{E+}04 = 1.23 \times 10^{+4} = 12,300$; $4.44\text{E+}05 = 4.44 \times 10^5 = 444,000$.

▶   *'Conservative assumption'*
Used in exposure and risk assessment, this expression refers to the selection of
assumptions (when real-time data are absent) that are unlikely to lead to under-
estimation of exposure or risk. Conservative assumptions are those which tend
to maximize estimates of exposure or dose – such as choosing a value near the
high end of the concentration or intake rate range.

▶   *'Worst case'*
A semi-qualitative term that refers to the maximum possible exposure, dose, or
risk to an exposed person or group, that could conceivably occur – regardless
of whether or not this exposure, dose, or risk actually occurs or is observed in

a specific population. Typically, this should refer to a hypothetical situation in which everything that can plausibly happen to maximize exposure, dose, or risk actually takes place. This worst case may indeed occur (or may even be observed) in a given population; however, since this is usually a very unlikely set of circumstances, in most cases, a worst-case estimate will be somewhat higher than occurs in a specific population. In most health risk assessments, a worst-case scenario is essentially a type of bounding estimate.

▸ *'Risk of 1 × 10^{-6} (or simply, 10^{-6})'*
Also written as 0.000001, or one in a million, means that one additional case of cancer is projected in a population of one million people exposed to a certain level of chemical X over their lifetimes. Similarly, a risk of 5×10^{-3} corresponds to 5 in 1,000 or 1 in 200 persons; and a risk of 2×10^{-6} means two chances in a million of the exposure causing cancer.

INDEX

Printed in the USA
CPSIA information can be obtained
at www.ICGtesting.com
LVHW021942161123
764134LV00001B/1